C0-AVH-907

Regional Development
in
Western Europe

Regional Development in Western Europe

Editor
HUGH D. CLOUT
Department of Geography, University College London

JOHN WILEY & SONS

London · New York · Sydney · Toronto

Library of Congress Cataloging in Publication Data:

Main entry under title:

Regional development in Western Europe.

1. Regional planning—Europe—Addresses, essays, lectures. 2. Europe—Economic conditions—1945- —Addresses, essays, lectures. I. Clout, Hugh D.

HT395.E8R4 309.2'5'094 75-11963

ISBN 0 471 16112 8 (cloth)
ISBN 0 471 16113 6 (pbk)

Printed in Great Britain by Unwin Brothers Ltd.
The Gresham Press, Old Woking, Surrey.

To

Our Colleges

List of Contributors

Mark Blacksell	*Department of Geography, University of Exeter, England*
Frederic Chiffelle	*Séminaire de Géographie, Université de Neuchâtel, Switzerland*
Hugh D. Clout	*Department of Geography, University College London, England*
Christopher Elbo	*Department of Geography, University College London, England*
François J. Gay	*Institut de Géographie, Université de Rouen, France*
James H. Johnson	*Department of Geography, University of Lancaster, England*
Russell King	*Department of Geography, University of Leicester, England*
John Naylon	*Department of Geography, University of Keele, England*
Fritz Schadlbauer	*Geographisches Institut, Hochschule für Welthandel, Vienna, Austria*
Kjell Stenstadvold	*Geografisk Institutt, Norges Handelshøyskole, Bergen, Norway*
David Thomas	*Department of Geography, St. David's University College, Lampeter, University of Wales*

Preface

This book owes its origin to the growing interest in Western Europe associated with the entry of the United Kingdom into the European Economic Community and the increasing problem-orientation of modern 'regional geography', which examines regional problems and the planning programmes that have been established to tackle them. Although the readily available literature on European regional planning themes is expanding, it is still patchy in coverage. The chapters of this book attempt to fill some gaps, solidify certain well-documented areas, and provide references for further student enquiry.

A broad spatial framework has been adopted, to include all the countries of non-communist Western Europe, although admittedly Iceland and Luxembourg receive only passing mention. Eleven geographers, teaching in universities in Britain and on the Continent, have contributed chapters on regional development in individual countries or groups of countries with which they are well acquainted through earlier research. Not surprisingly, their views and approaches are diverse and thereby mirror some of the very complex issues and attitudes that are subsumed under the deceptively brief title of 'regional development'. In no way have they attempted to give a 'complete' description of all the regions and regional problems of Western Europe.

The sixteen chapters are structured in two blocks. The first of these comprises four chapters which offer a West European overview of regional development, plus the attempts to establish a common regional policy for the Nine, and then summarize social and economic processes that have operated since 1945 to intensify regional problems. The second block examines these processes of change and the approaches to regional development for each of the Six original members of the E.E.C., for the three new members, and finally for the remaining nations of northern, southern and Alpine Europe that are linked to the Nine to a greater or lesser extent by virtue of politics, trade, tourism, labour migration, and in many other ways. A brief conclusion offers a number of spatial scenarios for Western Europe in the remainder of the twentieth century but, appropriately enough in a time of economic uncertainty, contains many more questions than answers.

A few words of explanation are required regarding the presentation of placenames. The appropriate national, or regional, language form has been used unless there is a familiar English equivalent. Thus 'København' and 'Bruxelles/Brussel' have not been used, but rather 'Copenhagen' and 'Brussels'. Names of official regions and provinces are always given in their local form, for example 'Oost-Vlaanderen' and 'Sicilia'; but where English versions exist for broad geographical areas (e.g. 'Flanders'), islands (e.g. 'Sicily'), rivers (e.g. 'the Rhine') and other geographical features they have been preferred.

As editor I must express my thanks to the contributors and especially to Continental colleagues who have tolerated my consolidation of flowing European prose and have kindly refined my translations of several chapters. Six members of the Cartographic Unit in the Department of Geography at University College London prepared almost all the illustrations, doing so with their customary skill and speed. The remaining maps, in Chapters 8 and 9, were drawn at Exeter and Lampeter respectively. I am indebted to my parents, for their patient checking of material and comments on my written use of English, and to my colleague Gerald Manners for encouragement throughout the project. Students at University College London, Birkbeck College and the University of Georgia have offered useful criticisms of European regional development over the years, and I extend my thanks to them. Finally, David Walker (University of Waterloo) and John Taylor (University of Oslo) kindly advised on particular points.

HUGH D. CLOUT

Contents

I ECONOMIC AND SOCIAL TRENDS IN WESTERN EUROPE

II REGIONAL DEVELOPMENT IN PRACTICE

I

Economic and Social Trends in Western Europe

1
Regional Development in Western Europe

Hugh D. Clout

Regional Diversity: a Matter of European Concern

The diversity of Western Europe, fashioned by the interaction of history and geography, has fascinated and challenged generations of social and physical scientists. Their findings, with respect to changing political boundaries, spatial variations in physical geography, appraisal and use of natural resources, development of transport networks and many other topics, contribute the essential background against which the contents of this book must be set. 'Regions', variously differentiated and defined, formed the usual canvases on which geographical views of socio-economic differences were painted and the scale at which they were presented. However, the theme of regional diversity has attracted more than academic interest in the past quarter century as policies for regional development have been devised and implemented by individual countries in an attempt to modify disparities in environment and in economic and social life that their territories displayed. Some politicians and social scientists have argued cogently for more of a pointilliste[1] approach, in which the problems of particular groups in society living in particular localities may be isolated and, hopefully, tackled. While appreciating the merits of such a realistic scale of enquiry and action, our present concern is for the higher level of spatial generalization that 'regional development' implies.

At the regional level, demonstrations of cultural identity and even of 'autonomist' sentiment have emanated from Basques and Bretons, Cornishmen and Corsicans, Walloons and Welshmen, and members of many other minority groups who are concerned about their economic wellbeing, their cultural survival and the nature of their relationship with central administrations (Héraud, 1973; Sérant, 1965). At a higher level, national governments have intervened increasingly in the operation of their respective economies. At first they needed to define priorities for the recovery of war-torn Europe but later began to introduce regional or spatial objectives to complement programmes for managing national economies. The numerous systems of regional planning that resulted have been concerned entirely with tackling the problems of individual countries. Rather different problems existed or, perhaps more accurately, were perceived to exist in each country. Resultant policies have operated in differing ways over varying lengths of time and have experienced varying degrees of success in meeting different objectives. By virtue of their diversity such policies defy presentation in summary form and will be examined in appropriate detail in later chapters.

In the last few years, problems of regional variation have shifted from simply being the concern of individual governments to occupy a point in the forefront of the international political arena, as the staff of the European Commission have attempted to define regional expressions of economic and social inequality in an objective fashion, to examine critically national policies for regional aid, and then devise, finance, and implement a Common Regional Policy for the nine member nations of the European Economic Community (E.E.C.). Their attempts will be outlined in the following paragraphs by way of introduction to regional inequalities in a Community of 253,000,000 inhabitants. Neighbouring West European countries outside the E.E.C. continue to refine and implement their domestic policies for tackling regional problems.

Search for a Common Regional Policy

The idea of correcting imbalances between developed and declining areas in Western Europe is far from new, having been discussed as early as 1956 in the report of a committee headed by M. Paul-Henri Spaak in connection with negotiations which led to the E.E.C. being created one year later. The preamble to the Treaty of Rome expressed the view that member states were 'anxious to strengthen the unity of their economies and ensure their harmonious development by reducing present differences between the various regions and by mitigating the backwardness of the less favoured' (Palmer and coworkers, 1968). In similar vein, Article 2 outlined the objective of promoting 'harmonious development of economic activities throughout the Community'.

A number of supranational organizations such as the European Investment Bank (E.I.B.) were established and have had important implications for regional development in the Common Market, but there was no direct reference in the Rome Treaty to a common regional policy as such. In subsequent years the E.I.B. loaned approximately £1,400,000,000 (1958–72), 75 per cent of which was allocated for regional development schemes, with Italy being the major beneficiary (Figure 1.1). Finance from the European Coal and Steel Community (E.C.S.C.) has contributed to the creation of 110,000 new jobs and has helped retrain and re-employ nearly 500,000 workers in the coal and steel industries of the Six. The European Social Fund (E.S.F.), in providing 265,000,000 units of account[2] for resettlement and training of workers, has also had an important regional impact, as has the guidance section of the European Agricultural Fund (F.E.O.G.A.) where 150,000,000 units of account have been spent on modernizing farming and providing higher living standards for the agricultural population.

The issue of a Common Regional Policy was raised on several occasions, as, for example, in 1965 when the European Commission sketched broad policy outlines for regional development. All types of region were to be involved but the implementation of regional schemes was to remain the responsibility of national authorities albeit with some kind of coordination from the Commission. Political commitment to coordination proved to be far from forthcoming in the years that followed and member nations continued to operate their own separate policies for regional development. Doubtless much was achieved, as individual nations and component regions attempted to outbid each other in their offers of aid to attract investors in order to create more jobs in manufacturing and in offices to help satisfy the need for employment for their growing workforces and to replace jobs lost in farming, coalmining, textile manufacture and other labour-shedding activities. An analysis of regional economic changes in the Six during the 1950s and 1960s showed that in spite, or perhaps because, of these measures the Rhinelands, Paris, northern Italy and smaller core areas in the E.E.C. continued to

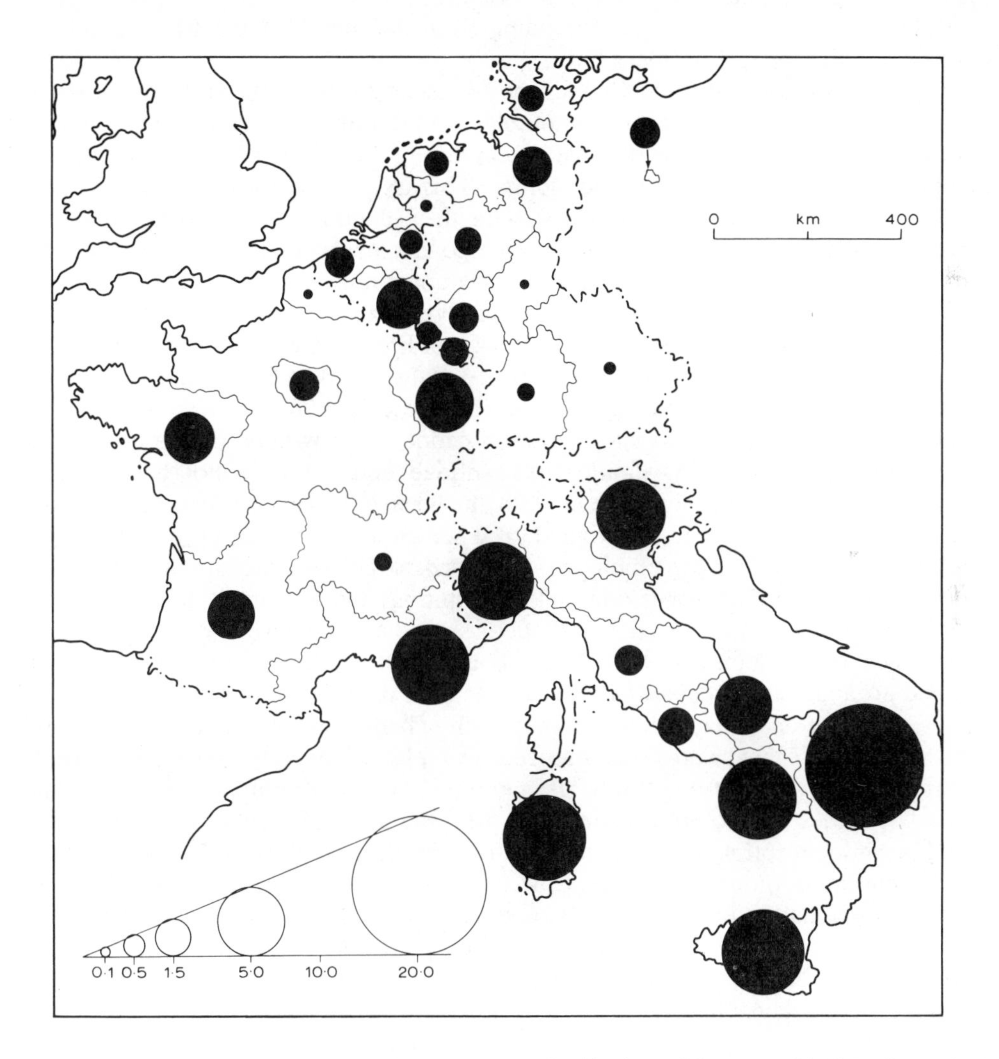

Figure 1.1. European Investment Bank: percentage distribution of loans to 31.12.70 (by region).

exercise their powerful attraction for new employment (Commission of the European Communities, 1971). In the light of this trend there is evidence that the 'gap' between rich and poor areas had become broader after the establishment of the E.E.C. Major urban areas outside the then Common Market were also highly attractive for new employment in offices and factories, with the result that southeastern England, Copenhagen and other economic cores in Western Europe functioned as national versions of the Rhinelands' 'Golden Triangle', the 'Lotharingian axis', or the 'hourglass' of economic strength extending from the mouth of the Rhine southeastwards across Europe into northern Italy.

Not until June 1971 did the Council of Ministers examine the idea of initiating a Common Regional Policy for the Community, and it took a further sixteen months for this to be recognized by heads of government during the course of the Paris summit in October 1972 for enlarging the Community. This issue of regional aid was of crucial importance to potential new members, with the Republic of Ireland being heavily dependent on agriculture and having low average incomes when measured against other parts of Western Europe, the United Kingdom containing mining and industrial areas that had received aid from domestic sources over many years, and western parts of Denmark having unemployment problems. Accordingly, it was decided that regional policy was to be given a new and high priority in the enlarged Community as it moved towards Economic and Monetary Union (E.M.U.) in 1980. It was declared that differences in living standards and in rates of economic growth between rich and poor regions in the Community should not continue and certainly not become more pronounced. Serious disenchantment with the idea of West European unity would undoubtedly threaten if wealth continued to be amassed in economic core areas where it was already plentiful. New employment needed to be installed urgently in mainly agricultural areas and in zones dominated by mining and declining industries to help resolve this problem. Congested conurbations should not, however, be forgotten. Their environmental problems — inadequate housing, overcrowded transport systems, polluted air and water — required greater attention than they had received in national schemes for regional management, since such schemes had been concerned mainly with rural areas that were characterized by relative poverty and high rates of unemployment and out-migration. Programmes for decentralizing economic and cultural activities were often implemented from the point of view of the benefits they would bring to reception areas rather than the relief they would give in congested core areas. Continuing progress towards E.M.U. would further emphasize the need for a Common Regional Policy since removing the opportunity for individual member states to devalue or revalue their currencies would accentuate regional differences.

Mechanisms therefore needed to be set up at the Community level to enable resources to be channelled from affluent areas to poorer ones, most of which were already assisted by member states. In more pragmatic terms, the international cooperation that would of necessity accompany the Common Regional Policy would benefit areas that straddled national boundaries inside the Community (Verburg, 1964). In addition, restriction of competition between member states in offering regional aid would allow the Community to harmonize action and, hopefully, also to reduce the overall costs of regional development. A Community Fund for regional investment and a Regional Development Committee were duly recommended as components of the proposed Common Regional Policy. To supplement the Regional Development Fund, finance would be allocated from the European Agricultural Fund to establish industrial jobs for people quitting farmwork in primarily agricultural regions. A regional development company was proposed to act as an information centre for European industrial expansion and a Community guarantee system was to back regional loans. Finally, all regions in the Nine were to be classified as 'peripheral' or 'central', with aid in 'central' regions to be subject to a ceiling of 20 per cent of investment costs.

Defining the Community's Problem Regions

In order to convert these proposals into realities, the European Commission was invited to analyse regional problems in the old and new member states and to offer suggestions for future action. Heads of state or government in the Nine also undertook to coordinate their regional policies and requested that the Regional Development Fund should be set up before the end of 1973, to be financed from the Community's own resources after the beginning of the second phase of E.M.U. Intervention by the Fund together with coordinated national aids for regional assistance were considered to be appropriate for tackling major spatial imbalances. Subsequent events have shown the immensity of the practical and political difficulties involved in trying to operationalize a Common Regional Policy. But before that stage could be approached a mammoth fact-finding operation had to be undertaken to define the E.E.C.'s problem regions.

This task was entrusted to a team working under the direction of Mr. George Thomson who had been appointed Commissioner responsible for regional policy. After tackling the formidable problem of collating and analysing regional data for all parts of the Community, he addressed a report on the nature of regional disparities in the Nine to the European Parliament in May 1973. His sources of information were surprisingly diverse, being produced independently by the authorities of each member state. Definitions of socio-economic categories, dates when information was collected and size of regional units by which they were organized varied considerably from country to country and did not help him in his task. These facts emphasize the need to exercise extreme caution in interpreting the final results of the *Thomson Report.* Some sets of information were simply not available at a regional level and so the analytical work in the *Report* was restricted to processing a fairly modest range of socio-economic indicators.

Fundamental decisions had to be taken for selecting which of those indicators should be retained and which critical values be considered appropriate for defining 'problem regions' at the Community scale rather than just in the context of each of the nine member countries. Eventually, regions deemed to be in need of help were defined as those with *per capita* gross domestic products below the Community average and with at least one of the following characteristics:

(i) more workers in farming than the Community average;
(ii) at least 20 per cent of the workforce in declining industries (e.g. coal mining and/or textile manufacture);
(iii) persistently high unemployment (at least 3·5 per cent, and over 20 per cent above the appropriate national average) or annual out-migration in excess of 1·0 per cent over a long period.

Evidence included in the *Thomson Report* represents a unique attempt to define regional disparities quantitatively whose rough outlines are well known and had been described many times before in verbal terms. But the apparent precision contained in the *Report* needs further comment. Generalized spatial information inevitably conceals a multitude of differing local conditions. In any case, apparently similar regional conditions (for example with respect to income levels or unemployment) may have been produced by the interaction of quite different sets of basic causes. Furthermore, social and economic conditions were largely treated in the *Report* as if they were static phenomena whereas, of course, they experience substantial change with the passage of time. Admittedly, the *Report* did stress that future changes should be monitored and used to redefine areas requiring aid.

The Evidence

Even a simple examination of population densities shows a fundamental contrast between heavily urbanized areas at the heart of the Community (supporting more than 450 people per square kilometre) and regions on its sparsely populated fringes for example in central France, Scotland and most of Ireland (with fewer than 50 people per square kilometre) (Figure 1.2). Such a core/periphery contrast will emerge time and time again when other indicators are scrutinized.

Financial evidence showed that Denmark was the most affluent member of the Nine in 1970, with an average *per capita* G.N.P. some 28 per cent above the Community average. West Germany (+ 23 per cent), France and Luxembourg (both + 18 per cent) and Belgium (+ 8 per cent) also fared well. By contrast, the Netherlands (— 3 per cent), the United Kingdom (— 12 per cent), Italy (— 31 per cent) and the Republic of Ireland (— 46 per cent) fell below the mean. According to the evidence of the *Thomson Report* each region of Italy, Ireland and the United Kingdom registered below-average figures. Five of the other countries in the Nine contained regions with values both above and below the Community mean. Only one country, tiny Luxembourg, was entirely above the average. Greater Paris was by far the most affluent region in the Nine, with an average *per capita* gross regional product of more than 4000 units of account, which was almost double the Community average (2469) and almost five times the values recorded for the poorest parts of southern Italy. That country undoubtedly contained the lowest values for any part of the Community but she also displayed a reasonable spread of regional figures, rising to more than 2000 units of account. By contrast, regional figures for the Republic of Ireland clustered around the 1000 mark indicating a more even distribution of poverty.

Rather different indicators of affluence were calculated by *The Economist*, which showed levels of personal income in 1973. Figure 1.3 confirms the relative poverty of the periphery of the Community, especially the Republic of Ireland and Italy, where average annual incomes were below £700. Northern England, Northern Ireland and central and northeastern Italy recorded incomes in the next class (£700–£900). All parts of Belgium, Denmark, France and West Germany were located toward the higher end of the affluence scale, with regional incomes of more than £1250 per head in 1973. The Six registered a 5·4 per cent average annual increase in G.N.P. between 1960 and 1970 (which represented a substantial rise in real living standards) but at the end of the decade the richest areas of the Six still had a *per capita* income about five times that of the poorest areas, just as they had at the beginning.

Jobs in farming, forestry and fishing in the Nine fell dramatically from 17,050,000 to 11,420,000 during the 1960s. But regions in the Republic of Ireland and southern

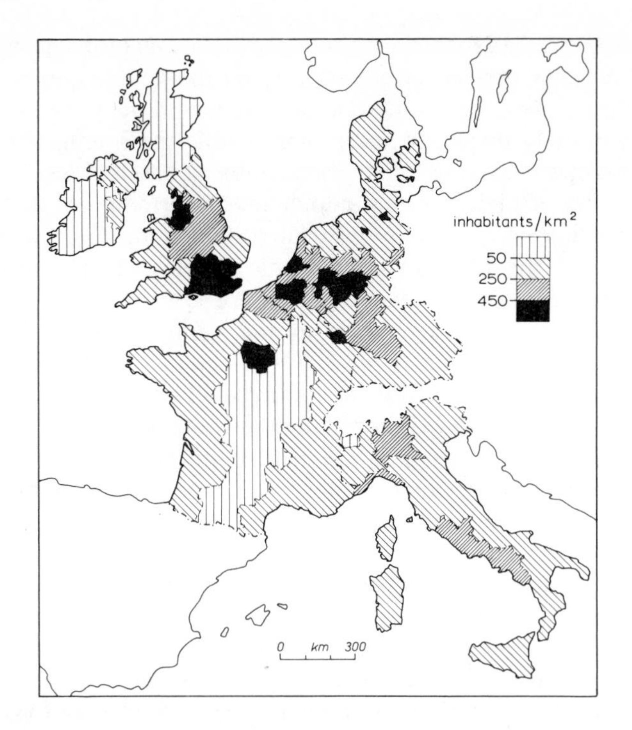

Figure 1.2. Population density in the Nine, 1971 (by region).

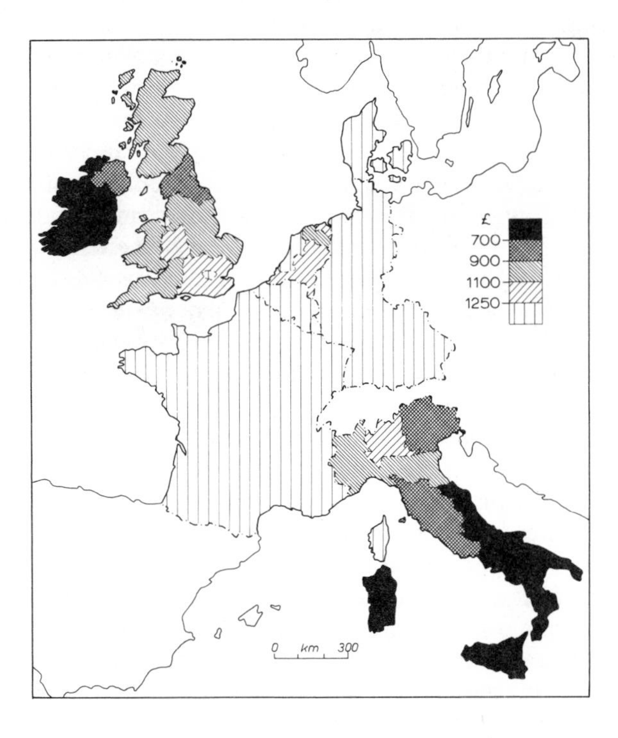

Figure 1.3. Annual average incomes in the Nine, 1971 (by region).

Italy were still very heavily dependent on farming at the end of the decade with over 30 per cent of their regional labour forces working on the land (Figure 1.4). Such rates were more than three times the Community average (9·8 per cent in 1971). Western France and much of Italy displayed more than double the average, with most of the remainder of those two countries plus considerable areas in West Germany having more than 10 per cent of their working population in farming. At the other extreme, nine of the eleven standard regions in the United Kingdom, along with greater Paris, parts of Belgium and the German Rhinelands had less than 5 per cent of their workforces in agriculture, forestry and fishing.

By contrast with this rapid decline in primary employment, the Community's industrial workforce increased during the 1960s, albeit at a relatively slow rate, rising from 43,450,000 to 45,530,000. Employment in manufacturing was above the average (43·9 per cent) in most of the United Kingdom, northern France, northern Italy, southern parts of the Netherlands, and much of Belgium and West Germany (Figure 1.4). Finally, numbers in tertiary employment underwent very important growth in the Nine during the past decade rising from 39,180,000 to 44,600,000, to rival the size of the manufacturing workforce. As one might expect, tertiary employment was in excess of the Community average (46·3 per cent) in regions with large cities and conurbations or well-developed tourist activities.

Indicators of declining industries highlighted the well-known pattern of coal mining regions plus a few other areas where quarrying was important and the perhaps less-familiar distribution of textile production (Figure 1.5). Taken together, these two maps pinpoint a scatter of small but often densely-populated areas with very important social, economic and especially environmental problems that were quite different from those of the rural fringes of the Community.

Unemployment conditions for a four-year period were collated from diverse national statistics and attempts at international comparison must therefore be made with caution. In spite of reservations about detail, the message is relatively clear. High unemployment affected both agricultural and industrial areas in the Republic of Ireland, southern and central Italy, southern Belgium, Denmark, Scotland, Wales and northern England (Figure 1.6). By contrast, unemployment rates were of slight significance in France, West Germany and much of the Netherlands.

Net out-migration in excess of 1·0 per cent per annum was a reflection of limited job opportunities in poor rural areas of southern Italy and western Ireland, while rates of 0·5–1·0 per cent involved not only rural areas but also smaller patches of contracting employment in mining areas and old industrial zones, for example in northern and northeastern France (Figure 1.6).

When all these sets of information had been collated according to the criteria defined by Community experts, the complete picture emerged of regions that were deemed to be in need of assistance and thus qualified for aid under the projected Common Regional Policy (Figure 1.7). Community problem regions included the whole of Ireland, most of Italy, western France, Scotland, Wales, northern England, together with smaller parts of Belgium, Denmark, West Germany, and the Netherlands. Some 40 per cent of the Community's population lived in these regions. In the words of the *Thomson Report*, 'the regional problem encompasses the whole range of economic evolution from excessively agricultural regions to areas that have experienced their first industrial revolution and are facing unemployment as a result of structural changes necessary to change those industries and bring them up to date'. Economic core areas

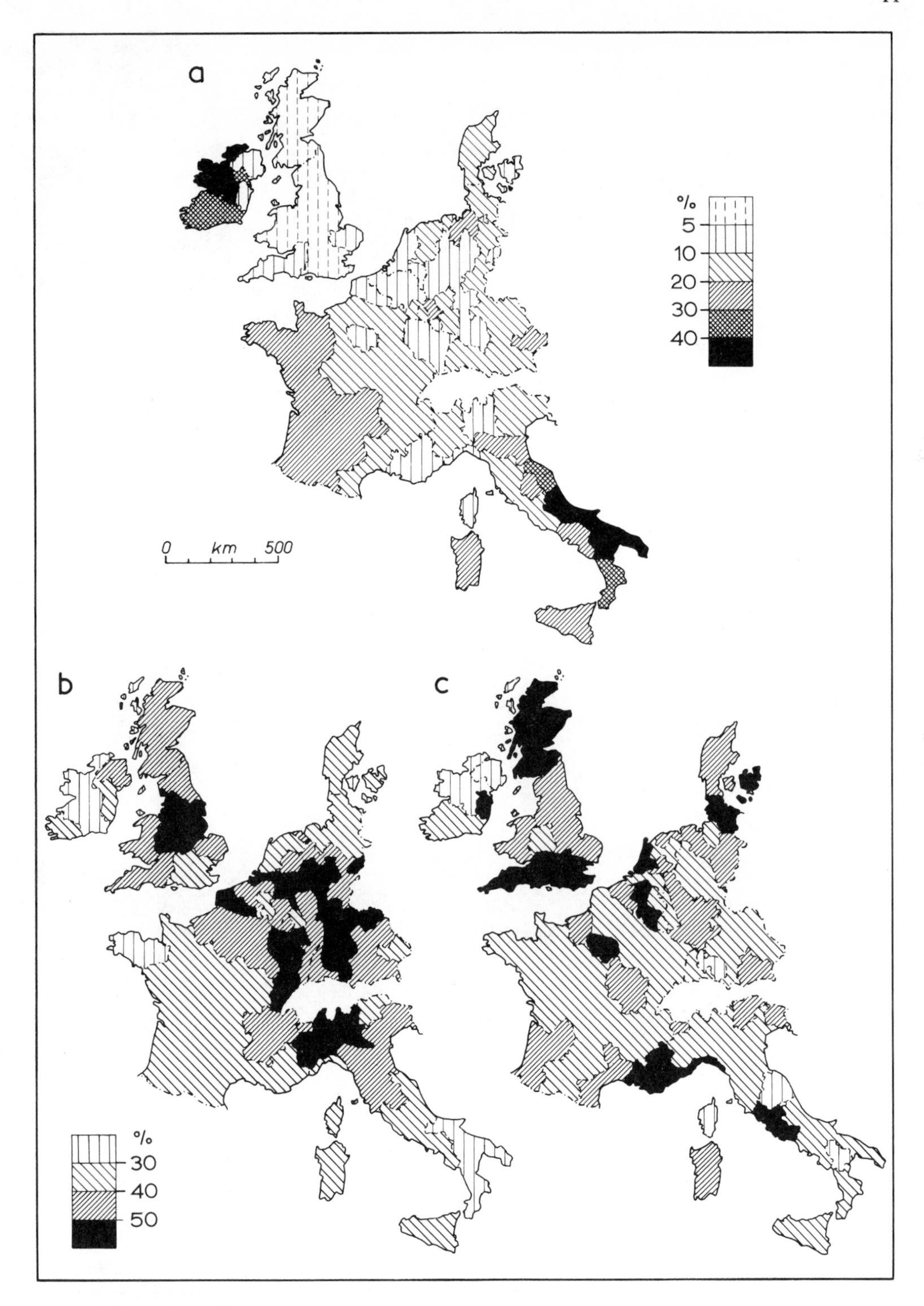

Figure 1.4. Proportion of the workforce in (a) primary, (b) secondary and (c) tertiary sectors in the Nine, 1971 (by region).

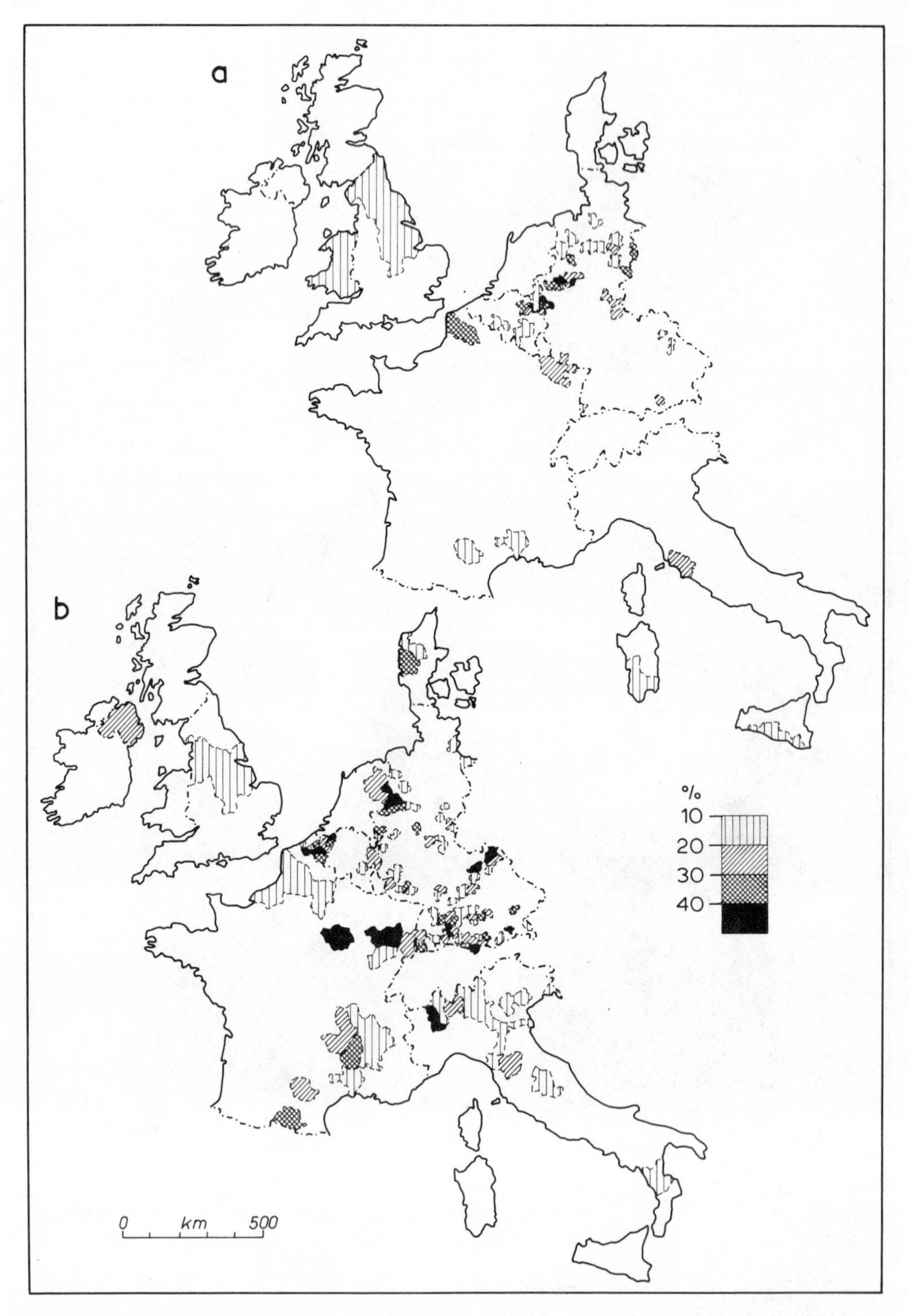

Figure 1.5. Proportion of the workforce in (a) coal mining and (b) textile manufacture in the Nine, 1971 (by region).

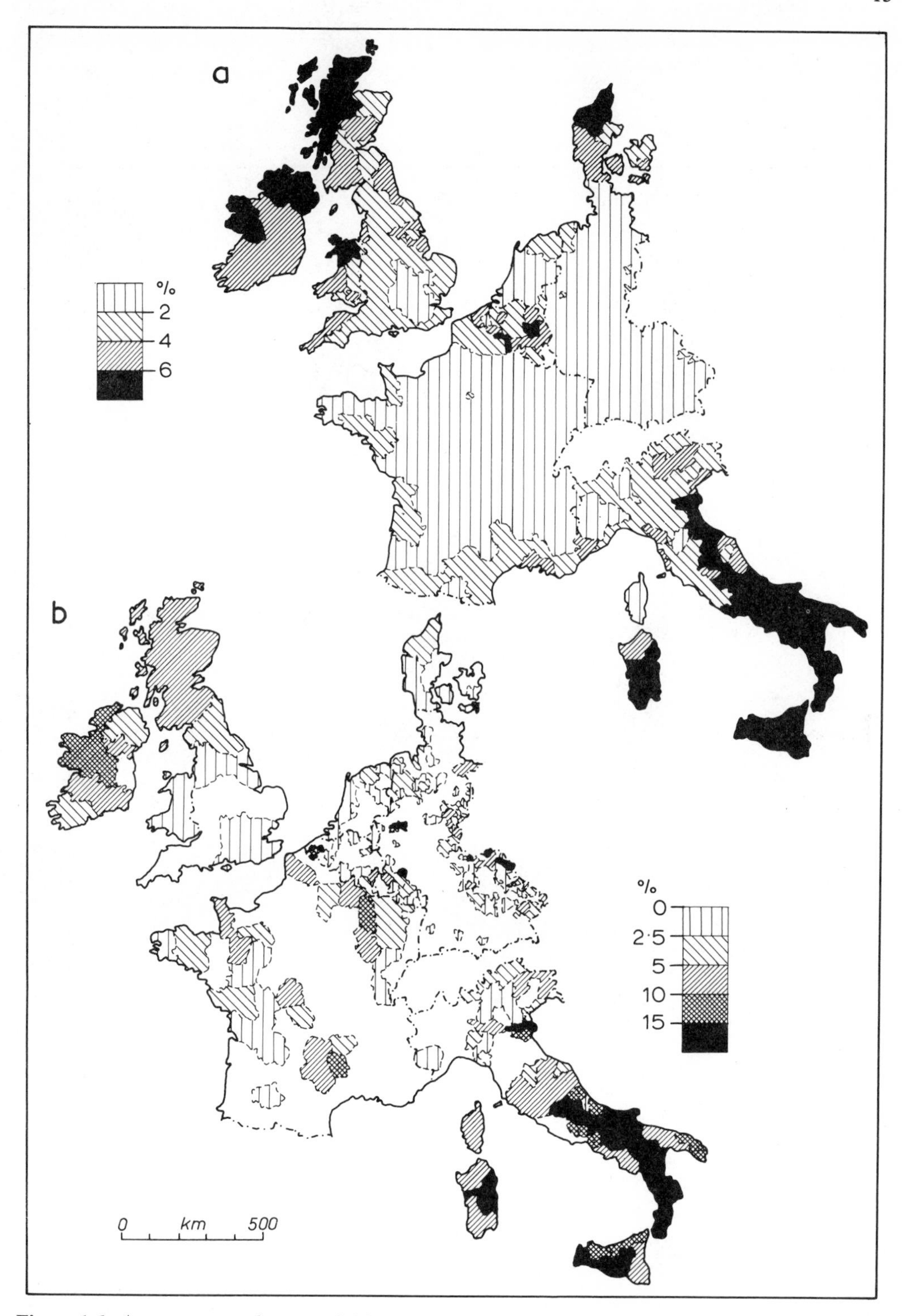

Figure 1.6. Average annual rates of (a) unemployment and (b) out-migration since 1960 in the Nine (by region).

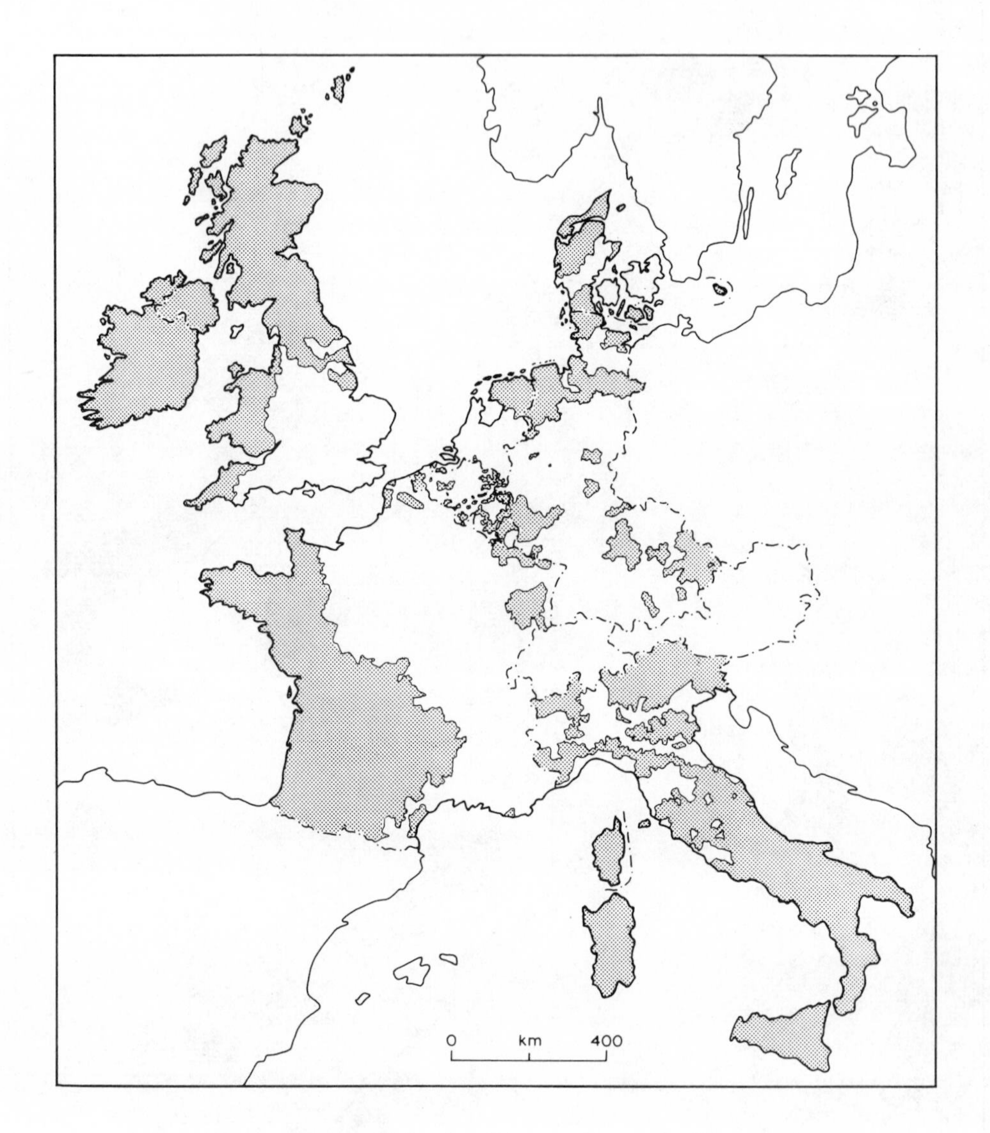

Figure 1.7. Regions qualifying for aid according to the *Thomson Report*.

did not emerge as 'problem regions' according to the above criteria, which picked out problems of poverty rather than of wealth. But the environmental difficulties that beset Western Europe's conurbations would not be ignored by the Common Regional Policy since they provided reasons for undertaking further decongestion operations that would benefit both core and periphery, provided that provincial industrialization could be undertaken with care and respect for relatively unspoiled environments.

Perhaps not surprisingly, the pattern of Community problem regions bears strong similarities to that produced by collating zones that already received assistance for regional development from domestic schemes operated by the member nations of the Nine (Figure 1.8). These similarities are more apparent than real. Member states exercised in various ways their initiative for devising schemes of regional aid which, when subjected to the degree of scrutiny attempted in later chapters, emerge as extremely diverse and complicated. The Common Regional Policy was intended to complement rather than to completely replace national systems of regional aid; but it would first be necessary to coordinate and harmonize existing schemes. In addition, the cost of the projected Regional Development Fund would have to be agreed upon and a means implemented for obtaining appropriate contributions from each member state. Such contributions would be used quite independently of any criterion of 'just return' in order to tackle the most serious and pressing regional problems in the Nine. Real transfers of resources from rich areas to poor ones would be absolutely essential. Member states would have to sacrifice a highly cherished portion of their national powers as they are being required to blend their own objectives and policies for regional development for the sake of a wider European goal. The *Thomson Report* noted that 'the acceptance of this principle will be an important test of Community solidarity'. The truth of these words has been proved many times by the heated discussions that took place in the chambers of European power in the closing months of 1973, the ritual 'stopping of the clock' before the old year ran out (by which time the Common Regional Policy should have been agreed), and the ensuing position of stalemate between member states. In order to begin to appreciate the background to this confrontation, brief attention will be paid to the diversity of existing national schemes for regional aid (which are examined in depth in later chapters) and the problems associated with financing a Community Regional Fund.

Diversity of National Systems of Regional Aid

National schemes for regional aid in Western Europe were devised under very different constitutional and political conditions, of which France, with a highly centralized administration, exemplifies one extreme (Rhodes, 1974). A system of national economic planning was introduced as early as 1947 from which objectives for regional development evolved under tight supervision and coordination from Paris (Despicht, 1971). The British approach has been rather more pragmatic and has developed under a different but still strongly centralized system of administration (McCrone, 1969). By contrast, West Germany lies at the other extreme of the spectrum with eleven individual states (*Länder*) having responsibility for regional development and other economic issues. However, legislation was passed in the last decade to strengthen links between the central *Bund* and the *Länder* so that by 1969 the task of establishing guidelines for regional economic policy had become a matter of joint responsibility. Decisions on sources of finance and broad policy are agreed between federal and state

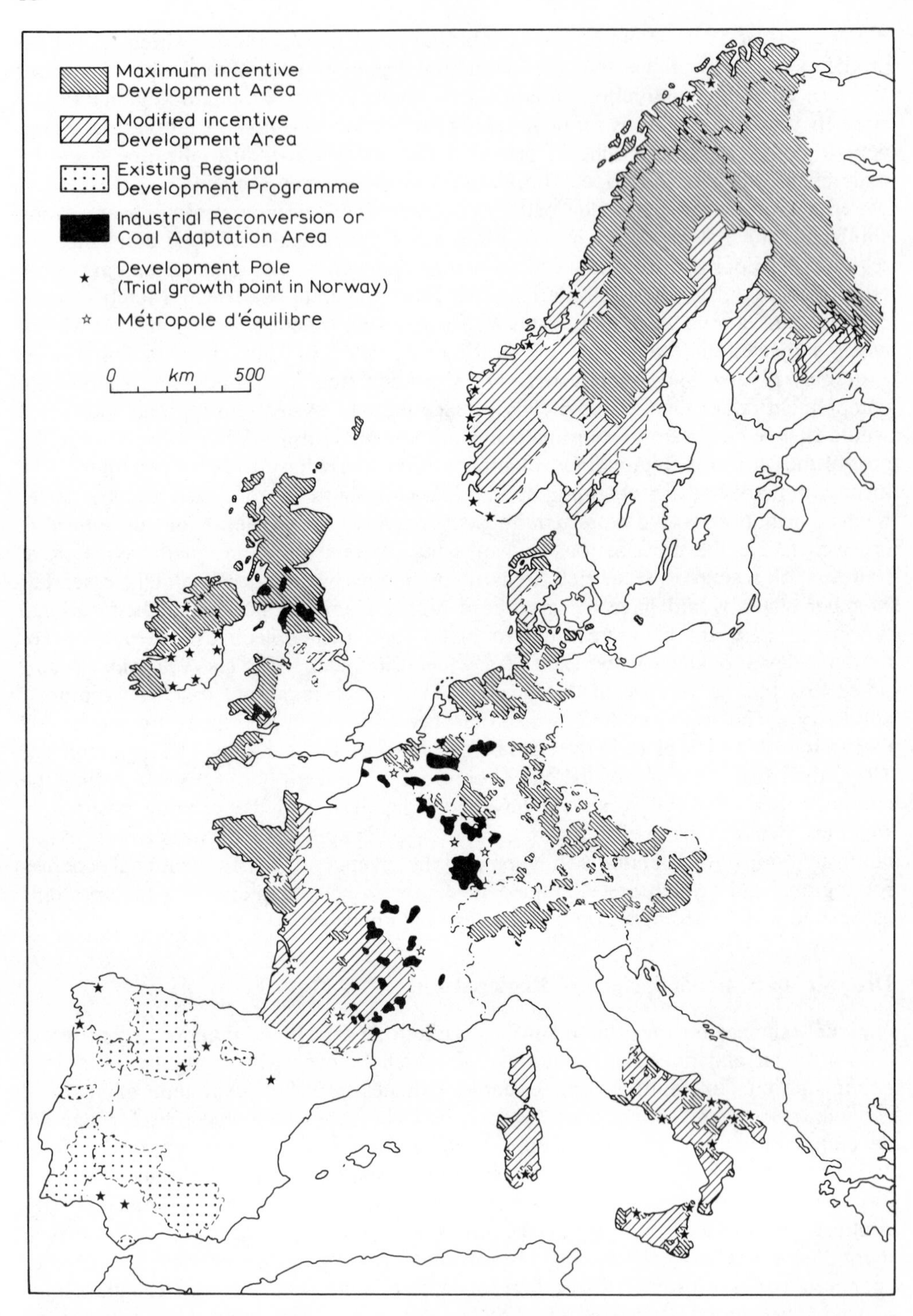

Figure 1.8. National systems of regional assistance in Western Europe, *c.*1970.

ministers but supervision of individual projects is left to ministers in the *Länder*. A variation on this theme of internal bargaining in regional policy is found in Belgium, where the problems and claims of the Flemish North and the Walloon South and the resultant policies have to be delicately counterpoised (Stephenson, 1972).

The nine existing systems of regional aid were started at different periods in time, in response to various social and economic problems, and have evolved in different ways as the problems themselves have changed. For example, British schemes began in 1934 as welfare measures to aid mining and industrial areas with high unemployment. Objectives have broadened since then to embrace the rational distribution and use of economic resources throughout the country. In Italy, regional development was stimulated by the political, economic and social problems that loomed in the rural South immediately after the Second World War. Assistance has been operationalized by the Cassa per il Mezzogiorno, which started work in 1950, and by other organizations. Objectives have broadened from initial concern with agricultural improvement to the development of manufacturing and tourism. The regional problems of several other countries have been perceived in terms of a basic contrast between capital cities and the provinces. Copenhagen, for example, contains 35 per cent of the Danish population, and southeastern England houses 31 per cent of the U.K. total on 11 per cent of the national surface. However, the classic formulation of a core/periphery contrast is in J-F. Gravier's study of *Paris et le désert français* (1947). In the past quarter century, regional objectives in France have broadened progressively from reducing the disparity between Paris and the provinces, to providing industrial jobs in the rural West, and renovating old industrial and mining areas in the North and North-East. In a similar way, greater attention has been devoted to the problems of mining areas in Belgian and Dutch regional policy since the mid-1960s.

Delimitation of assisted areas has varied from country to country and has also, of course, evolved through time. Figure 9.1 records the main changes in the British situation from the isolated Special Areas of the 1930s to the much broader and graded system of financial aid in the 1970s, ranging from maximum incentives for regional industrial development in Special Development Areas, through moderate aid in Development Areas, to minimum help in Intermediate Areas and controls on further development in southeastern and central England (Chisholm, 1974). A roughly similar graded system operates in France, although the precise nature of financial aid and fiscal relief differs considerably from the British counterpart (Figure 6.4). By contrast with such large areas qualifying for assistance in France, the United Kingdom and Italy, regional aid in West Germany, Belgium and the Netherlands has been concentrated in much smaller areas. The growth pole concept has been accepted in many regional development strategies but in varying degree. At one extreme are the eight *métropoles d'équilibre* that have been designated to counterbalance the predominance of Paris in French national life, and the large growth centres for industrializing southern Italy in which functional linkages between enterprises have been carefully planned. By contrast, much smaller development foci have been favoured in West Germany, the Republic of Ireland and the Netherlands.

Finally, the nature and value of regional assistance varies both within and between the members of the Nine. Each country offers inducements in the form of capital grants, cheap state loans or guarantees, tax relief, aid for installation and equipment of factory estates, and finance for training and retraining labour, so that work may be taken to the workers in depressed and declining regions. In addition, the United

Kingdom pays a Regional Employment Premium on manufacturing jobs created in development areas; and in the Netherlands a premium of over £1000 per worker is paid to firms that move to reception areas. Labour premiums of this kind are highly controversial, since they actively distort competition between regions as they bid for new employment.

To complement such inducements in development areas, a few countries operate controls in their congested areas with varying degrees of strictness. Ever since 1948 the United Kingdom has required firms to obtain Industrial Development Certificates for creating or extending factories above certain thresholds and beyond the assisted areas. At present a Certificate is required for any new industrial building or extension of more than 1400 m^2 outside the development areas and over 1000 m^2 in London and the South-East. This has been complemented since 1965 by the need to obtain Office Development Permits, currently for any new office or extension exceeding 1000 m^2 in London and the South-East. Control of industrial and tertiary growth has also been attempted in Paris and in 1973 the Dutch government tightened up on planning permission for new development in its western regions where employment growth is to be slowed down in the future. Relocation of administrative offices and state-owned factories away from national capitals operates in the United Kingdom, France and the Netherlands. In Italy 40 per cent of total public sector investment is now allocated to the Mezzogiorno each year. Rapid rates of employment growth in tertiary activities and further stagnation of some industries during the 1960s suggest that in the future much greater attention will have to be paid to the very difficult task of steering office jobs away from economic core areas to assisted zones.

When seen from the Community viewpoint these various systems of regional aid undoubtedly appear complicated and unwieldy. But it is important to recall that they have each been nurtured and implemented by member countries in order to tackle their *own* specific regional problems, relying primarily on domestic sources of finance. As recent events have shown, such a weighty legacy of policy and practice will not be easily coordinated under the mantle of a Common Regional Policy operated from Brussels. Matters of finance lie at the heart of the matter.

The Question of Finance

In addition to coordinating national schemes, an effective Common Regional Policy would require the transfer of financial resources between member states according to the principles of greatest ability to contribute and of greatest need of assistance. West Germany, with relatively few areas qualifying for help according to the Community's criteria, would be the major net contributor to the proposed Fund, while poorer members of the Nine would be net beneficiaries. As a result, the precise extent of assisted areas, the underlying criteria used for definition, the resultant size of the Regional Fund, and the ratio between likely contributions and receipts have proved to be issues of the most acrimonious debate.

At the end of November 1973 the European Commission succumbed to French pressures by insisting that the nine member countries should agree to new restrictions on national aid to backward regions by the end of the year. In spite of considerable opposition from France all the Development Areas and Special Development Areas in the United Kingdom were eventually classified as 'non-central' and thus remained outside the E.E.C. ruling that limited aid to central areas to 20 per cent of the value of any investment. But at the Copenhagen summit in December it was evident that West Germany had set itself against the Thomson Plan with its proposal for 2,250,000,000 units of account (*c.* £900,000,000) to be invested by the E.E.C. in regional projects over the years 1974–76. The West German government proposed a much smaller fund of only 600,000,000 units of account (*c.* £250,000,000) over the three years. Such a reduction would render meaningless the criteria that had been outlined by Thomson and force a re-writing of the list of regions that would stand to benefit. Moderately poor parts of the Nine might be deprived of help and the volume of assistance available to the very poor parts of the Community, such as Ireland and southern Italy, would also be reduced. After the European clock had been stopped at the end of 1973 the stalemate solidified. West Germany made it clear that it was not willing to contribute the volume of resources favoured by the United Kingdom; and France stressed that it was not prepared to go along with proposals for a small fund with the bulk of the money being channelled to the countries in greatest need, namely the Republic of Ireland and Italy.

Eleven months later, at the Paris summit, the Nine cleared a major obstacle on 1 December 1974 by agreeing that a relatively small joint regional development fund would be set up by 1 January 1975. It is expected that operations will begin in the summer and money could start flowing in the autumn. The Fund will involve 1,300,000,000 units of account (*c.* £540,000,000) to be shared among the poorer regions of the Community over the next three years. Some 300,000,000 units of account will be spent in 1975, and 500,000,000 units in each of the succeeding two years. Details of how the money is to be used in each member state remain to be worked out in 1975 by the Commission in a series of working groups with officials of the member states and finally by the Council of Ministers. The Commission is also working on a new system for categorizing development areas. However, by virtue of the reduced size of the Fund the number of areas eligible for help may have to be scaled down but the general pattern is not expected to change substantially. The Fund will be available for both private and public projects. According to the Commission's original proposals the Fund could contribute up to 15 per cent of the total cost of a private project, and up to 30 per cent of public expenditure on new infrastructure. More generally, the Fund can contribute up to 50 per cent of the total national regional aid in each case.

One of the tasks for 1975 will be the appointment of a committee to play a watchdog role on the administration and spending of the Fund. This will include officials of the governments of the Nine together with representatives of the Commission. The Fund will be shared out in the following proportions: Italy, 40 per cent; United Kingdom, 28 per cent; Republic of Ireland, 6 per cent (these three are net beneficiaries); France, 15 per cent; West Germany, 6·4 per cent; the Netherlands, 1·7 per cent; Belgium, 1·5 per cent; Denmark, 1·3 per cent (mainly for Greenland); Luxembourg, 0·1 per cent (these six countries are net contributors to the Fund). Clearly this scheme for dividing up the Fund on a national basis is an important new principle that will demand further reappraisal of the map of eligible regions that had been defined in the *Thomson Report.* At the time of writing (February 1975) the Common Regional Policy and most other E.E.C. schemes are exposed to bitter debate and the question of whether the United Kingdom should continue being a member is to be submitted to the nation in a referendum. The chapters that follow are based on firmer ground, examining salient social and economic trends in North-West Europe in the past quarter of a century, and then investigating the nature of regional problems and the schemes that have been implemented to tackle them in individual countries or in groups of nations.

References

Adamson C., 1973, Regional policies as a strategy for the development of Europe, *Chartered Surveyor,* **106**, 109–114.

Allen, K., 1974, European regional policies, in Sant (1974, pp. 87–99).

Cassidy, M., 1973, Community rationalization: regions, transport and energy, *Built Environment,* **2**, 205–209.

Chisholm, M., 1974, Regional policies for the 1970s, *Geographical Journal,* **140**, 215–244.

Commission of the European Communities, 1971, *Regional Development in the Community; analytical survey,* Brussels.

Commission of the European Communities, 1973, *Report on the Regional Problems in the Enlarged Community* [the *Thomson Report*], Brussels.

Despicht, N., 1971, From regional planning to regional government, in Kalk (1971, pp. 65–73).

Ezra, D. J., 1972, The conditions for an increase of investments of Community enterprises in the less developed regions of the Community, in *Towards a European Model of Development,* Venice, 213–253.

Ezra, D. J., 1973, Regional policy in the European Community *National Westminster Bank Quarterly Review,* 8–21.

Gravier, J-F., 1947, *Paris et le Désert Français,* Flammarion, Paris.

Froment, R. and F. J. Gay, 1970, *L'Europe Occidentale d'Economie Libérale,* Sirey, Paris.

Hall, P. G., 1970, *Theory and Practice of Regional Planning,* Pemberton, London.

Héraud, G., 1973, *Contre les Etats, les Régions d'Europe,* Presses d'Europe, Paris–Nice.

Kalk, E. (Ed.), 1971, *Regional Planning and Regional Government in Europe,* International Union of Local Authorities, The Hague.

McCrone, G., 1969, *Regional Policy in Britain,* Unwin, London.

Organization for Economic Cooperation and Development, 1969, *The Industrial Policies of Fourteen Member Countries,* Paris.

Organization for Economic Cooperation and Development, 1970, *The Regional Factor in Economic Development,* Paris.

Palmer, M. and coworkers, 1968, *European Unity: a survey of the European organizations,* George Allen and Unwin, London.

Rhodes, R. A., 1974, Regional policy and a 'Europe of Regions': a critical assessment, *Regional Studies,* **8**, 105–114.

Sant, M. E. C. (Ed.), 1974, *Regional Policy and Planning for Europe,* Saxon House, Farnborough.

Sérant, P., 1965, *La France des Minorités,* Laffont, Paris.

Stabenow, W., 1974, Regional policy in the E.E.C., in Sant (1974, pp. 71–86).

Stephenson, G. R., 1972, Cultural regionalism and the unitary idea in Belgium, *Geographical Review,* **62**, 501–523.

Thirlwall, A. P., 1974, Regional economic disparities and regional policy in the Common Market, *Urban Studies,* **11**, 1–12.

Verburg, M. C., 1964, Location analysis of the common frontier zones in the E.E.C., *Regional Science Association Papers,* **12**, 61–78.

Viot, P., 1971, Through regional planning towards regional administration, in Kalk (1971, pp. 83–100).

Notes

1. The term 'pointilliste' is borrowed from Dr. Edwin Brooks of the University of Liverpool.
2. Each unit is the equivalent of $1 U.S. before devaluation.

2

Population and Urban Growth

Hugh D. Clout

Introduction

By contrast with stagnation in the 1930s and devastation during the Second World War, Western Europe has been characterized by substantial economic growth and demographic vitality over the greater part of the past three decades. Urbanization has taken on new forms and has acquired a new 'mix' of functions. Historic cities have been outpaced by rapidly expanding suburbs, and urban activities have become more strongly associated with providing services than with manufacturing goods. Economic recovery has been paralleled by the rise of mass car ownership after 1950 and new dimensions of residence and recreation have been offered to growing numbers of Western Europeans. At the same time, inherited systems of settlement and communication have been subjected to serious and previously unimagined strains. Employment centres have exerted broadening control of their surrounding space, with individual towns and cities becoming bound together by roads, railways and pipelines to give rise to dynamic city regions and axes of movement and economic activity. This kind of transformation has, in itself, given rise to serious regional problems. Remote areas appear to have become even more deprived, lacking the stimulus of good communications to encourage employment growth. But such areas still retain the real or latent advantage of having relatively unspoiled environments. City regions, for all their apparent affluence, are beset by problems of congestion and pollution, not only in coal-based industrial areas that are now losing their *raison d'être*, but also in industrial and urban zones that experienced important growth earlier in the present century but are now equally plagued by spiralling costs of housing and service provision.

Population Change

Spatial variations in job opportunity and environmental attractiveness, as well as factors relating to population itself (such as age structure and natural increase or decrease), have interacted to produce a highly complicated pattern of population movement over the last quarter century. Changing political conditions have been of crucial importance in shaping population flows. For example, in 1950, refugees from the eastern provinces, from the Sudetenland and from German minority areas in Eastern Europe comprised 16 per cent of the total population of what was to become West Germany, with regional proportions rising to 21 per cent in Bayern, 27 per cent in Niedersachsen and 33 per cent in Schleswig–Holstein (George, 1972). During the 1960s 2,000,000 migrants moved from East Germany to the West.

Western Europe's population grew by 17·5 per cent from 277,500,000 in 1950 to reach 326,500,000 in 1971. Two-thirds of this increase of 49,000,000 involved only

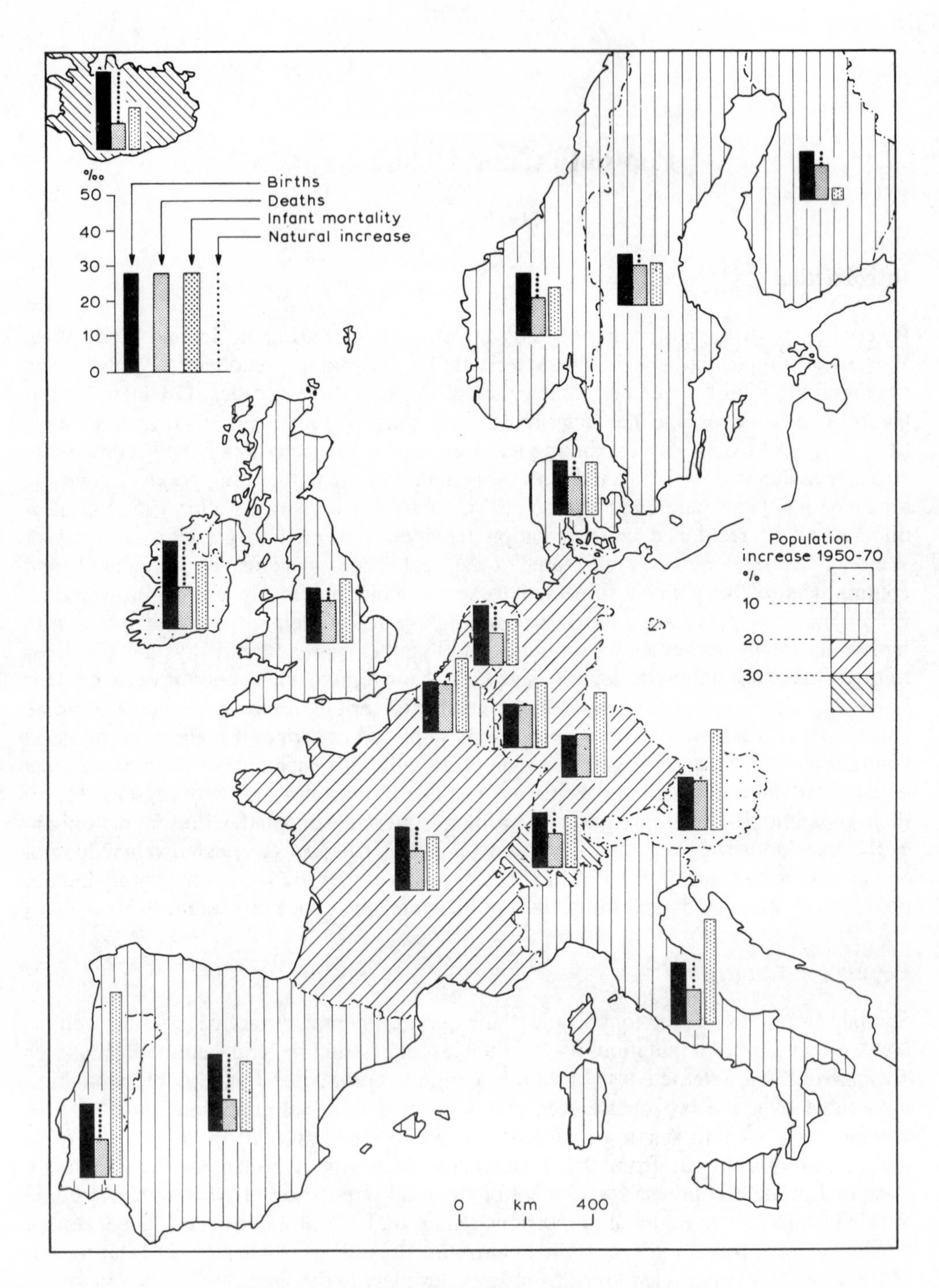

Figure 2.1. Demographic variables in Western Europe, 1972.

four countries (West Germany, France, the United Kingdom, Italy), with each experiencing population growth under somewhat different socio-economic and political conditions. The German 'economic miracle' was paralleled by an increase of 11,000,000 to give a national population of almost 62,000,000 in 1971, with the results of natural increase being inflated by the presence of refugees from the East and by labour migrants primarily from Mediterranean countries. Repatriates from North Africa and from other former French territories together with labour migrants from Mediterranean Europe and French-speaking parts of Africa contributed to the 9,000,000 increase in France, which had 51,000,000 inhabitants in 1971. Migrants from the Republic of Ireland and from Commonwealth countries helped swell the population of the United Kingdom by 5,000,000 to reach 56,000,000. By contrast, the 7,000,000 increase in Italy (54,000,000 inhabitants in 1971) was largely the result of relatively high rates of natural increase in the South, since important emigration continued to operate in the post-war years as in earlier times. The thirteen other West European countries shared the remaining 17,000,000 increase. In relative as opposed to absolute terms, Switzerland (+ 34 per cent) experienced the strongest growth, largely as a result of immigration. As we have seen, high rates of increase in West Germany (+ 24 per cent) and France (+ 22 per cent) were aided by the same process. However, rapid growth in the Netherlands (+ 29 per cent) was linked primarily to high rates of natural increase over the greater part of the period, with repatriation and immigration playing smaller roles in the overall increase.

Following the post-war baby boom and a second peak in births in the mid-1960s, when that generation reached marriageable age, rates of natural increase have slackened in varying degrees in all parts of Western Europe. This trend has developed in response to a wide and complicated range of variables which include age structure, value of family allowances, religious observance and attitudes to family size and birth control. When compared with conditions in the late 1940s, birth rates are now particularly low in West Germany (11·5 per thousand), Luxembourg (11·8 per thousand), Finland (12·7 per thousand) and Austria, Belgium and Sweden (each with 13·8 per thousand) (Figure 2.1). Rapid decline has also affected countries characterized by relatively high fertility in the past. For example, average birth rates in the Netherlands plummetted from 19·2 per thousand in 1969 to 16·1 per thousand in 1972 and from 16·6 to 14·9 per thousand in the United Kingdom.

In 1972 the outer periphery of Western Europe formed a zone of relatively high natural increase, comprising Iceland (13·6 per thousand), Spain and the Republic of Ireland (11·2 per thousand) and Portugal (9·6 per thousand). Medium rates characterized not only Italy (6·7 per thousand), with a 'dual economy' in which the affluent North contrasted with the impoverished South, but also rich industrialized countries, such as the Netherlands (7·6 per thousand), France (6·3 per thousand) and Switzerland (5·7 per thousand). Elsewhere, rates of natural increase were slight, with deaths actually exceeding births during 1972 in West Germany and Luxembourg. Earlier estimates of future population numbers have had to be revised downwards in the light of these reduced rates of natural increase. For example, current estimates for the population of the Netherlands in the year 2000 are in the order of 15–16,000,000, in contrast to a figure of 20,000,000 proposed in 1964 (Tinbergen, 1972). Nevertheless, substantial population growth will continue in Western Europe during the 1970s, with the 1980 total estimated at 347–350,000,000, some 20,000,000 greater than in 1971 and representing an increase equivalent in size to the present population of Belgium plus Portugal.

The national trends of change that have been discussed so far are undoubtedly of vital importance but much more detail is required for appreciating the population background to regional development. Maps published by the French Institut National d'Etudes Démographiques offer more precision than the evidence of the *Thomson Report* and broaden the picture to include Iberia, Norden and the Alpine states as well as the Nine. They form the framework for the paragraphs that follow.

Western Europe's economic core regions exhibited the greatest volumes of population increase during the 1960s but rates of growth in excess of 30 per cent occurred around Barcelona, Bilbao, Copenhagen and Oslo, away from the familiar axis running from southeastern England through the Rhinelands to northern Italy (Figure 2.2). The declining residential function of many large European cities was not directly in evidence at this scale; however, the modest rates of increase in inner London were certainly a result of that tendency. Absolute decline of population affected rural regions on the periphery of Western Europe from southern Italy, through large parts of Spain and Portugal, Ireland and Scotland, to northern Sweden and Finland. In addition, the Massif Central and the Po Valley stood out as 'islands' of population loss.

The demographic decline of these peripheral areas was predominantly due to strong out-migration since many of the same areas had relatively youthful age structures, high birth rates and significant volumes of natural increase (Figure 2.3). A second type of region displayed an important surplus of births over deaths. This comprised major employment centres which attracted large numbers of young workers of marriageable age. By contrast, deaths exceeded births in severely depopulated parts of the Pyrenees, Massif Central and upland Wales, where decades of out-migration had distorted the age structure of the resident population, and also in retirement areas such as Sussex and the Alpes Maritimes. In each case the local age structure was characterized by large numbers of elderly residents.

The demographic reservoirs of Western Europe's periphery were tapped by labour-hungry economic core areas with important flows of migrants operating both within and across national boundaries (Figure 2.4). For example, large numbers of temporary migrants abandoned the limited employment prospects of Spain and Portugal in order to find work in the factories, mines, transport industries and service activities of France and other affluent West European neighbours. Out-migration involved not only rural areas but also old industrial and mining zones with declining employment possibilities (such as northeastern France) and urban areas, such as Bremen and Hamburg, where the number of residents in inner cities is declining.

The diversity of man/land relationships in Western Europe as expressed by population density is outlined in Figure 2.5. This forms a fundamental working document against which national approaches to regional development should be appraised. On the outer margins of Western Europe broad expanses of central Spain, the Scottish Highlands and the greater part of Norden support fewer than 20 inhabitants per square kilometre. The economic cores correspond roughly to areas with over 200 inhabitants per square kilometre, but densities rise to more than 1000 inhabitants per square kilometre in heavily urbanized areas, such as Lancashire, London and the Ruhr.

Whilst the general pattern of urban areas may be deduced from the density map it is virtually impossible to arrive at a precise definition of the degree of urban development in all parts of Western Europe. This difficulty arises from three main problems. First,

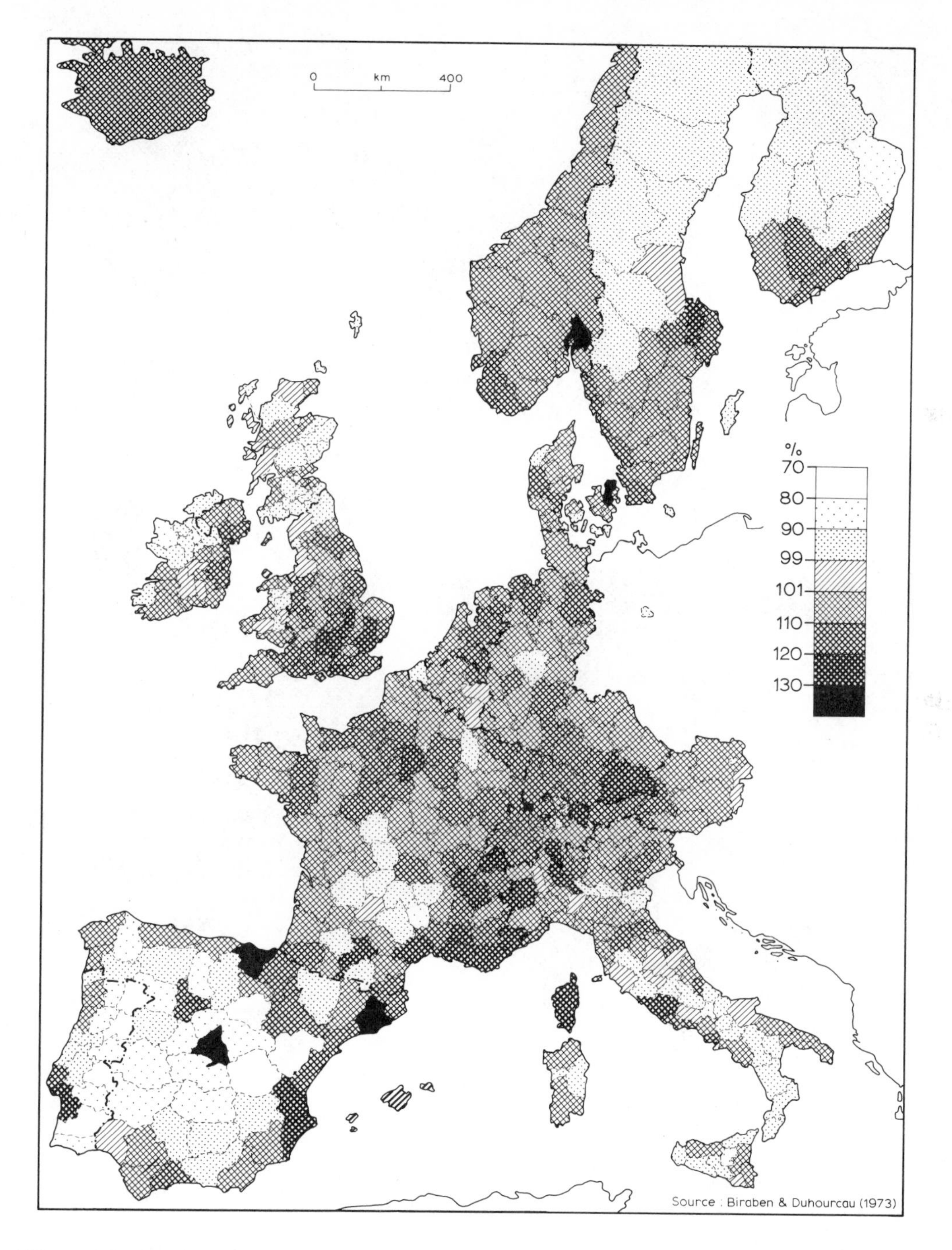

Figure 2.2. Population change in Western Europe, 1960–70. (Reproduced by permission of Institut National d'Etudes Demographiques and J. N. Biraben and F. Duhourcau.)

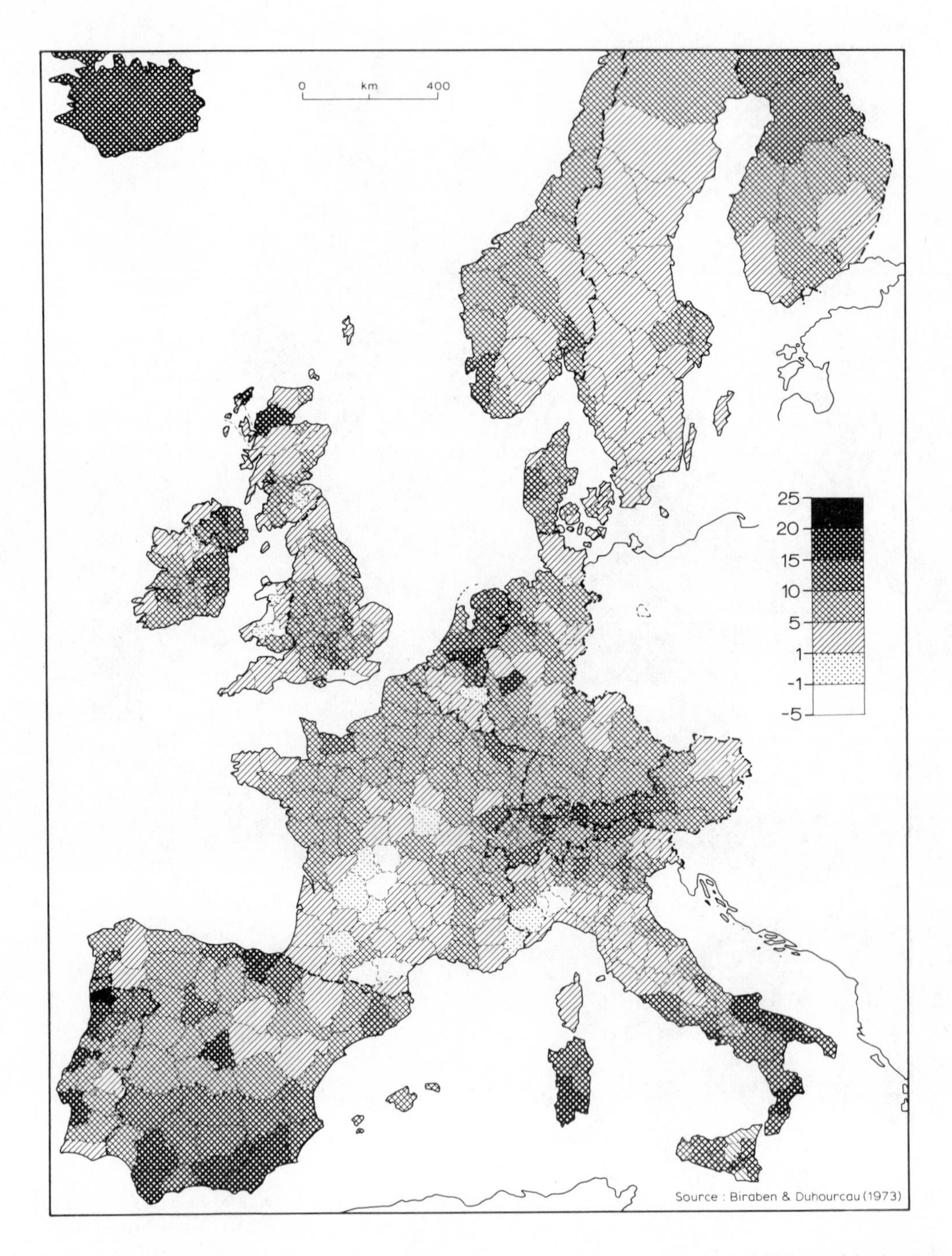

Figure 2.3. Mean natural change in Western Europe, 1960–70. Calculated as $\frac{\%\ \text{Natural increase}}{\text{P60}}$.
(Reproduced by permission of Institut National d'Etudes Démographiques and J. N. Biraben and F. Duhourcau.)

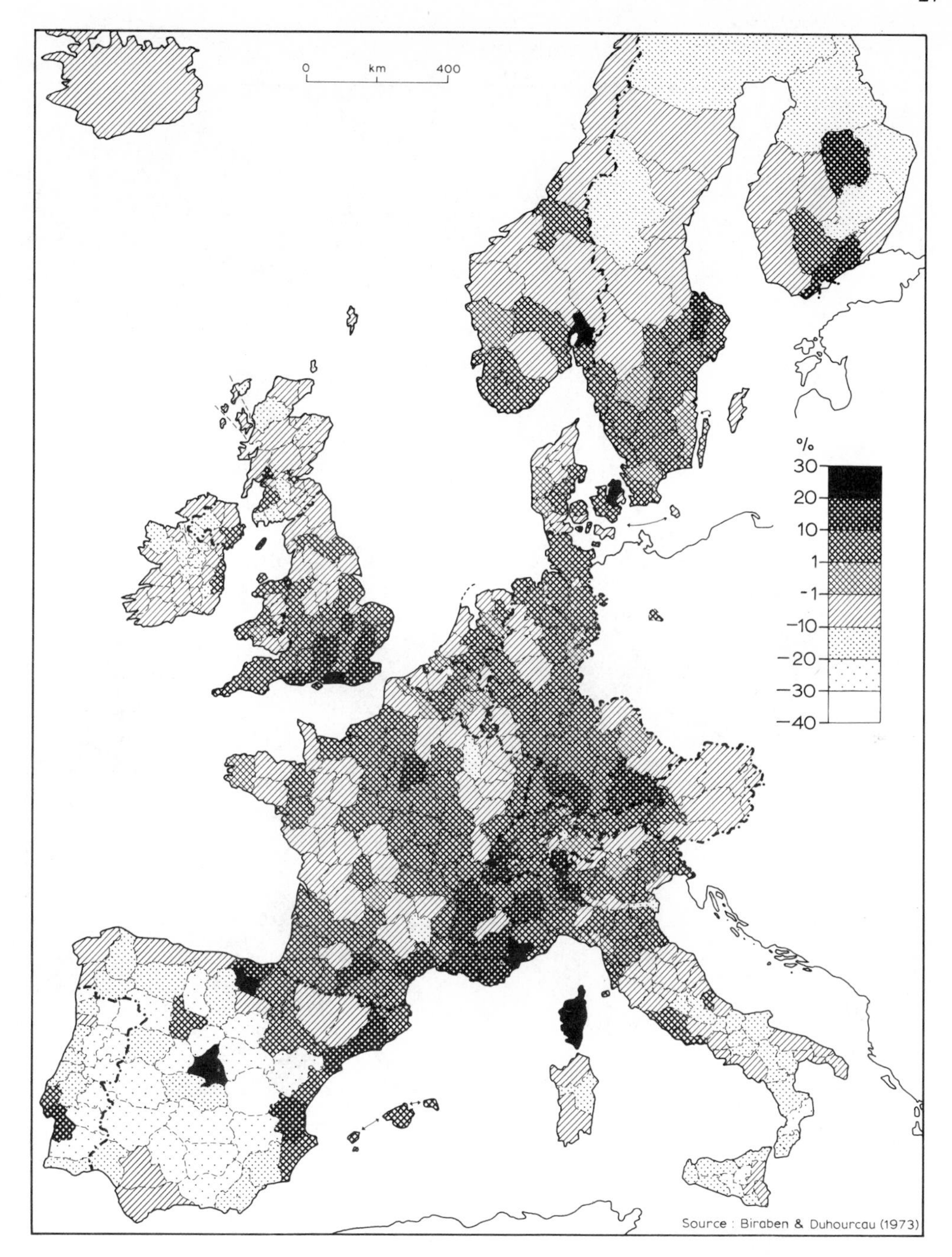

Figure 2.4. Net migratory change in Western Europe, 1960–70.

Calculated as $\frac{\text{P70} - \text{P60} - \text{Natural increase}}{\text{P60.}}$

(Reproduced by permission of Institut National d'Etudes Démographiques and J. N. Biraben and F. Duhourcau.)

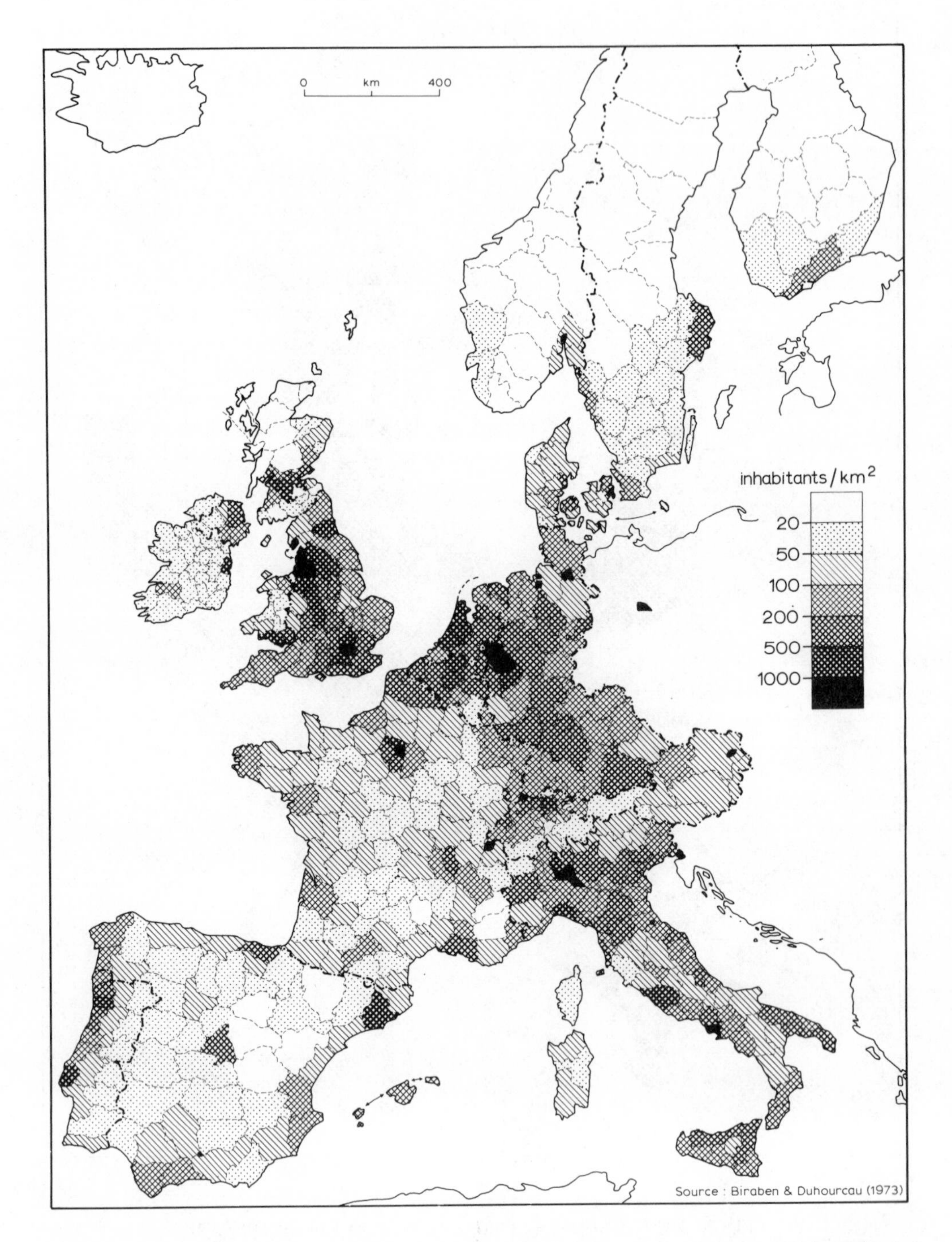

Figure 2.5. Population density in Western Europe, 1970. (Reproduced by permission of Institut National d'Etudes Demographiques and J. N. Biraben and F. Duhourcau.)

there is no standard definition of 'urban' settlement, since each country employs its own criteria. Second, basic administrative units vary in size and rarely correspond with any accuracy to built-up areas plus their surrounding suburbs. Finally, there is the conceptual problem of setting an outer limit to an urban area since commuting by public and private transport carries the 'dispersed city' far into areas that may be recognized visually as countryside. By virtue of these problems and the functional changes in central city areas it is possible for official statistics to register a decrease in 'urban' population even though outer suburban areas in districts recognized as 'rural' for administrative purposes are increasing their residents.

Bearing in mind these reservations, the officially defined 'urban' population of Western Europe appears to have increased from 158,000,000 in 1950 to 187,000,000 twenty years later (+ 18 per cent). Rather curiously, 57 per cent of the increased total population lived in 'urban' settlements at both dates. If outer suburban residents were to be included in the official statistics the proportion would undoubtedly be higher for 1970. At that time, six countries were highly urbanized (Belgium, France, the Netherlands, Sweden, the United Kingdom and West Germany), with more than two-thirds of their populations living in towns and cities (Figure 2.6). From 1950 to 1970, the number of urban agglomerations with more than 100,000 inhabitants apiece grew from 200 to 300, and the population living therein rose from 75,000,000 (27 per cent of the West European total) to 116,000,000 (35 per cent) (Figure 2.7). Cities of more than 100,000 were particularly important in three countries, housing 53 per cent of the population of the United Kingdom, 42 per cent in the Netherlands and 41 per cent in France.

In short, Western Europe's population growth over the past thirty years has primarily involved urban areas. A multitude of problems have arisen, in relation, first, to remodelling inner cities to perform new functions and cater for vastly increased flows of traffic and, second, to channelling the future growth of metropolitan regions along desired lines. It is surely curious that in defining 'development regions' relatively little attention has been paid by national governments to the problems of Western Europe's major cities.

Changing Imperatives in Urban Planning

The first task facing city planners in post-war Europe was to make good the damage to housing and urban facilities incurred as a result of hostilities. About 4,000,000 buildings had been destroyed throughout Western Europe during the Second World War, with more than 1,000,000 lost in Germany, 450,000 in France and 200,000 in the United Kingdom (George, 1960). In cities such as Dunkirk, Le Havre and Rotterdam, more than 90 per cent of the central areas had been flattened, and over half of the inner parts of Amiens, Caen, Frankfurt and Mannheim had been razed to the ground.

Reconstruction made relatively slow progress until 1950, with two differing principles being employed. In some cities, such as Caen and Le Havre, the urban pattern was substantially modified as land-ownership plots were consolidated and large new blocks with broad streets were laid out. Other cities, such as Saint Malo and many German centres, were faithfully restored to their previous appearance. Although they may look like 'urban museums' their internal functions have often been modified and vehicles have been banned in some cases (Claval, 1972). By way of example, the main shopping street of Cologne, the Hoherstrasse, had been wholly demolished but has

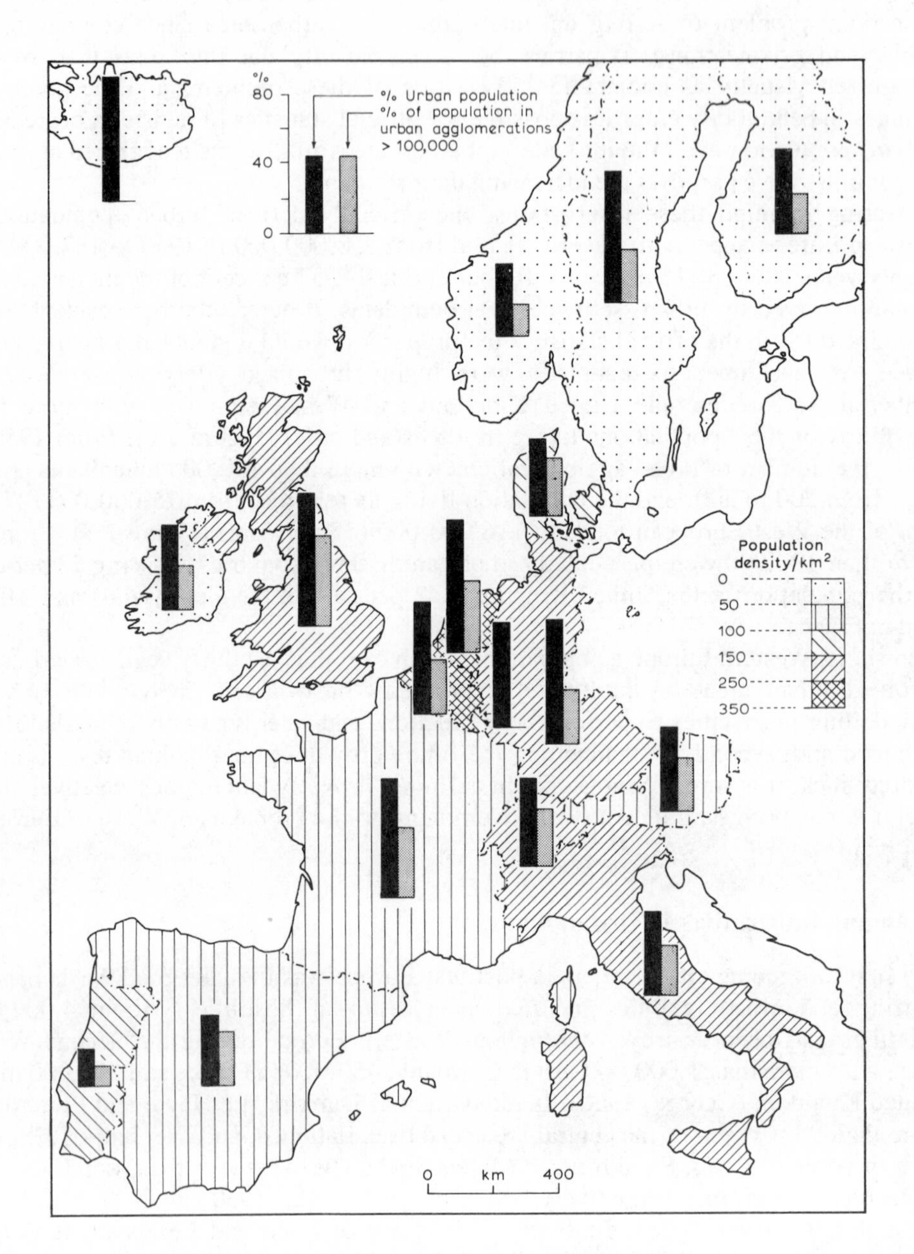

Figure 2.6. Urbanization in Western Europe, 1970.

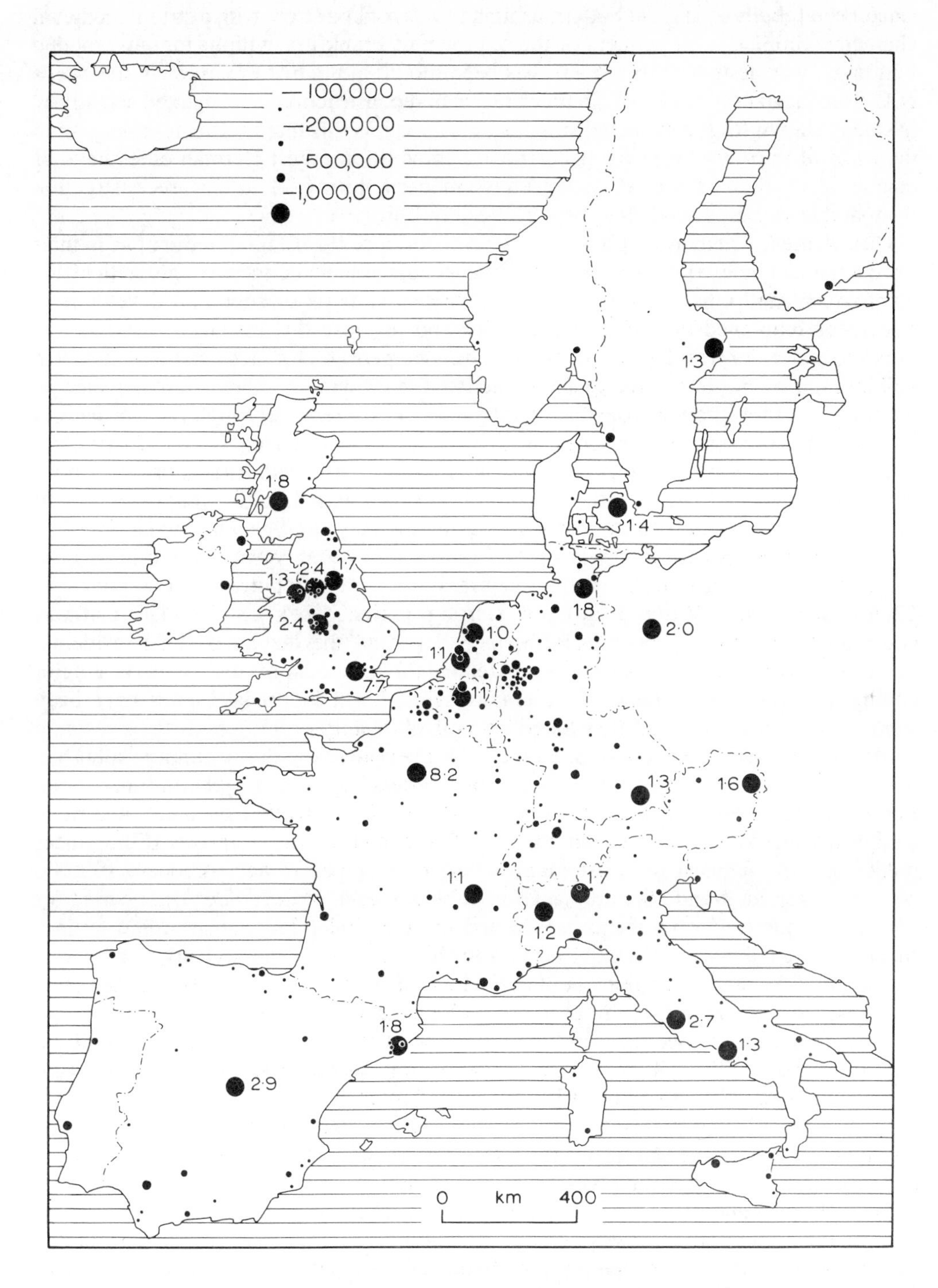

Figure 2.7. Urban agglomerations of more than 100,000 inhabitants in Western Europe in 1970 (population of 'million' cities indicated).

since been rebuilt on the old pattern as a narrow, crooked street with a strong medieval character. Similarly, 80 per cent of the old town of Frankfurt, with its famous wooden buildings, was destroyed; but it also has been rebuilt along historic lines. In the words of C. Buchanan (1963, p218), 'in much German reconstruction . . . marked sensitivity has been shown for the medieval street pattern, and it is this that is often re-emerging in the form of narrow streets for pedestrians'. In addition, West German cities have of course, had their share of completely modern forms of urban reconstruction comparable to those found elsewhere in Western Europe.

City centres, whether rebuilt or not, have experienced substantial reductions in their residential and industrial functions and have become associated increasingly with office employment and other tertiary activities. Various patterns of suburban development have been employed in war-damaged cities and in those that managed to escape, in order to cater for residential 'overspill', to compensate for a stagnation in house building during the 1930s and to accommodate the growing number of urban residents produced by in-migration from the countryside and by relatively high rates of natural increase in many outer urban areas. National traditions of housing have been strongly reflected in the morphology of post-war suburban growth, with estates of one-family houses that surround British cities standing in contrast with massive areas of apartment blocks that are characteristic of post-war housing in many Continental city regions.

The largest apartment complex (*grand ensemble*) in France was started in 1955 at Sarcelles, some 11 km to the north of central Paris, and now houses 50,000 people in 12,000 apartments. After a difficult starting period when commercial, cultural, educational and employment facilities lagged behind the demands of its residents, Sarcelles now contains 200 shops in a dozen neighbourhood units and supports a major commercial centre with local office employment. Light industrial estates have been started and it is anticipated that Sarcelles will ultimately employ half of its resident workforce and also provide jobs and services for people living in surrounding suburban areas. At present, however, the majority of workers commute to jobs in inner Paris. Large and impressive as Sarcelles undoubtedly is, it is dwarfed by other apartment complexes elsewhere in Western Europe. For example, the 1800 ha Bijlmermeer polder to the southeast of Amsterdam is being developed to accommodate 100,000 people, largely in 9-and 16-storey apartment blocks. Half of the polder will be used for housing, a quarter for recreation space and the remainder for industrial and tertiary functions (McClintock and Fox, 1971). A further example of massive urban growth is provided by the satellite settlement of Perlach, to the east of Munich, which is designed to house 80,000 residents in apartment blocks and will also play an important sub-regional role, providing jobs for 40,000 workers and commercial and cultural facilities for up to 350,000 people, to be drawn from surrounding rural areas and from inner Munich to which it will be linked by rail and motorway.

In addition to such planned forms of suburban or satellite growth that have been grafted on to most large European cities, two almost spontaneous forms of urbanization have occurred. The first involves shantytowns that have mushroomed around many cities in France and elsewhere on the Continent. They provide rudimentary housing mainly for immigrant workers, many of whom are employed (ironically) in the construction industry. The shanties are ghettos of deprivation. Official policies aim at rehousing their residents in better conditions but many shanties still remain. By contrast with the poverty of the shantytowns, access to public, and increasingly to private, transport has encouraged the growth of commuter villages and expanded towns beyond continuously built-up suburbia throughout Western Europe. For example, rising output of compact cars and motor-scooters has made the north Italian one of the most mobile of West European workers, so that the daily urban system of Milan now covers a much broader area of the western Po Basin than twenty years ago (Turton, 1970). Similarly in Belgium and the Netherlands, motor-assisted bicycles have been instrumental in broadening daily commuting hinterlands (Dickinson, 1957). In some cases, ex-farmers continue to live in their farmhouses and work their plots on a part-time basis while holding down a full-time urban job.

In most West European countries, positive attempts have been made to steer new employment, housing and cultural facilities away from the major economic core areas of the inter-war years to provincial urban centres. Recent rapid growth in many such city regions must be seen as the outworking of these policies. In West Germany, capital-city functions that had formerly been concentrated in Berlin were not transferred simply to Nordrhein–Westfalen, which is the largest concentration of population. Instead, Bonn became the seat of national government, but other vital national organizations were dispersed and rehoused in Frankfurt, Karlsruhe, Munich, Nuremberg and elsewhere (Gravier, 1972). Commercial functions of national significance moved to a number of cities, with, for example, the newspaper industry flourishing in Frankfurt and Hamburg, the advertising industry developing rapidly in the same two cities plus Düsseldorf, and branches of the fashion clothing trade becoming established in Munich and Düsseldorf (Merritt, 1973). Munich is now the leading example of a post-war provincial success story. Its population had dropped from 800,000 in 1938 to 400,000 in 1944 but thereafter rose to 1,300,000 in 1970. The Siemens electrical firm which moved from Berlin in 1948 has been instrumental in stimulating job opportunities and now employs more than 40,000 of the inhabitants of the *Stadt der Lebensfreunde* (Krümme, 1970; Pourtier, 1967). Its role is bolstered by other 'migrant' firms that have transferred from Berlin and from other cities that are now in East Germany. In any case, provincial cities in West Germany have a strong legacy of administrative, educational and industrial functions which has been built upon in recent years. By contrast, provincial cities in highly centralized countries such as France have been less well-endowed. Decentralization under such circumstances has proved a formidable challenge to regional planners.

One particularly distinctive feature in the urban growth of Western Europe since 1945 has been the establishment of 'new towns'. These have been designed to meet various objectives and some are completely new creations whilst others represent substantial grafts of new housing and services on to existing urban cores. Likewise, their degree of self-sufficiency varies considerably, with some new towns forming parts of the daily urban systems of metropolitan areas almost like outer suburban estates. Legislation in 1946 heralded their designation in the United Kingdom with the main objective being to decant population from congested (and often war-torn) cities. By 1950, fourteen new towns had been founded, including eight in the South-East, two apiece in northeastern England and the Scottish lowlands, and one near Cardiff, together with Corby which had been upgraded to 'new town' status. These new settlements have been built on the basis of rigorous land-use zoning. They vary greatly in size and pattern, but neighbourhood units of 5000–10,000 residents apiece are familiar components. Now some 2,000,000 people live in over 30 settlements with 'new town' status, which provide over 700,000 jobs. Roughly half of their total population comprised families that had moved in following designation. The bulk of the new housing was constructed by Development Corporations, with a much smaller quantity being built by local authorities and other agencies. Unlike their Continental counterparts, most dwellings in British new towns are single-family houses.

Until the acceptance of the new town (or, perhaps more appropriately, new city) formula in France in the mid-1960s, there were few new settlements in mainland Western Europe that were planned to function as complete towns with official new town status. However, Emmeloord (13,000 inhabitants in 1971) and Lelystad (15,000 in 1974) on the polderlands of the former Zuyder Zee are notable examples. Lelystad is planned to reach 100,000 inhabitants by 2000 A.D. More generally, isolated housing estates were constructed to fulfil particular housing needs at newly-developed mining or industrial sites such as Mourenx in the gasfield of southwestern France, the iron mining centres of Luossavaara–Kiiruna in northern Sweden and Salzgitter in the German Harz mountains, and the coal mining settlement of Marl, north of the long-exploited section of the Ruhr. A number of satellite estates in Northern Europe, such as Tapiola and Vällingby, are sometimes referred to as 'new towns' but they are intimately bound to their central cities for employment, with 80 per cent of Tapiola's workforce having jobs in Helsinki and 50 per cent of those living in Vällingby commuting to jobs in Stockholm (Waller, 1963). In these and many other satellite settlements in Scandinavia and Randstad Holland, more than 80 per cent of the total dwellings are provided in apartment blocks.

Managing Metropolitan Regions

Above the level of individual towns and cities, schemes for managing whole metropolitan regions have brought a new dimension to urban planning in Western Europe (Hall, 1967). Management on this scale has proved necessary because of the changing spatial organization of modern cities. Many are experiencing a reduction in number of residents, as a result of urban renovation programmes and increasing emphasis on tertiary functions. For example, thirteen of the twenty cities in West Germany with more than 250,000 inhabitants recorded a decline in residential population during the 1960s (Gravier, 1972). In inner Amsterdam the number of residents has declined as a result of out-migration and natural decrease (Gastelaars and Cortie, 1973) and a similar story could be told for many other cities. The traditional urban tissue of historic centres in Western Europe is undoubtedly breaking up; and this process is being accelerated by urban renewal schemes, as in Stockholm where a plan was approved in 1962 to demolish a considerable part of the business section and devote a quarter of the cleared area to new streets, a quarter to multi-level parking garages, and the remaining half to new commercial buildings (Sidenbladh, 1965).

This kind of operation is being complemented by important growth at the edge of present built-up areas and at greater distances along communications axes. Such vigorous outward expansion is in response to a range of complementary factors. Multi-generation households are much rarer than in the past, since married children prefer to live away from their parents. Migrants to metropolitan areas are often relatively young with young families and therefore need housing space that may not be available close to the city centre. Accommodation in suburban areas may be cheaper away from inner city land prices. Single-family houses, as opposed to apartment blocks, are usually more readily available in outer suburbia. Finally, rising standards of living and increasing rates of car ownership permit greater flexibility in choosing where one wants to live, allowing a broad area of desirable commuter settlements to fit in the daily urban system. Metropolitan regional plans are therefore absolutely essential in order to direct the future pattern of outer residential growth and ensure an adequate provision of public transport, local service centres and employment opportunities.

The plans that have been produced for metropolitan areas vary considerably in content and emphasis. This is in response to five main groups of reasons: varying site conditions, patterns of past urban development, transport facilities, impact of legislation for urban planning, and land ownership. Site conditions impose a number of immediate constraints on metropolitan regional plans. Schemes for Copenhagen, set at a 120° angle on Zeeland island, or for Stockholm, with its islands and inlets, would necessarily differ in pattern from those for London or for Paris, which are both set in structural basins and bisected by major waterways. Monocentric metropolitan areas such as these have, in turn, produced patterns of outward growth that are quite different from those found in the polycentric urban regions such as the Dutch Randstad or the Ruhr. Differences in transport provision affect the degree of flexibility or innovation that may be included in plans for metropolitan regions. For example, both Paris and Stockholm have new railway systems under construction which permit completely new directions for urban building, but this is not necessarily so easy in other major cities which lack new transport links. Past legislation for city development may impose important constraints on future expansion, as in the form of London's Green Belt incorporated in the Greater London Plan (1944). Recent residential developments in the form of new towns and commuter settlements have leapfrogged that belt of protected land. By contrast, an alternation of urban areas (plus communications axes) and wedges of open space has been preferred in planning the Ruhr and the Randstad. Variations in land ownership offer differing possibilities for future urbanization. The best example of foresight in land acquisition is provided by Stockholm where, in the early twentieth century, the city council purchased extensive tracts of farmland and forest beyond the city limits to accommodate future expansion (Davies, 1973).

Most metropolitan plans have favoured developing urban sectors along lines of communication that lead from central cities (Hall, 1967; 1974b). Thus future urban growth in Randstad Holland is being steered away from the remaining 'greenheart' of the metropolis along axes leading south to the Deltalands, north into Noord-Holland and east into the drained polders. The identity of individual cities is being preserved as much as possible in this polycentric metropolitan region, so that by the end of the century it will be possible to identify two centres of more than 1,000,000 inhabitants apiece (Amsterdam and Rotterdam), three cities with over 500,000 and five with 250,000–500,000 residents.

Around Copenhagen two 'fingers' of urban growth are being developed to the west and southwest of the city. Each of these is planned to contain 100,000 dwellings (for about 250,000 residents) and will be equipped with a wide range of services and employment in order to reduce reliance on central Copenhagen (Diem, 1973). Linear growth has been approved along railway lines leading from central Stockholm, with particular attention being paid to conserving lake shores and water surfaces for recreation and to zoning residential districts, with densities decreasing with distance from stations along the main railway lines (Merlin, 1971). In a rather different way, two axes of growth running roughly NW–SE on the north and south banks of the river Seine are accommodating new cities, a new international airport and other major urban facilities for greater Paris. By contrast, the continued existence of London's Green Belt prohibits massive development along radial growth axes for the immediate future at least.

Each of these metropolitan plans has required modification, if not significant readjustment, in the light of changing social and demographic circumstances. In the past, it was usually a matter of raising target populations because of high rates of natural increase and inmigration. Now declining rates of natural increase may provoke readjustments in the opposite direction. By way of example, the Finger Plan for Copenhagen was drawn up in 1948 and had envisaged the city's population growing from 1,100,000 to 1,500,000 by the end of the century. However, that figure was already reached by 1960 and had to be revised upwards to *c.* 2,500,000 for 2000 A.D. Likewise, the 1,500,000 target population for greater Stockholm in 1990 was already reached in 1970 and was revised to 2,200,000 for the year 2000 (Stäck, 1971). Detailed components of Stockholm's city plan have had to be adjusted in response to rapid immigration. Originally, each satellite neighbourhood had been planned for *c.* 10,000 residents apiece. This level proved to be too small to support an adequate range of shopping facilities and cultural activities and hence neighbourhoods, such as Vallingby to the west and Farsta to the south, have been planned to receive up to 25,000 inhabitants (Sidenbladh, 1965). The P.A.D.O.G. (1960) plan, which envisaged restraining the future growth of Paris, was superseded in 1965 by a far more expansionist programme. But, in the last few years, reduced rates of natural increase in many countries of Western Europe have called into question many of these metropolitan population targets. In Amsterdam, doubts are being expressed about the desirability of and the underlying need for massive urban expansion to the north of the North Sea Canal. Instead, pleas have been made for renovating existing housing stock in inner parts of the city.

The Transport Revolution

Each of these metropolitan schemes has recognized, admittedly in varying degree, the impact of rising car ownership on daily travel patterns and also the need to provide adequate public transport between city centres and satellite communities. Rapidly rising rates of car ownership have led to something of a 'transport revolution' in Western Europe which made its real impact after 1950 (two decades later than in the U.S.A.). In 1972 there were 69,000,000 private cars on the roads of Western Europe, of which 59,000,000 were registered in the Nine. The total had doubled since 1960. However, there were significant national differences in rates of ownership. For Western Europe as a whole there were 210 cars for every thousand inhabitants, but this figure was greatly exceeded in the wealthier nations. In Sweden there were 219/1000 (compared with 42/1000 in 1950), France 256 (39), West Germany 253 (13) and Switzerland 236 (32). By contrast, the Republic of Ireland had only 139 (31), Spain 84 (2) and Portugal 54 (7). A further 8,500,000 commercial vehicles and 22,000,000 motorcycles and mopeds were in use. As a result of these increases the number of passenger kilometres travelled in private cars in the Nine doubled between 1963 and 1970 to reach a million million passenger kilometres.

Such a trend brought the well-known advantages of flexibility and convenience in travel and permitted new features in urbanized society, such as residence in car-based commuter suburbs, driving for leisure by a large proportion of the population, and shopping in out-of-town commercial centres and hypermarkets, rather than in city centres. The Main–Taunus–Zentrum 12 km from central Frankfurt provides a fine example of this kind of phenomenon. The site commands good access by motorways and main roads, having a potential clientele of 1,500,000 people less than 30 minutes' driving time away. Free parking is provided for 5000 cars and bus routes serve thirty destinations (Brunet and Pinchemel, 1972). More than 500 out-of-town shopping centres are in existence in West Germany and they are numerous in many other countries of mainland Western Europe (Smith, 1972–3).

However, the increasing orientation of modern life to car-based mobility has brought serious problems of congestion, pollution and consumption of land to the city regions of Western Europe. In addition, 60,000 people are killed and 1,500,000 injured on the roads of the Nine every year. Many cities have restricted car access to their innermost shopping areas in order to reduce congestion and pollution. In West Germany alone, 28 cities introduced car-free zones between 1967 and 1972 and such examples have been emulated in historic cities in many other parts of Western Europe that are keen to preserve their heritage. General evidence points to an increase in retail trade when shopping streets become pedestrian areas. For such schemes to operate successfully, adequate parking space is vital on the margins of inner areas and thus multi-storey and underground car parks are familiar features of the modern urban pattern. Some cities, such as Bremen and Gothenburg, have introduced schemes to discourage non-essential traffic from using central areas, for example by dividing them into quadrants which allow local traffic to circulate but prevent through-traffic flows. Another negative effect of increased car ownership has been the reduction in demand for, and ultimately in provision of, public transportation, especially buses, because of diminishing financial returns. In such a situation the car-less become even more seriously deprived. In addition, urban motoring is a highly inefficient and expensive use of energy resources. Between 1961 and 1971 the share of expenditure in the Six on transport and communications increased from 8·5 per cent of private consumption to 11·0 per cent and has now advanced steeply following the rise in oil prices.

Whilst an increasing volume of road traffic brings chaos and consternation in historic city centres, construction of motorways and improvement of main road links add a new degree of ease and flexibility for the movement of goods and passengers. The various countries of Western Europe have approached the task of road building with varying degrees of enthusiasm. High densities of motorway now serve the Rhineland economic core in West Germany and the Netherlands and important construction has been completed in Italy (Figure 2.8). Most other countries are still quite poorly served. West Germany inherited a considerable legacy from the inter-war years with over 2000 km in existence in 1940. By 1970 this had been extended to 4500 km and a total of 13,000 km is planned for 1985. Motorway construction has been far more recent in Italy where the national total increased from 480 km to 4000 km during the 1960s and by the late 1970s is scheduled to reach 6700 km.

Impressive though these achievements have been, the motorways were built (like their predecessors, the railways and main roads) primarily as systems of national rather than international communication. They radiated out from national capitals, major ports and economic core areas but often failed to reach international boundaries until very recent years. Attempts had been made as early as 1950 to establish a system of Europe-wide highways; but governments were slow to comply. Fifteen years later the E.E.C. recommended that attention should be devoted to filling the gaps in the motorway system to facilitate access between, as well as within, member states of the

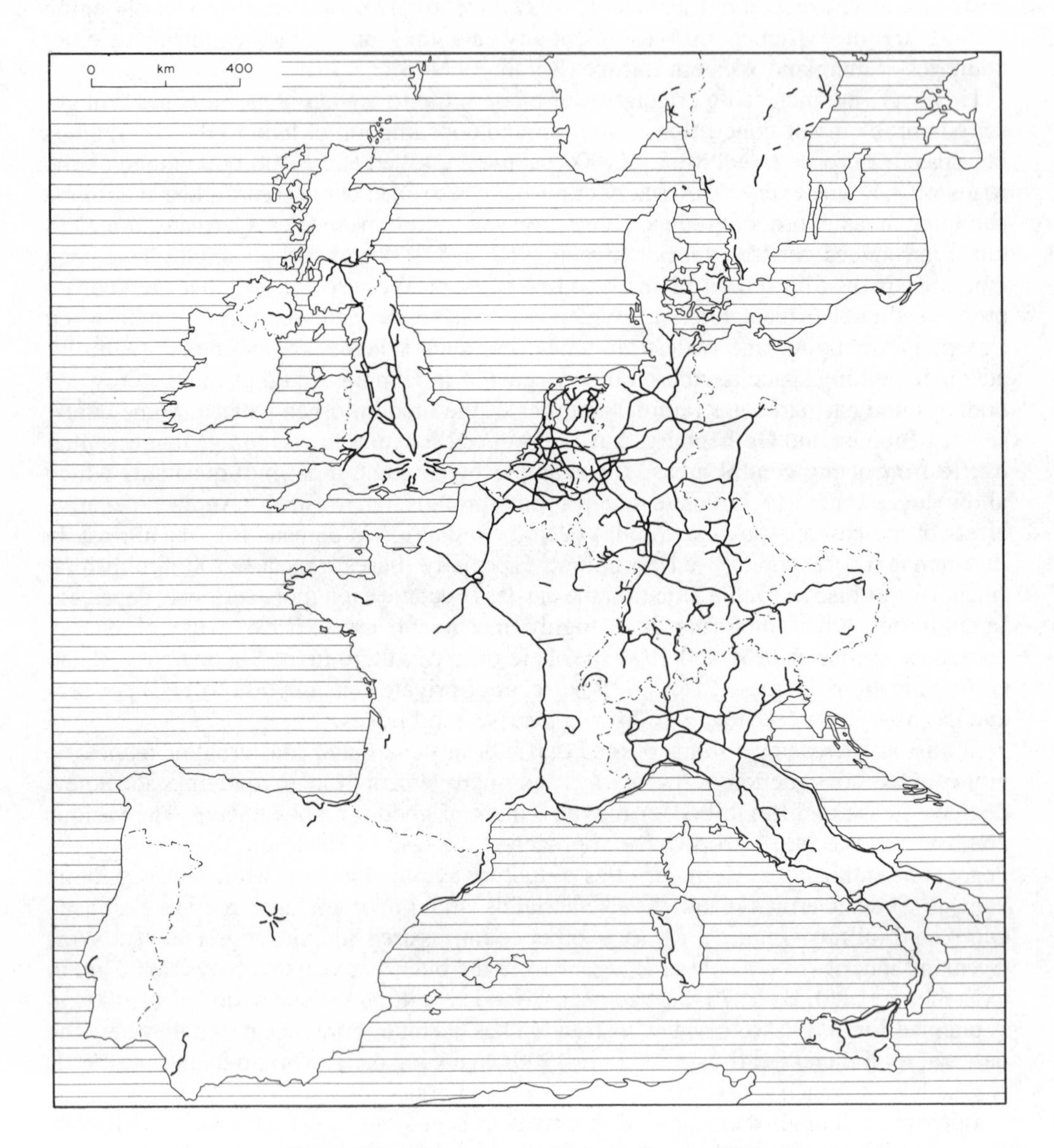

Figure 2.8. Motorways in Western Europe, 1974.

Six. The absence of links between France and Belgium, between France and Italy (along the Mediterranean coast and through the Alps), and between Italy and West Germany (via the Brenner Pass) posed particularly serious problems. Loans from the E.I.B. have been instrumental in allowing progress to be made on many of these cross-border links, including the Brenner and Val d'Aosta motorways, parts of the axis between Paris and Brussels, the Franco-Italian link on the Mediterranean, and the motorway between Metz and the Saar.

In recent years added attention has also been paid to the need to improve inter-city rail links because of the numerous environmental problems that arise from projects for building inter-urban motorways and new airports in proximity to centres of population, and also because trains are more efficient consumers of energy than either cars or planes for inter-city travel on the West European scale. Significant progress has been made already on many lines and most fast inter-city links now operate at 90–120 km/h. Key sections of the German inter-city system, inaugurated in 1971, run at 120–140 km/h with twelve lines linking 73 cities. Higher speeds are envisaged on lines from London to Edinburgh and Glasgow, between Hamburg and Dortmund, Paris and Bordeaux, and on much of the Italian network. Speeds of 200 km/h are a practical limit on existing systems but new lines to take speeds of 250 km/h are contemplated along heavily-used axes from the Channel coast to Paris and thence to Lyons, plus several inter-urban sections in Belgium, West Germany and Italy.

Further integration of Western Europe's railway networks demands extensive modernization along saturated sections. These include the link between Barcelona and Narbonne, and the trans-Alpine routes between Turin and Chambéry, between Munich and Verona (Brenner Pass) and between Basle and Milan (using the Saint Gotthard and Simplon links). The Channel Tunnel scheme has been abandoned, but key elements for the improvement of international transport links in Western Europe include duplicates for the Brenner and Saint Gotthard tunnels, a tunnel between Denmark and Sweden, and a bridge across the Straits of Messina linking Sicily to the Italian mainland.

Further improvements in transport are of vital importance in an age of high mobility in urbanized Western Europe. Completion of missing links will assist the flow of goods and people between member states in the Nine and will strengthen European economic integration. Some manufacturing activities are still orientated towards mineral reserves or markets but, as Clark and coworkers (1969, p. 197) have stressed, 'the majority of industries can now be described as footloose'. In other words, there is a theoretically wide range of sites where a firm might reasonably locate so far as transport of materials or of products is concerned. Nonetheless, it is the ever-widening metropolitan regions of Western Europe (with high population densities, large and diversified labour forces, specialized services, and good access to information facilities and government departments) that are retaining their grasp on growth industries.

The role of rail links in stimulating and servicing industrial sites is now being complemented by motorways which are increasingly becoming the industrial boulevards of Western Europe and are incorporating new areas into her broadening metropolitan zones. More than six hundred new factories were established between 1968 and 1973 at sites along the Italian Austostrada del Sole and this example may be replicated in many other parts of Western Europe (Pacione, 1974). Access to metropolitan areas through good surface communications is clearly a vital factor affecting opportunities for old industrial and mining regions to revitalize their economies and for predominantly rural regions to attract the industrial enterprises of which they are in dire need in order to diversify their employment base and raise living standards. Nevertheless, increased reliance on road transport is not without its dangers as recent increases in oil costs for Western European consumers have all too painfully shown.

References

Biraben, J-N. and F. Duhourcau, 1973, La rédistribution géographique de la population de l'Europe occidentale de 1961 à 1971, *Population,* **28**, 1158–1169.

Bouchez, G. and A. Dubau, 1968, Etude pour un habitat non ségrégatif, *Urbanisme,* **106**, 54–57.

Boudeville, J. R., 1974, European integration, urban regions and medium-sized towns, in Sant (1974, pp. 129–156).

Brunet, P. and P. Pinchemel, 1972, Grandes opérations d'urbanisme en République Fédérale allemande, *Annales de Géographie,* **81**, 555–578.

Buchanan, C., 1963, *Traffic in Towns,* Penguin, Harmondsworth.

Buchanan, R. H., 1969, Towards Netherlands 2000: the Dutch national plan, *Economic Geography,* **45**, 258–274.

Clark, C., F. Wilson and J. Bradley, 1969, Industrial location and economic potential in Western Europe, *Regional Studies,* **3**, 197–212.

Claval, P., 1972, La grande ville allemande, *Annales de Géographie,* **81**, 538–554.

Clerc, P., 1967, Grands ensembles: banlieues nouvelles, *Cahiers de l'Institut National d'Etudes Démographiques,* **49**.

Davies, M. L., 1973, The role of land policy in the expansion of a city: the case of Stockholm, *Tijdschrift voor Economische en Sociale Geografie,* **64**, 245–250.

Dickinson, R. E., 1957, The geography of commuting: the Netherlands and Belgium, *Geographical Review,* **47**, 521–538.

Dickinson, R. E., 1967, *The City Region in Western Europe,* Routledge and Kegan Paul, London.

Dieleman, F. M. and R. B. Jobse, 1974, An economic spatial structure of Amsterdam, *Tijdschrift voor Economische en Sociale Geografie,* **65**, 351–367.

Diem, A., 1973, The growth and planning of Copenhagen, *Revue de Géographie de Montréal,* **27**, 41–51.

Gastelaars, R. V. E. and C. Cortie, 1973, Migration from Amsterdam, *Tijdschrift voor Economische en Sociale Geografie,* **64**, 206–217.

George, P., 1960, Problèmes géographiques de la reconstruction et de l'aménagement des villes en Europe occidentale depuis 1945, *Annales de Géographie,* **69**, 1–14.

George, P., 1972, Questions de géographie de la population en R.F.A., *Annales de Géographie,* **81**, 525–537.

Goddard, J. B., 1974, The national system of cities as a framework for urban and regional policy, in Sant (1974, pp. 101–127).

Gravier, J. F., 1972, *Paris et le Désert Français en 1972,* Flammarion, Paris.

Hall, P. G., 1966, *World Cities,* Weidenfeld and Nicolson, London.

Hall, P. G., 1967, Planning for urban growth: metropolitan area plans, *Regional Studies,* **1**, 101–134.

Hall, P. G., 1974a, The containment of urban England, *Geographical Journal,* **140**, 386–418.

Hall, P. G., 1974b, *Urban and Regional Planning,* Penguin, Harmondsworth.

Hellen, J. A., 1974, *North-Rhine Westphalia,* Oxford University Press.

Johnson, J. H., (Ed.), 1974, *Suburban Growth: geographical processes at the edge of the Western city,* Wiley, London.

Kosinski, L., 1970, *The Population of Europe,* Longman, London.

Krümme, G., 1970, The inter-regional corporation and the region, *Tijdschrift voor Economische en Sociale Geografie,* **61**, 318–333.

Labasse, J., 1972, L'aéroport et la géographie volontaire des villes, *Annales de Géographie,* **81**, 278–297.

Lawrence, G. R. P., 1973, *Randstad, Holland,* Oxford University Press.

Lewan, N., 1969, Hidden urbanization in Sweden, *Tijdschrift voor Economische en Sociale Geografie,* **60**, 193–197.

Lewin, J. and J., 1970, A specimen of the timber industry and town growth in Finland, *Geography,* **55**, 129–145.

McDonald, J. R., 1969, Labor immigration in France, 1946–65, *Annals of the Association of American Geographers,* **59**, 116–134.

McClintock, H. and M. Fox, 1971, The Bijlmermeer development and the expansion of Amsterdam, *Journal of the Royal Town Planning Institute,* **57**, 313–316.

Merlin, P., 1971, *New Towns,* Methuen, London.

Merritt, R. L., 1973, Infrastructural changes in Berlin, *Annals of the Association of American Geographers,* **63**, 58–70.

Pacione, M., 1974, Italian motorways, *Geography,* **59**, 35–41.

Pourtier, R., 1967, Munich: croissance démographique et développement industriel, *Annales de Géographie,* **76**, 129–151.

Riquet, P., 1972, Conversion industrielle et réutilisation de l'espace dans la Ruhr, *Annales de Géographie,* **81**, 594–621.

Salt, J., 1973, Job finding in a united Europe, *Geographical Magazine,* **45**, 768–770.

Sant, M. E. C. (Ed.), 1974, *Regional Policy and Planning for Europe,* Saxon House, Farnborough.

Sidenbladh, G., 1965, Stockholm: a planned city, *Scientific American,* **213**, 106–121.

Smith, B. A., 1972–3, Retail planning in France, *Town Planning Review,* **44**, 279–306.

Stäck, J., 1971, Le plan régional du grand Stockholm, *Urbanisme,* **126**, 18–23.

Steigenga, W., 1972, Randstad Holland: concept in evolution, *Tijdschrift voor Economische en Sociale Geografie,* **63**, 149–161.

Thomas, W. S. G., 1974, Gastarbeiter in Western Germany, *Geography,* **59**, 348–350.

Thompson, I. B., 1973, *The Paris Basin,* Oxford University Press.

Tinbergen, J., 1972, Population, in R. Mayne (Ed.),*Europe Tomorrow*, Fontana, London, 64–84.

Turton, B. J., 1970, The western Po Basin in Italy: a study in industrial expansion and the journey to work, *Town Planning Review,* **41**, 357–371.

Waller, E., 1963, Vällingby et Farsta: essais de solution aux problèmes urbains de Stockholm, *Revue de Géographie de Lyon,* **38**, 33–46.

3

Energy and Regional Problems

Hugh D. Clout

Phases in Energy Supply

Clark's (1969) analysis of industrial location in Western Europe highlights one group of processes, subsumed under the heading of transport costs, that have operated in the recent past to reduce the earlier correlation between manufacturing sites and mining towns or major markets. 'A general description of what is happening in the modern industrial world is that the macro-location of industry and population is tending towards an ever-increasing concentration in a limited number of areas[1]; their micro-location, on the other hand, towards an ever-increasing diffusion or ''sprawl'' ' (Clark and coworkers, 1969, p. 197). However other important processes are at work which include a wide range of policy measures for decentralizing industry away from flourishing economic cores and for stimulating manufacturing activities in less-developed, predominantly rural regions. Such measures vary considerably in detail and are best examined in their appropriate national contexts in later chapters. In spite of these various trends for weakening resource-based manufacturing locations, changes in energy supply and consumption should not be underestimated as factors contributing to regional problems and potentials in Western Europe.

The post-war period was heralded by a reconstruction phase, when the disrupted coal industry received massive investments to regain and then surpass pre-war production levels. Coal was vital for industrial revival and general economic recovery. Then a long and complicated period followed in which coal output continued to rise at first but was cut back later as cheaper energy supplies became available in the form of imports of crude oil and coal (from the U.S.A.). At the same time, European energy sources, such as hydro-electricity and natural gas, were exploited more intensively. From simply being experiments, nuclear power and tidal energy became practical realities. In Jensen's (1967, p. 192) words, 'we moved from a situation in which the economy of almost the whole of Europe depended on one single source of energy—coal, to what can fairly be described as a multi-fuel economy'. Coal mines were closed by the hundred and miners made redundant by the thousand in the name of rationalization, as governments indulged in what Gordon (1970) has called 'the reluctant retreat from coal'. Western Europe became increasingly dependent on overseas supplies of cheap oil. Her economies and the living standards of her people thus became vulnerable to increased costs of imported energy. The latest phase of energy supply has seen the era of cheap oil come to an end with the 'energy crisis' being unleashed in 1973. Nevertheless, North Sea natural gas and oil offer a European energy bonanza for the future—but not the immediate future! Coal resources in old mining areas are being reappraised in the light of altered economic circumstances, and important new deposits have been discovered. Rising energy costs have implications that strike right at the roots of Western Europe's economy and affect the future prosperity of all types of region—developed and less-developed, rural and urban. Such broad speculations are inappropriate in the present discussion which will concentrate on the changing production and consumption of energy and their immediate regional impact in Western Europe.

The Reluctant Retreat from Coal

On the eve of the Second World War coal had supplied more than 90 per cent of Western Europe's primary energy requirements. Wartime devastation was accompanied by disruption of coal supplies from Poland, Germany and the United Kingdom. Oil imports declined and many refineries were destroyed, with the crude oil consumption of France, Italy, the Netherlands and West Germany declining from 10,560,000 tons in 1938 to 3,790,000 tons in 1946. Recovery of output from the coal industry was surprisingly rapid, with pre-war levels being achieved in many European countries by mid-1947. Such success was linked to government intervention in the industry, such as nationalizing mining activities, in France and the United Kingdom for example, elevating coal to a prime position in national economic priorities and investment schemes, and establishing a supranational organization (the E.C.S.C.) for coordinating coal production.

The E.C.S.C. was the brain child of the French Foreign Minister, M. Robert Schuman, who proposed in May 1950 that the coal and steel industries of France and Germany should be pooled, and offered invitations for the United Kingdom and other countries to join. Hard bargaining ensued and in March 1951 the leaders of France, Germany, Italy, Belgium, Netherlands and Luxembourg agreed to set up a common market for coal and steel. This was a deliberate effort to increase production and also to make the major internal source of energy in Western Europe less expensive. In fact, consumer costs continued to vary between countries, being lowest in Germany, with the Netherlands, France and Belgium coming in ascending order. A complex system of price adjustment was brought into operation with, for example, high cost Belgian mines being assisted from a perequation fund that was financed by a levy on low cost German and Dutch producers. Rationalization was coupled with mechanization which reduced labour demands. Nevertheless, hard coal output from West Germany, France, Belgium and the Netherlands (plus Spain and the United Kingdom outside the E.C.S.C) rose from 336,000,000 tons in 1946 to 467,000,000 tons in 1955 when these countries were extracting 29 per cent of world output. In addition, output of lignite from West Germany, Austria, France and Spain almost doubled from 57,000,000 tons to 101,000,000 tons, largely owing to recovery of the West German brown coal industry where production rose from 52,000,000 to 91,000,000 tons.

By the mid-1950s, growing energy demands assumed such proportions that the coal industry was unable to meet the requirements of energy-hungry Western Europe. Imported oil was used to satisfy demands and crude oil consumption in Western Europe almost doubled between 1950 and 1955, thus making a massive breakthrough on the energy market. Imported supplies of coal, especially from the U.S.A., were essential for maintaining Western Europe's high level of industrial output. The Suez political crisis in 1956 provoked uncertainties about Middle East oil supplies and placed further emphasis on American coal. Favourable transatlantic contracts for shipping coal were signed and oil remained subsidiary to coal for some time.

By the late 1950s the absolute rule of 'King Coal' had come to an end and the two-fuel economy began to make its appearance. From being a net producer of energy, Western Europe was converted into a net consumer. This shift was associated with such advances as progress in refinery technology, and the operation of supertankers and the impact of the pipeline revolution to reduce transport costs. Recent events have demonstrated with startling clarity the importance of political and economic agreements contracted during this phase between oil companies and the Middle East countries to guarantee supplies of cheap oil. In addition, mild winters in the late 1950s and slackened industrial activity (1957–59) weakened the relative position of coal. After 1958 the industry changed emphasis away from attempting to maximize production and towards maintaining output of specific types of coal that satisfied market requirements. As a result of reduced demand, pithead stocks of coal piled up, increasing from 16,000,000 tons at the end of 1957 in the five main coal producing countries to 38,000,000 tons at the end of 1964, after having reached 67,000,000 tons in 1959. Total output from the six main producing nations in Western Europe was cut from 467,000,000 tons (1955) to 279,000,000 tons (1972) when Western Europe contributed only 14 per cent of world hard coal production.

Faced with this situation, production was reduced dramatically. Appraisals of the energy situation by the O.E.C.D. Energy Committee in 1966 assumed that rationalization in the coal industry would continue; oil would be available at cheap prices; natural gas consumption would increase; and nuclear power would be competitive with conventional sources of energy by 1980. This made grim reading for Western Europe's mining regions. Reductions in coal output from 1955 to 1972 ranged from 12 per cent in Spain and 23 per cent in West Germany, to 45 per cent in France and the United Kingdom, 65 per cent in Belgium and 77 per cent in the Netherlands. By contrast, lignite production continued to rise from 101,000,000 to 120,000,000 tons in the four main West European producers.

The regional implications of declining demand and falling production of coal have been suggested in Chapter 1. Key aspects in regional employment have been cut back and in some cases have disappeared completely. Complex social problems have ensued, involving retraining and re-employment of ex-miners. The harsh environmental legacy of mining landscapes and patterns of settlement and communication that were geared to coal but to little else needs to be remedied.

Agencies of the E.C.S.C. and the European Social Fund have taken action to ease coal mining and old-established steel producing regions in the Six (and now the Nine) through this difficult period of adaptation, and, in so doing, disbursed £750,000,000 up to the end of 1972 to help more than 400,000 workers, four-fifths of whom were ex-coalminers. Workers have qualified for four types of assistance from the E.C.S.C., but the details vary because of different national policies for social security. First, money has been provided for up to 12 months after initial redundancy to encourage workers to accept alternative jobs that may be less well paid than their former employment. This is particularly true for Continental coalminers who had received relatively high rates of pay. Second, financial help has been available to help men take on jobs in other regions, to cover costs of retraining and to guarantee wages during the retraining period ranging from 85 to 100 per cent of their former salaries. Third, workers receive compensation for costs incurred in moving their families to new employment areas and, if this is not possible, may claim transport expenses incurred in daily travel or weekend visiting. Finally, assistance is available to employees awaiting new jobs.

Further modernization of the coal and steel industries also qualifies for financial aid and more than half of the E.C.S.C. loans in this category were paid to West Germany (53 per cent), with the single region of Nordrhein–Westfalen (the Ruhr) receiving 40 per cent. Mining areas in northern and northeastern France and southern Belgium were also important beneficiaries. Loans for constructing and modernizing workers' housing were distributed similarly, with 47 per cent going to Nordrhein–Westfalen (Figure 3.1). By contract, E.C.S.C. assistance for creating new jobs was distributed differently, with 19 per cent going to Nordrhein–Westfalen and the Dutch Zuid region (southern Limburg) respectively, 15 per cent to Belgian Wallonia, 11 per cent to Eastern France, and 7 per cent to Sardegna. As well as disbursing finance, the High Authority of the E.C.S.C. is responsible for studying the problem of providing new jobs in old industrial regions, can make firm recommendations for action, and can act as a high-level forum to coordinate schemes initiated by national governments or public agencies. The European Social Fund also has responsibilities for retraining and resettling workers and maintaining their wage levels between jobs. Its field of intervention extends over a much wider range than just the coal and steel industries. More than 1,000,000 European workers have received help, with Italian workers accounting for about half of the total.

Planning Problems in Coalfield Areas

Coalmining areas throughout Western Europe have shared common problems resulting from rationalization in the industry, but these problems have varied in intensity from region to region. Production has been terminated in some areas but continues in other mining zones, albeit with a much reduced labour force. The local significance of jobs lost in mining will depend on the range and future viability of alternative employment. Some areas, such as the Belgian Borinage, were almost exclusively involved with coal extraction; but in other regions mining formed just one of a number of activities. Regrettably, many of the industries located on or close to coalfields have also shed labour in considerable quantities. These include textile production in northern France, iron-mining and steelmaking in Lorraine, shipbuilding in northeastern England and many others. Variations in the size of mining areas and the number of workers released complicate the problem. But simple numbers cannot adequately describe the intensity of social and economic distress. The closure of a single pit can be a devastating blow to the inhabitants of a small community for whom coal was literally their life. In addition, the physical legacy of mines, spoil-heaps, communications and workers' housing varies greatly between regions. Nineteenth century housing in mining villages in County Durham or southern Belgium contrasts with *cités-minières* in northern France that were reconstructed after the two World Wars and with miners' housing of Dutch Limburg or the Belgian Kempenland that dates only from the present century.

The *Thomson Report* emphasized that relatively small parts of Western Europe were involved in coalmining but stressed the intensity of their adaptation problems because they supported high densities of population (Figure 1.4). Two contrasting areas (southern Limburg and the Ruhr) exemplify the kinds of problem that have to be tackled. The concealed coalfield of southern Limburg began to be mined early in the present century. Rationalization was started in the late 1950s and the number of miners

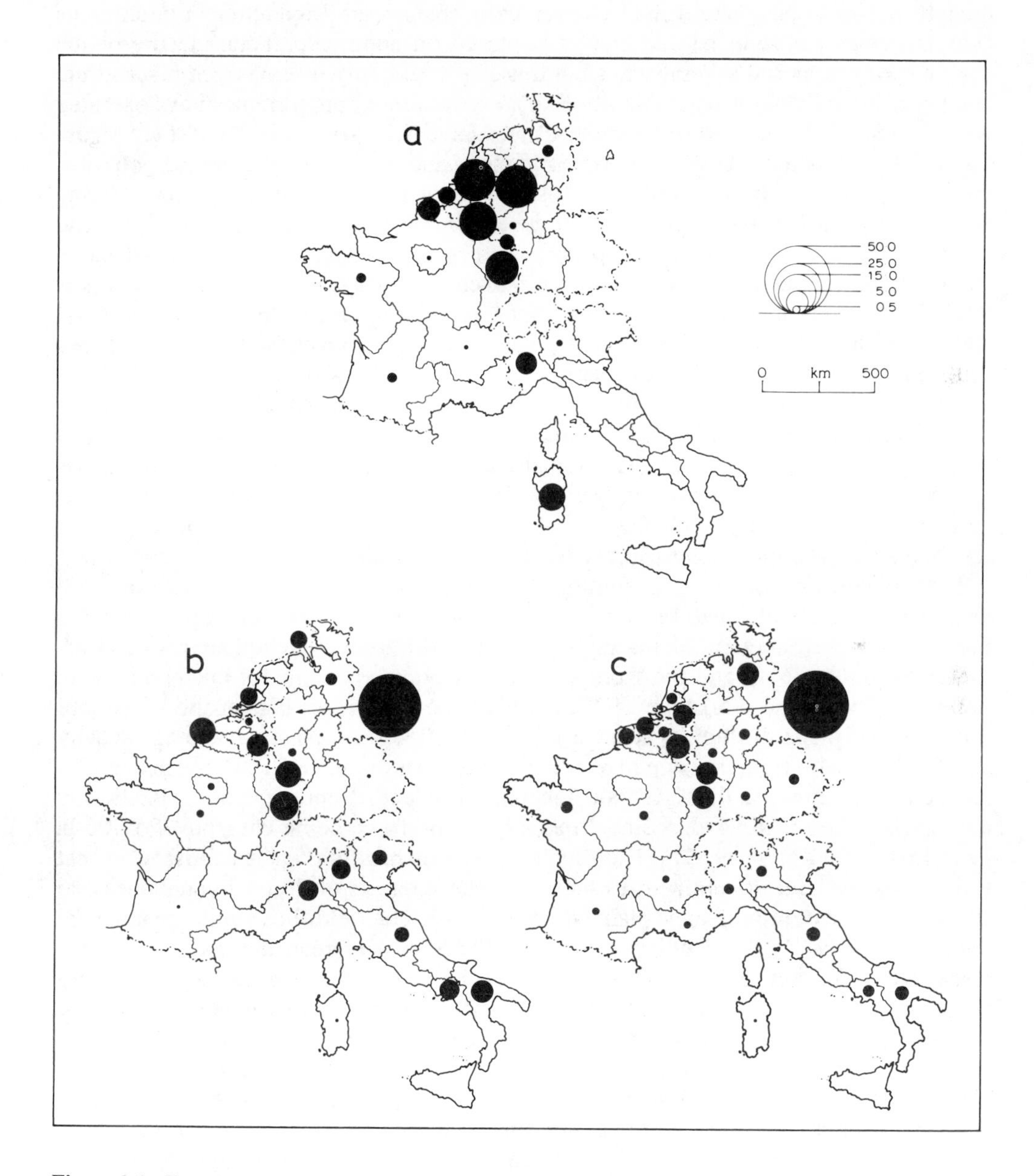

Figure 3.1. European Coal and Steel Community assistance to 31.12.70 (by region): (a) for developing new industries; (b) for investment in the coal and steel industries; (c) for workers' housing.

was reduced by 1500 per annum until 1965 when only 45,000 workers remained. Labour release was then accelerated to 6000 per annum so that a further 30,000 mining jobs were lost by 1971. Operation of the new Beatrix state mine was postponed (even though shafts had been sunk) and was eventually abandoned. Production of bituminous coal and coke was reduced and emphasis placed on anthracite. Total closure of the coalfield was proposed several times but this step could only be envisaged if substitute employment could be made available. Extensive retraining programmes have operated with E.C.S.C. finance and new factory estates have been opened in the last ten years. Careful coordination between releasing miners and providing alternative jobs has meant that most collieries could be closed without causing unemployment. Mining organizations have been instrumental in developing industrial activities that make use of coal byproducts, for ammonia, plastics and fertilizers. Newly installed industries include light engineering, food preparation and car assembly, with the D.A.F. works at Born starting production in 1967 and now employing more than 5000 workers. Old-established chemical industries using gas from coke ovens have been converted to consume byproducts from oil and natural gas. New motorways and canal links have eased access between Limburg and other parts of the Rhinelands core; and the mining environment has been improved as shafts have been filled and new housing and shopping facilities constructed. The social and economic conversion of Limburg has been something of a success story. Nevertheless, white collar jobs are short, as in other mining regions, and attention needs to be paid to providing local employment for school leavers with increasingly high educational standards.

The problems encountered in southern Limburg and the approaches used to tackle them have been replicated in varying degree in mining areas throughout Western Europe. However, such problems are of an entirely different magnitude in regions such as the Nord, Wallonia and the Ruhr. In spite of grave economic problems in recent years, the Ruhr—the workshop of Western Europe—produces half of the Six's hard coal and nearly one-third of its steel. It houses 5,500,000 people at an average density of 1230/km^2, with figures rising to over 3000/km^2 in its urbanized heart which contains a dozen cities of more than 100,000 inhabitants apiece. Important rationalization in mining has taken place with the coal industry's workforce being cut from 325,400 in 1964 to 123,000 in 1969. The Ruhrkohle A.G. organization was established in that year by the Federal government to take over the interests of 26 coal companies and allow the future of mining to be planned for the whole coalfield. In a similar fashion the steel industry has been shedding labour and the Ruhr has been the major target for assistance from E.C.S.C. coffers. Many problems have been tackled but more replacement jobs are needed for male workers and the regions's 'image' needs to be boosted among industrialists since footloose industries tend to be drawn to more attractive environments, especially in southern Germany. Improvements in communications, housing, shopping and cultural facilities, including the construction of four universities where there was none pre-war, have played a substantial role in helping the Ruhr escape from its history of coalmining, steel and pollution.

The region's planning achievements are of interest in other respects. In spite of administrative fragmentation it has been managed since the First World War by the Ruhr Planning Authority (the Siedlungsverband Ruhrkohlenbezirk). Initially, the S.V.D. was concerned with workers' housing but later it embraced a wide range of environmental issues and much has been achieved in this respect. Green wedges have been designated to separate urbanized zones. Special subsidies encourage owners of woodland to preserve their trees. Recreation spaces have been established in the form of 'free-time parks' and the 900 km^2 Naturpark Hohe Mark with its heaths and woodland. Land reclamation by the S.V.D. has involved more than 500 ha of pit heaps being restored in the last twenty years, together with more than 200 ha of embankments (along canals, roads and railways) and 500 ha of miscellaneous derelict land. Over 600 ha of reclaimed land have been put to good use for housing, industrial estates or formal recreation space. The need to restore the environment, improve communications and provide social and cultural facilities is at a greater scale in the Ruhr than in any other mining region in Western Europe, but such tasks are essential elsewhere if employers are to be encouraged to open factories and offices in old industrial regions and thereby diversify local economies.

Growing Dependence on Oil

By contrast with the declining position of coal, oil has risen to prime importance in Western Europe's energy balance. In 1960 it accounted for 29 per cent of the total energy consumed in the countries that were to make up the Nine, but in 1972 it had risen to 60 per cent. Looking at Western Europe more broadly, 58 per cent of the energy consumed in 1972 was in the form of oil. Production at present is of slight importance, with no more than 16,000,000 tons being derived from West European wells, of which more than half is produced in West Germany (Figure 3.2). Western Europe has therefore become heavily dependent on overseas countries for energy supplies. By 1971 60 per cent of the energy consumed by the Nine was imported, having grown from 31 per cent in 1960. Both Denmark and Luxembourg relied on imports for more than 99 per cent of their energy consumption in 1971. Belgium (83 per cent), the Republic of Ireland and Italy (both 81 per cent), and France (76 per cent) were also heavily reliant, with West Germany (53 per cent) and the United Kingdom (50 per cent) importing roughly half of their energy. By contrast, only 16 per cent of the Netherlands' energy supplies were imported by virtue of its rich resources of natural gas.

Pipelines for crude oil and refined products traverse the Continent, adding an important innovation to patterns of commodity flow and transporting oil from maritime ports to coastal and inland refineries (Figure 3.3). Western Europe now has an annual refining capacity of 750,000,000 tons (1972), some 27 per cent of world capacity. Once again there have been rapid increases, with growth from 152,000,000 tons in 1957 to 350,000,000 tons in 1964. At present it is Italy (174,000,000 tons), that has the largest refining capacity among the West European countries, followed by West Germany (126,000,000 tons), France (122,000,000 tons), the United Kingdom (121,000,000 tons), and the Netherlands (78,000,000 tons). Imports of crude oil have greatly boosted the volume of cargo handled by major ports such as Marseilles–Fos, Le Havre, Dunkirk, Genoa, Antwerp and Rotterdam—the largest port in the world.

Over one-third of the port fabric of Rotterdam was destroyed during the Second World War. Its trade hinterland in the Ruhr was in ruins in 1945 and trade with the East Indies was seriously curtailed (Whittick, 1967). Within two years the city council had approved a plan for reconstruction and enlargement of the port to handle 65,000 ton vessels at Botlek and incorporate space for industrial development on reclaimed land alongside deepwater terminals. Facilities were vastly increased in later years and Rotterdam capitalized on the congestion of rival West European seaports. Now a chain

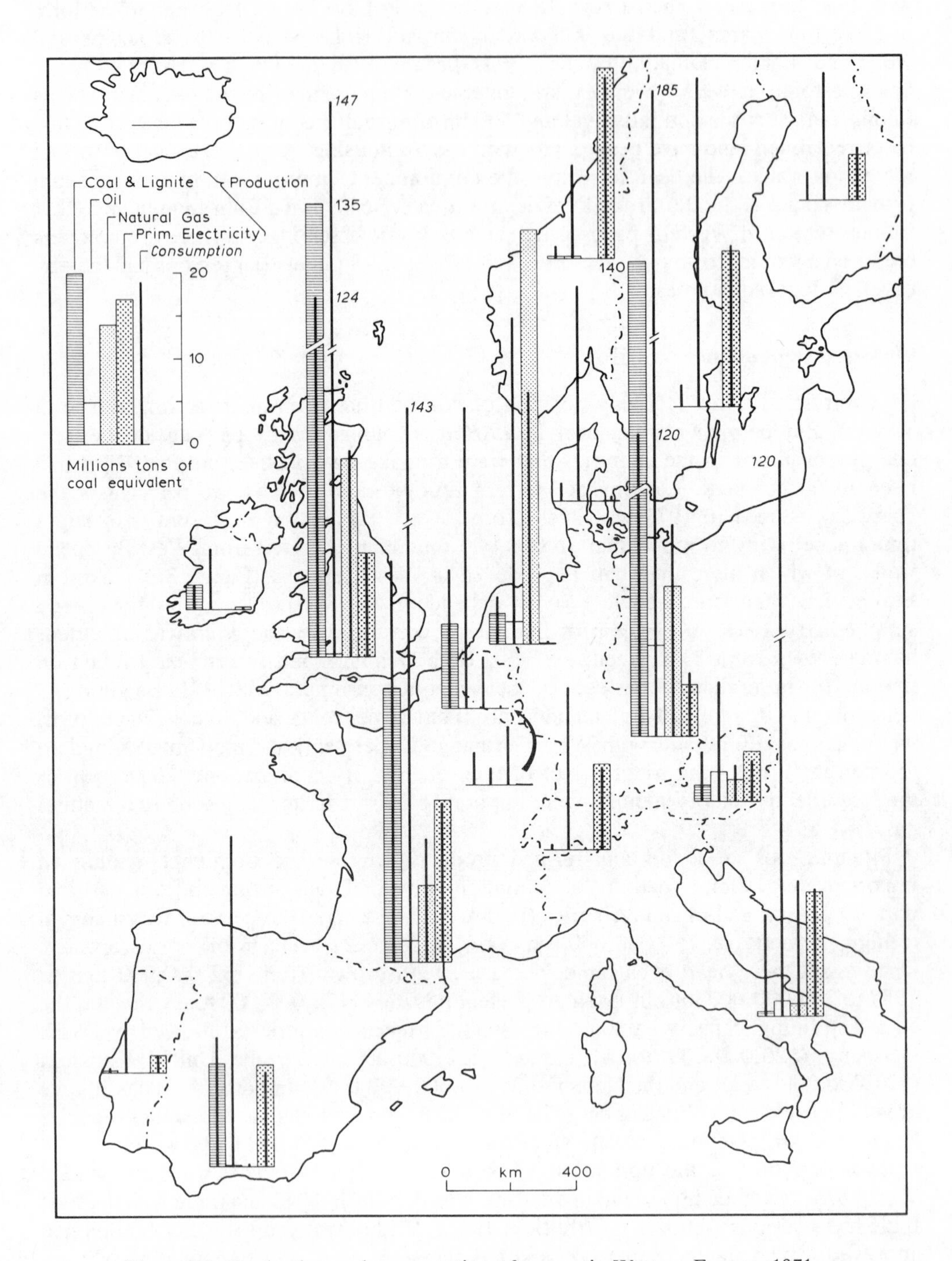

Figure 3.2. Production and consumption of energy in Western Europe, 1971.

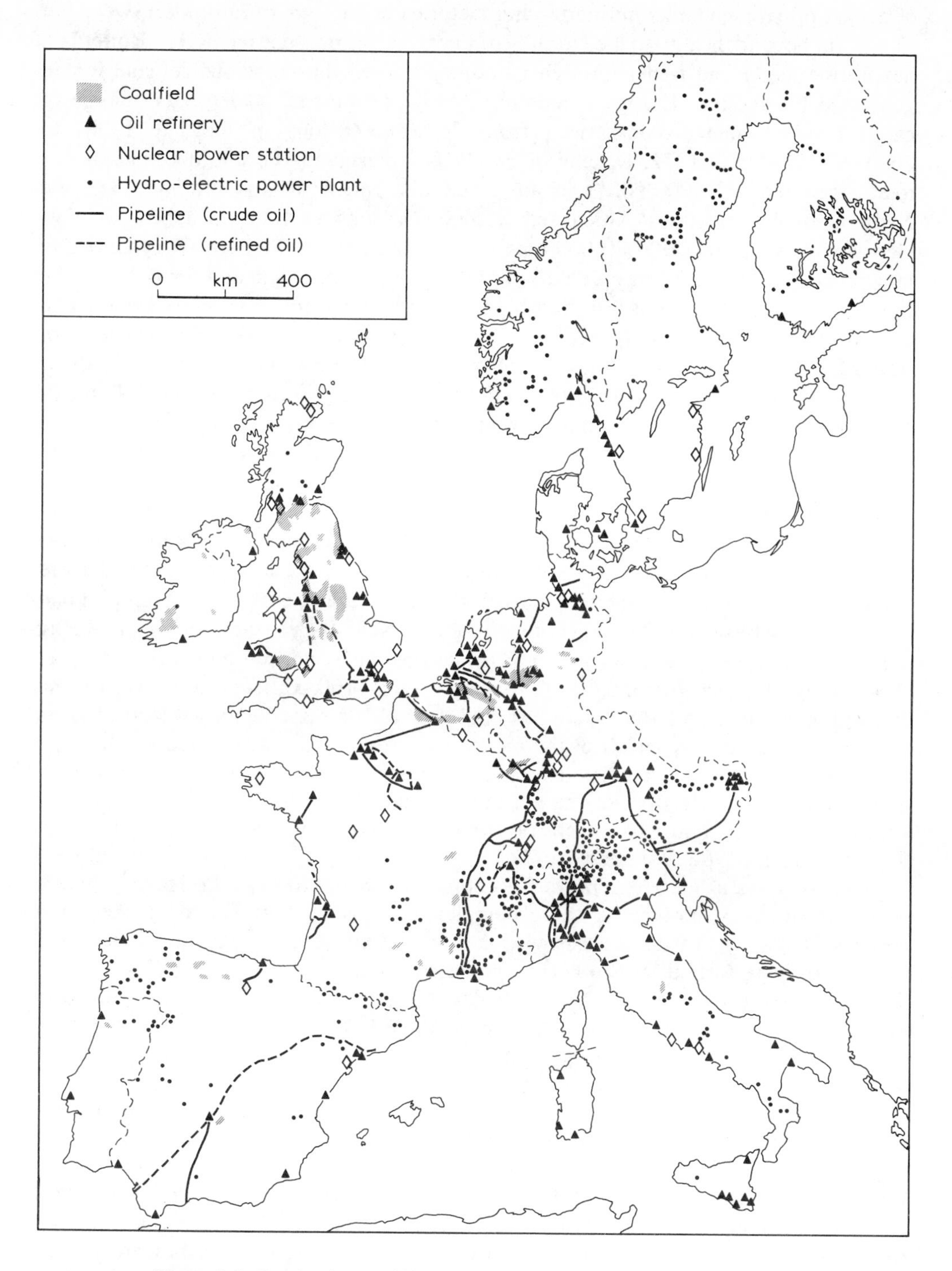

Figure 3.3. Production and distribution of energy in Western Europe.

of docks, oil-storage tanks, refineries and factories in the form of Europoort sweep out for 15 km beyond the city to the North Sea. For over two decades the port of Rotterdam has grown larger and richer through its ability to feed the voracious demand for oil generated by Western Europe's booming post-war industrial society. Its volume of cargo handled annually has risen from 36,000,000 tons in the late 1930s to 262,000,000 tons in 1972. Oil and oil products represented 188,000,000 tons of this total, including 130,000,000 tons of crude oil. Much of this promptly left the Netherlands by pipelines which fed 25,000,000 tons to four refineries in West Germany and a further 20,000,000 tons to four refineries in Belgium. West Germany now receives a quarter of its crude oil imports through Rotterdam and for Belgium the figure is one-third. Crude is also transhipped in Rotterdam for refining in Scandinavia and the United Kingdom. In addition, 80,000,000 tons of crude oil were kept for refining in Europoort's four massive refineries or at Amsterdam. Refined products were destined for Dutch use (28,000,000 tons) and for re-export (52,000,000 tons). Europoort alone contains one-tenth of Western Europe's refining capacity and can store 20,000,000 tons at any one time. Its refineries supply oil-based industries, such as plastics, chemicals, paints, synthetic rubber and a host of others, which line the river Maas and make up the 14,000 ha of port and industrial facilities of Europoort.

Rotterdam functions as the 'oil jugular' of Western Europe pumping in more than one-fifth of the oil it needs every day. But two-thirds of these imports were derived from Arab sources. Events in 1973 and 1974 revealed that the wealth of both port and city were established on a precarious footing. Nevertheless new areas are being reclaimed from the North Sea, out on the Maasvlakte beyond Europoort, and dredgers are digging channels for 350,000 ton supertankers—100,000 tons larger than those handled at present. But Rotterdam has paid dear for its massive growth and obvious prosperity based on expansion of the oil industry. Fumes and waste products from its refineries and factories have polluted the atmosphere to such an extent that local opinion, and especially that of the agricultural industry in the Westland, is strongly opposed to further industrial expansion if more environmental degradation is likely to follow (Kuipers, 1962; Weigend, 1973).

Rotterdam has at least one powerful competitor in the form of Le Havre where a deepwater offshore harbour is being completed to handle 500,000 ton tankers and perhaps ultimately to take vessels twice that size. Tankers using Le Havre would not have to navigate particularly crowded and narrow sections of the English Channel and the port also offers 10,000 ha of reclaimed estuarine land for new industrial installations. Plans have been prepared for pipelines to be built across northern France to feed southern Belgium and the Ruhr. If constructed, these pipelines must surely reduce Rotterdam's chances for massive expansion in the future.

Certainly public opinion, from quarters as diverse as the oyster fishermen of Brest and the vinegrowers of the middle Rhône, has been raised against the threat of further air and water pollution from new or enlarged oil refineries at many points in Western Europe. The possibility of pollution spreading from Liège in Belgium to the southern Netherlands brought opposition from the Dutch government to a Belgian scheme. However, landscaping and pollution control can be quite effective if they are strictly enforced. A good example would seem to be Milford Haven, which is Western Europe's second largest oil port. But if oil resources in the Celtic Sea are exploited commercially and million-ton tankers become standard, the deepwater facilities of

Milford Haven will be in much greater demand and the environment of the national park that surrounds it may be seriously threatened.

Tidewater sites have been of crucial importance not only for locating oil refineries and associated petrochemical industries but also for installing steelworks using imported ores and, in many cases, imported coking coal as well (Fleming, 1967; Warren, 1967). On the eve of the Second World War there was no major integrated works at a coastal location in Western Europe, but a coastward trend had started with the Dutch developing IJmuiden at the mouth of the North Sea Canal and the Italians deciding to build a works to the west of Genoa. Increasing reliance on imported minerals and fuel has made coastal sites all the more attractive in recent years. Tidewater steelworks have proliferated and now include Bremen, Dunkirk, Fos, Taranto and Zelzate (Wittmann and Thouvenot, 1972). Strangely enough, neither Rotterdam nor Le Havre has managed to attract a major steelworks to complement its port, refinery and other installations, but this has not been for want of trying (Malézieux, 1971).

The rise of oil refining has not just brought prosperity and the chance of industrial expansion to coastal sites in Western Europe. Inland refineries near major demand centres have multiplied since 1960 in response to pipeline construction. Changes in refinery location in West Germany exemplify trends that have occurred elsewhere. Before the Second World War, refining was concentrated near Hamburg, which dealt with imported crude, and on domestic oilfields near Hanover. During the war refineries were built in the Ruhr to distil oil from coal products. Those installations that were not subsequently dismantled were available for conversion to conventional oil refining. Since 1945 three phases of change have occurred, raising refining capacity in West Germany from 2,500,000 tons (1947) to 126,000,000 tons (1972). Up to 1955 there was roughly equal expansion of capacity in the North Sea ports and in the Ruhr region which was fed by Rhine barges loaded with crude oil after transhipment in Rotterdam. In the latter half of the 1950s the advantages of centrally placed refineries in the Ruhr and at Cologne were boosted by crude being piped from Wilhelmshaven and Rotterdam. Hamburg's importance as a refining centre declined, being peripheral to main demand centres in West Germany and having its hinterland truncated by the Iron Curtain. During the 1960s refineries were constructed in southern Germany, whilst expansion in the Ruhr and the North progressed more slowly. High energy prices had previously been a drawback to industrialization in the South, hence refining was encouraged by the Bavarian government, making use of crude oil piped from Marseilles to Karlsruhe/Mannheim (1967) via the Rhône valley and Alsace, or by the Central European Pipeline (1966) from Genoa, or by the Trans-Alpine Pipeline (1967) from Trieste to Ingolstadt. The net result has been greatly advantageous for southern Germany, reducing energy costs and encouraging industrial development. Broadly comparable inland locations for oil refining near major demand centres include the Paris Basin, northern and northeastern France, the Lyons region, southern Belgium and northern Italy (Figure 3.3).

Electricity and Natural Gas

Generation of primary electricity, mainly in the form of hydro-electric power but also nuclear and tidal energy, accounts for 21 per cent of all energy produced in Western Europe, being particularly significant in Norden and the Alpine states (Figure 3.2) but of limited significance in the Nine, where primary electricity accounts for only 11 per cent of domestic energy production. The proportion of energy consumed in the form of primary electricity is even smaller, representing 4·5 per cent for the Nine and 9 per cent for the whole of Western Europe. The pattern of hydro-electricity generation conforms to regions with suitable terrain conditions and river flow. Local environmental impact is considerable (in the form of valley drowning) but transmission of electricity to existing industrial demand centres has reduced the economic impact of H.E.P. in production areas. For example, six 400 kV grid lines bring power from northern Sweden to the cities and industries of the South and also Denmark, Finland and Norway. Tidal generation of electricity is restricted to the Rance estuary in Brittany, and nuclear energy has been accepted with varying degrees of enthusiasm. The first nuclear power plant put into operation in Western Europe was at Calder Hall in 1956. The *Target for Euratom* report (1957) pointed to the vulnerability of oil as an energy source and stressed the importance of developing nuclear power. The United Kingdom adopted an ambitious approach to nuclear power but progress was slower elsewhere in Western Europe than predicted in the late 1950s because of strong competition from cheap oil. However several schemes for constructing nuclear power stations have received E.I.B. loans including the Franco-Belgian power station at Tihange near Liège. Rising oil prices have caused the role of nuclear power in Western Europe's energy supplies to be reappraised and the French have announced an ambitious programme for constructing over a dozen nuclear power stations.

Natural gas exploitation has made great progress accounting for 19 per cent of Western Europe's energy production in 1972 and 29 per cent of that produced in the Nine. Italy (14,142,000,000 m^3), West Germany (17,700,000,000 m^3), the United Kingdom (24,995,000,000 m^3) and the Netherlands (58,420,000,000 m^3) were the leading producers, but the situation is rapidly changing as North Sea resources are exploited. Industrial use of natural gas was established in the Po Valley in northern Italy in the 1930s and, since the Second World War, further exploitation has occurred in Sicily and in southern and central parts of the mainland. Relatively small reserves were also discovered on the eve of the Second World War at Saint Marcet in southwestern France but the major discovery was at Lácq in 1951. This gas was of high pressure with a high sulphur content. It was corrosive and required purification before piping. Sulphur recovery works were opened in 1957 and pipelines built to serve demand centres in many parts of France. Further discoveries were made near Pau in 1965 and supplies from southwestern France now account for 60 per cent of French consumption of natural gas. The remaining 40 per cent is piped from the Netherlands and is brought in liquefied form by tanker from Arzew in Algeria to Le Havre. Similar shipments are made to Canvey Island for the British gas grid. In addition, piped supplies from the U.S.S.R. to France will be available after 1975.

Trial drilling for oil in the northern Netherlands during the late 1930s revealed natural gas at Schoonebeek and further discoveries were made in the 1940s and 1950s, including a major reserve at Slochteren near Groningen in 1959. Exploitation began in 1963 but Dutch demand was insufficient to consume total gas output. It was in the commercial interests of the two major companies and of the Netherlands to go for a high-price/limited-volume strategy to bring high returns on investment and raise foreign exchange earnings. Contracts were arranged to export gas to Belgium, northern and northeastern France, Luxembourg and the Ruhr at prices which would give competing fuels little chance of maintaining their markets. In addition, a 1350 km pipeline is to be built to take natural gas from Annerveen in the northern Netherlands to the industrial region of northwest Italy. Plentiful supplies of natural gas have, of course, caused important changes in energy consumption in the Netherlands, where consumption of fuel oil was 30 per cent less in 1973 than in 1969.

A North Sea Bonanza?

The virtually perfect geological characteristics of the Groningen oil and gas field came to light during the 1960s and interest was stimulated in the neighbouring North Sea as a potential reserve of energy. Surveys were started and prospecting flourished as it had done off California and Texas half a century earlier. Mineral rights under the waters of the North Sea were divided among bordering countries in 1964 according to the Geneva Convention of 1958, with the lion's share going to the United Kingdom. The riches of the North Sea oil and gas province have been revealed by successive discoveries. The first exploration licences were issued in 1964 when investigations started off the coasts of Yorkshire and East Anglia. Gas discoveries in the West Sole field in 1965 were the first reserves suitable for exploitation and gas came ashore two years later. Five major fields are now in production (Figure 3.4). As a result, the British gas industry has undergone major reorganization, switching from coal-derived gas to natural gas which now satisfies over 90 per cent of national consumption. The Anglo-Norwegian Frigg field will be producing by 1976 and the British Gas Corporation has negotiated for supplies from that source. By contrast, gas from the Norwegian Ekofisk field is to be piped via Emden to the Ruhr where the gas corporation is willing to pay twice as much as the British Gas Council. Belgian, Dutch and French customers will also receive Ekofisk gas. In any case, Norway is unable to use large quantities of gas domestically. It could only be burned in power stations as an expensive alternative to hydro-electric power. Norway will benefit mainly from tax and royalty payments rather than directly from cheaper energy supplies.

Of course the energy riches of the North Sea are not restricted to natural gas. The first important oil discoveries were made in the Norwegian Ekofisk field in 1969 and in the British Forties field in 1970. Many other fields have been declared commercially viable. By 1980 the North Sea may supply two-thirds of the United Kingdom's oil needs. By contrast with the other North Sea nations Norway is in the unique position of not being in the E.E.C. and not being obliged to make her oil available to the highest Common Market bidder. Oil (and of course gas) from the Norwegian Ekofisk field cannot be landed in Norway because of the presence of the Norwegian Trench preventing large-diameter pipelines being installed. Instead, the Norwegian government has consented for oil to be landed near Middlesbrough.

The ecological dangers that might arise from North Sea oil exploitation have been highlighted in many quarters. Wave scour could damage pipelines. Harsh weather conditions have already delayed their installation and may pose serious problems for production. Sites in Scotland have been sought for constructing production platforms and much controversy has resulted on social as well as ecological grounds. Put bluntly, the resources of the North Sea have the potential to bring development and prosperity to Scotland but, at the same time, may also involve social disruption and environmental disaster.

Informed opinion varies considerably on the likely impact of North Sea oil and natural gas. As Odell (1973a) has shown, Western Europe certainly offers a highly

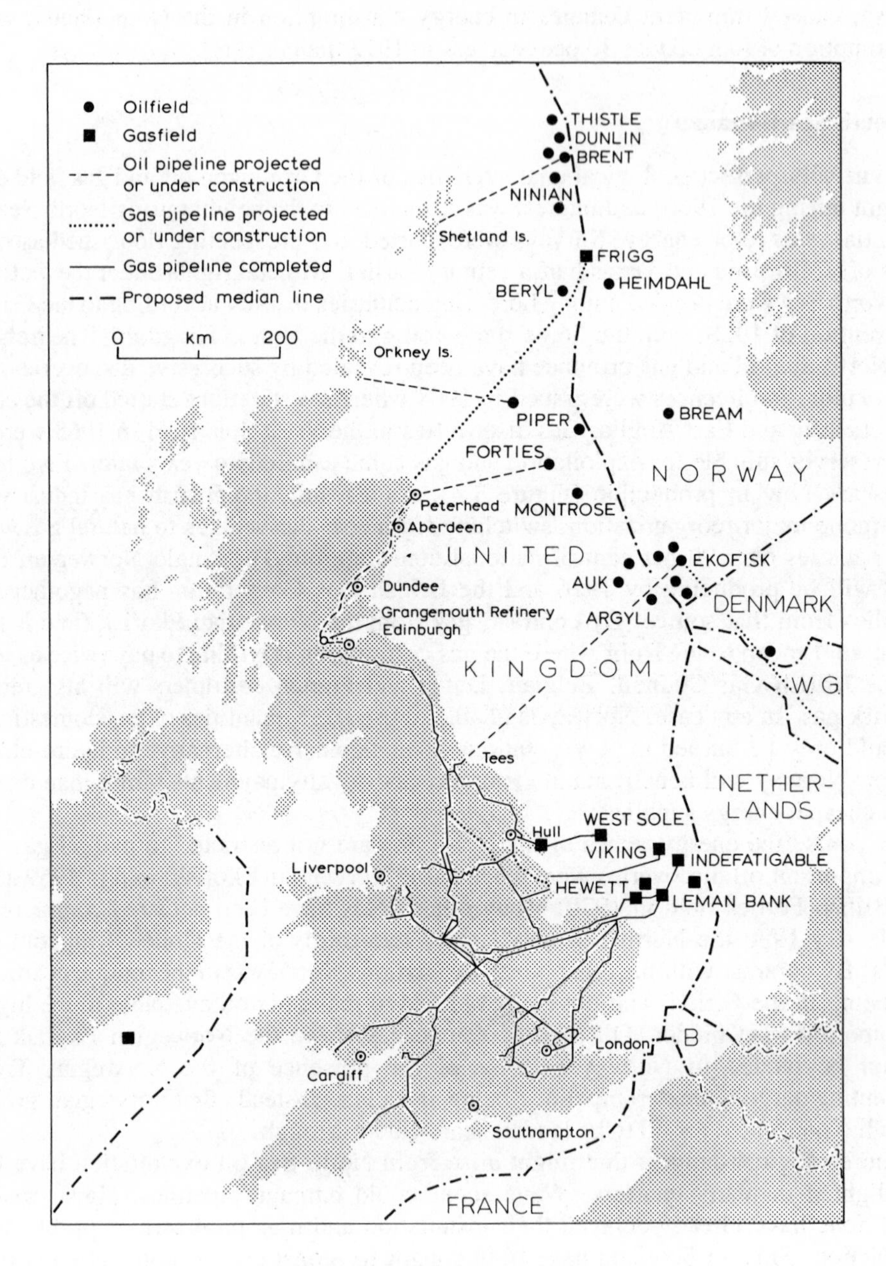

Figure 3.4. North Sea energy.

favourable location for their development. First, there is a powerful geography of demand, from highly industrialized economies of surrounding countries. Second, natural gas offers the advantage of being clean and pollution-free. Third, there is a considerable availability of European funds and skill, even though American intervention in North Sea development is highly important. Finally, West European supplies of oil and natural gas may reduce reliance on supplies from politically unstable parts of the world and also ease the balance of payments situation.

North Sea oil and natural gas resources are of crucial importance to national economies in Western Europe, but it is difficult to envisage the precise regional impact they may have. Important new industrial development on the shores of the North Sea has been envisaged by some experts, which might remove fears stemming from over-dependence on fishing and marginal agriculture, for example, in eastern Scotland and western Norway. Examination of what has happened in southwestern France and the northern Netherlands, where oil and gas deposits have been exploited, would hardly confirm that interpretation. Development of the Dutch Schoonebeek oilfield brought only *c.* 1000 new jobs, of which 600 are on the oilfield and the remainder at headquarters in Assen. The real regional impact has come in the form of new roads and additional taxes for local authorities. It is true that an aluminium smelter and chemical works have been opened in Groningen province in order to make use of relatively low cost gas, but the majority of gas supplies in the northern Netherlands and southwestern France have been piped to demand centres elsewhere . Admittedly important light industries, such as electrical goods, have been established in the northern Netherlands but they were not attracted by cheap energy. At present direct employment in the North Sea oil and gas industry is relatively slight, but onshore rig construction and provision of equipment and services have raised wage levels and inflated the demand for housing in such cities as Aberdeen, where land values have risen sharply. More than 200 firms associated with North Sea exploration have moved into the city, providing 4000 jobs related directly to the oil industry and helping indirectly to support many more. Construction of refineries and pipelines will demand supplies of temporary labour but once installed oil refining is a highly automated operation requiring relatively little in the way of labour. In addition, one should not ignore the possible negative impact of North Sea supplies on the existing oil refinery and import centres, for example at the mouth of the Rhine. Exploitation of North Sea oil will be to the benefit of refineries in Scotland and northeastern England, plus perhaps some new refineries (e.g. Shetland), but may lessen the rates of expansion in Rotterdam. An important share of oil imported from non-European sources may be directed to Le Havre for local refining or piping to interior refineries rather than being shipped through the congested Straits of Dover.

The end of the era of cheap imported oil plus the discovery of North Sea oil and natural gas have brought profound changes to Western Europe's industrial economies; but such changes are more likely to be national and international in impact rather than purely regional. Ultimately, the continued operation of policies for regional development in Western Europe will depend on economic growth continuing at a satisfactory level. Unfortunately, the full benefits of North Sea oil supplies will not be experienced until the early 1980s and serious energy problems threaten in the immediate future. In the light of these conditions, existing energy resources have had to be radically reappraised. Potential sites for future H.E.P. development have been re-examined in Norden, and France has embarked on a massive programme of nuclear power production. Western Europe's coal industries now have brighter prospects than at any time in the past two decades. For example, after the closure of 905 pits in the United Kingdom since 1947, leaving only 255 open with a colliery workforce of 240,000, eight major expansion schemes have been approved involving selected pits in the Midlands, Nottinghamshire, Fife, Yorkshire, Wales and northeastern England. Thick coal seams beneath Selby are to be exploited from which it is hoped to produce an annual output of more than 2,000,000 tons of coking and power station coal before 1980. Similar reappraisals are taking place in the hard-coal and lignite fields of other parts of Western Europe. One key problem is that the time gap between starting work on a new pit and the pit reaching full capacity production is eight to ten years, in the United Kingdom at least. Recent decisions to open new pits will not be realized until the early 1980s, by which time North Sea oil should be fully onstream. For this reason, the question of more extensive opencast mining, with appropriate remedial landscaping, has been reopened in an attempt to meet immediate demands. Undoubtedly the coal industry of the future will be highly automated and should not produce such harmful environmental effects as resulted during earlier phases of exploitation. Survival, and indeed possible revival, of selected components of Western Europe's coal industry adds a new and intriguing element to her problems of regional management.

References

Barr, J., 1970, Planning for the Ruhr, *Geographical Magazine*, **40**, 280–289.

Burtenshaw, D., 1972, Regional renovation in the Saarland, *Geographical Review*, **62**, 1–12.

Burtenshaw, D., forthcoming, *Saar–Lorraine*, Oxford University Press.

Clark, C., F. Wilson and J. Bradley, 1969, Industrial location and economic potential in Western Europe, *Regional Studies*, **3**, 197–212.

Clout, H. D., 1971, Regional revival in the Nord region of France, *Norsk Geografisk Tidsskrift*, **25**, 145–158.

Clout, H. D., 1972, Nord coal miners prepare for 1983, *Geographical Magazine*, **44**, 398–406.

Clout, H. D., 1975, *The Franco-Belgian Border Region*, Oxford University Press.

Elkins, T. H., 1972, Life on the European growth axis, *Geographical Magazine*, **44**, 375–380.

Fleming, D. K., 1967, Coastal steelworks in the Common Market, *Geographical Review*, **43**, 48–72.

Gordon, R. L., 1970, *The Evolution of Energy Policy in Western Europe: the reluctant retreat from coal*, Praeger, New York.

Guérin, P., 1973, Le gaz naturel en France, *L'Information Géographique*, **37**, 36–42.

Hauer, J. and coworkers, 1971, Changes in the industrial geography of the Netherlands during the sixties, *Tijdschrift voor Economische en Sociale Geografie*, **62**, 139–156.

Hellen, J. A., 1974, *North Rhine-Westphalia*, Oxford University Press.

Jensen, W. G., 1967, *Energy in Europe, 1945–80*, Foulis, London.

Johnson, J. A., 1967, Developments in the Swedish iron ore industry, *Geography*, **52**, 420–422.

Krümme, G., 1970, The inter-regional corporation and the region: a case study of Siemens' growth characteristics and response patterns in Munich, West Germany, *Tijdschrift voor Economische en Sociale Geografie*, **61**, 318–333.

Kuipers, H., 1962, The changing landscape of the island of Rozenburg (Rotterdam Port Area), *Geographical Review*, **52**, 362–378.

Lambert, A. M., 1971, Dutch steelmaking: past, present and future, *Geography*, **56**, 241–243.

Lawrence, G. R. P., 1971, The changing face of South Limburg, *Geography*, **56**, 35–39.

Malézieux, J., 1971, Signification géographique d'un projet d'investissement industriel: un centre sidérurgique dans la Maasvlakte de Rotterdam, *Annales de Géographie*, **80**, 428–438.

Martin, J. E., 1968, New trends in the Lorraine iron region, *Geography*, **53**, 375–380.

Martin, J. E., 1974, Some effects of the canalization of the Moselle, *Geography*, **59**, 298–308.

Michel, A. A., 1962, The canalization of the Moselle and West European integration, *Geographical Review*, **52**, 475–491.

Odell, P. R., 1972, *Oil and World Power,* Penguin, Harmondsworth.
Odell, P. R., 1973a, Indigenous oil and gas developments and Western Europe's energy policy options, *Energy Policy,* **1**, 47–64.
Odell, P. R., 1973b, The future of oil: a rejoinder, *Geographical Journal,* **139**, 436–455.
Perry, N. H., 1967, Recent developments in the West German oil industry, *Geography,* **52**, 408–411.
Riley, R. C., 1965, Recent developments in the Belgian Borinage, *Geography,* **50**, 261–273.
Riley, R. C., 1967, Changes in the supply of coking coal in Belgium since 1945, *Economic Geography,* **43,** 261–270.
Riquet, P.,. 1972, Conversion industrielle et réutilisation de l'espace dans la Ruhr, *Annales de Géographie,* **81**, 594–621.
Scargill, D. I., 1973, Energy in France, *Geography,* **58**, 159–163.
Sid Ahmed, A., 1973, Le gaz naturel dans le monde, *L'Information Géographique,* **37**, 11–36.
Spaak, F., 1973, An energy policy for the European Community, *Energy Policy,* **1**, 35–37.
Swann, D., 1972, *The Economics of the Common Market,* Penguin, Harmondsworth.
Tamsma, R., 1972, The northern Netherlands, *Tijdschrift voor Economische en Social Geografie,* **63**, 162–179.
Thomas, T. M., 1966, The North Sea and its environs: future reservoir of fuel? *Geographical Review,* **56**, 12–39.
Thomas, T. M., 1968, The North Sea gas bonanza, *Tijdschrift voor Economische en Sociale Geografie,* **59**, 57–70.
Thompson, I. B., 1965, A geographical appraisal of recent trends in the coal basin of northern France, *Geography,* **50**, 252–260.
Toonen, W. P. G., 1972, The economic restructuring of South Limburg between 1965 and 1971, *Tijdschrift voor Economische en Sociale Geografie,* **63**, 180–189.
Waller, P. P. and H. S. Swain, 1967,Changing patterns of oil transportation and refining in West Germany, *Economic Geography,* **43**, 143–156.
Warren, K., 1967, The changing steel industry of the European Common Market, *Economic Geography,* **43**, 314–332.
Warren, K., 1973, *North East England,* Oxford University Press.
Weichart, G., 1973, Pollution of the North Sea, *Ambio,* **2**, 99–106.
Weigend, G. G., 1973, Stages in the development of the ports of Rotterdam and Antwerp, *Geoforum,* **13**, 5–15.
Wever, E., 1974, Seaports and physical planning in the Netherlands: some comments on policy considerations relating to industrial activities in seaports, *Tijdschrift voor Economische en Sociale Geografie,* **65**, 4–11.
Whittick, A., 1967, The largest port in the world: Rotterdam rebuilt, *Town and Country Planning,* **35**, 353–358.
Wittmann, M., and C. Thouvenot, 1972, *La Mutation de la Sidérurgie,* Masson, Paris.

Note

1. 'Metropolitan regions' would perhaps be a more useful term.

4

The Rural Residuum

Hugh D. Clout

Introduction

Beyond the nodes and axes of urban development straddling Western Europe are found a wide range of rural areas with varied physical resources and socio-economic characteristics but normally distinguished by relatively low densities of settlement, stable or declining population numbers (largely due to out-migration to urban centres), relatively low average incomes (resulting from agricultural employment), a shortage of alternative jobs, and poorer provision of services and facilities than in urban areas. For these and other reasons the component parts of the rural residuum have long been recognized as 'problem areas' and this fact is reflected in the delimitation of zones for financial assistance undertaken by national governments and by the E.E.C. (Figures 1.7 and 1.8). But, on the other hand, rural areas form relatively unspoiled alternative environments for weekend recreation and holidaymaking by city dwellers. In addition, they represent 'reservoirs' of land from which space will be abstracted for constructing housing, factories, roads, airports and other material features of urban civilization. Planning future uses of Western Europe's countryside demands that a delicate balance be established between conserving valuable resources and permitting sufficient change and economic development for the living standards of country dwellers to be improved.

In spite of its relatively 'natural' appearance, the rural residuum is far from immutable and is undergoing important changes in land use and socio-economic composition as it experiences both the direct and indirect effects of urbanization. The concept of 'rurality' is too intangible for the dimensions of Western Europe's countryside to be known precisely, but roughly five-sixths of the total land surface is devoted to farming, forestry, economically unproductive land and small settlements, which are all recognized popularly as 'rural' land uses.

The Role of Agriculture

Agriculture represents by far the most extensive rural use of land in Western Europe and farming conditions vary enormously over such a large and internally diverse section of the Earth's surface. In 1971 44 per cent of the total area was devoted to agricultural production (exclusive of timber) which represented 149,000,000 ha (Figure 4.1). Spain (36,400,000 ha of farmland) and France (33,000,000 ha) together accounted for almost half of the total. National proportions of land under agriculture reflect the quality of physical conditions and the relative importance of other land-use components. It is remarkable that 78 per cent of the United Kingdom was still in agricultural use in spite of extensive urban development, but this was mainly because of the country's small timber cover. Very high proportions of land were also devoted to

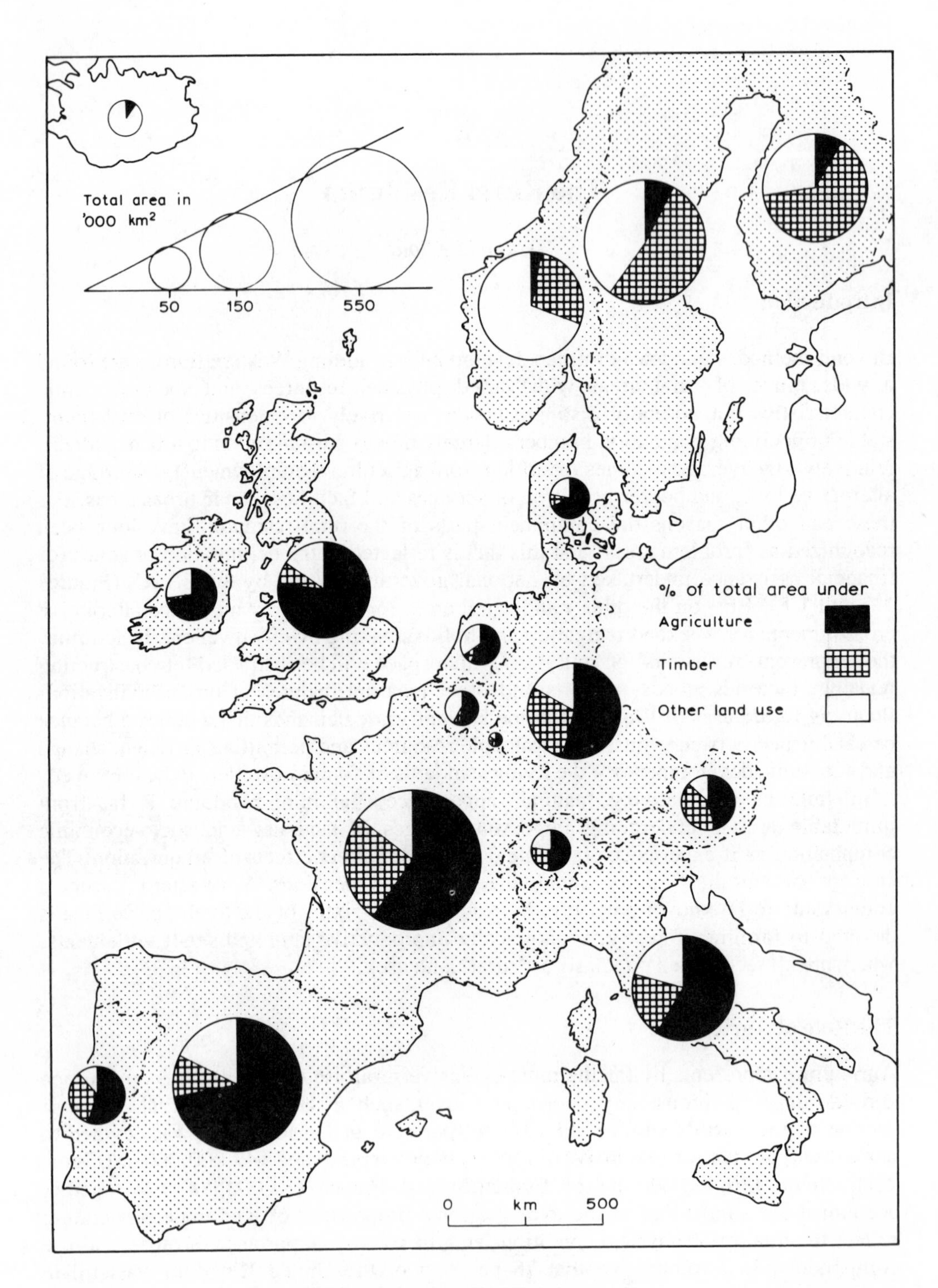

Figure 4.1. Land use in Western Europe.

farming in Spain (72 per cent) and the Republic of Ireland (70 per cent). Not surprisingly, the harsher physical environments of northern and Alpine Europe have only small proportions of their land under agriculture.

In 1971 168,000,000 people were employed in farming, forestry and fishing in Western Europe—12·6 per cent of the total workforce. Some 12·1 per cent of the Six's labour force was in these three primary activities (of which farming was by far the most important) but the proportion declined to 9·9 per cent when the three new member states of the Common Market were included. By contrast, primary employment was of far greater importance in West European countries outside the E.E.C., involving 21·8 per cent of the total workers of Iberia, Norden and the Alpine states. At the national level primary employment varied from a mere 2·7 per cent of the workforce in the United Kingdom to 29·1 per cent in Spain and 31·0 per cent in Portugal (Figure 4.2). Percentages convey only part of the picture and one must remember that Italy (3,600,000), Spain (3,600,000) and France (2,800,000) together contained three-fifths of Western Europe's farmers and farmworkers and represent sizeable political interests in Western European affairs (Franklin, 1971). Nevertheless, agriculture has shed large numbers of personnel in the past quarter century. In the Six farming employed 28·0 per cent of the workforce in 1950, 21·0 per cent in 1960, but only 12·1 per cent in 1971. To take a single example, 5,000,000 Italians left the land between 1951 and 1971, cutting the agricultural proportion of the workforce from 42·0 per cent to 18·9 per cent. By 1980 this is expected to fall to 10·0 per cent.

In spite of the large primary labour force and the high proportion of land given over to farming and woodland, agriculture, plus forestry and fishing contribute only a small proportion of Western Europe's G.D.P. In the Nine only 5.3 per cent is derived from these sources, falling to 2·9 per cent in the United Kingdom and 3·5 per cent in West Germany, but rising to 9·8 per cent in Italy and 16·4 per cent in the Republic of Ireland (Table 4.1). Primary activities are still of great importance in the national economies of Spain (13·5 per cent of G.D.P.), Finland (13·7 per cent) and Portugal (16·2 per cent). A rough indication of the relative efficiency of national primary activities is shown in Table 4.1. In Belgium, Denmark, the Netherlands and the United Kingdom the percentage of the workforce in the primary sector is approximately equal to the proportion of the G.D.P. derived from it, but elsewhere in Western Europe large agricultural labour forces make only small contributions to the national economy.

In spite of marked regional differences in the physical and human aspects of Western European farming there have been important increases in both production and productivity in almost all branches since 1950. For example, the volume of agricultural output from the Six rose by 3–4 per cent per annum between 1963 and 1970. Such advances reflect the technical and biological progress of recent decades. Highly productive or resistant crop strains have been developed, new breeds of livestock perfected, new fertilizers, agricultural chemicals and machines manufactured, all combining to reduce the volume of land and labour required to produce a given quantity of foodstuffs. Nevertheless, large areas of land drainage and irrigation have been completed, sometimes with financial backing from the E.I.B. as in the case of irrigation schemes in Languedoc and Sicily. Similar technical achievements have occurred outside the E.E.C., as in Spain where a further 41,000 ha were irrigated during 1972 to bring the national total to more than 2,600,000 ha. Improved land

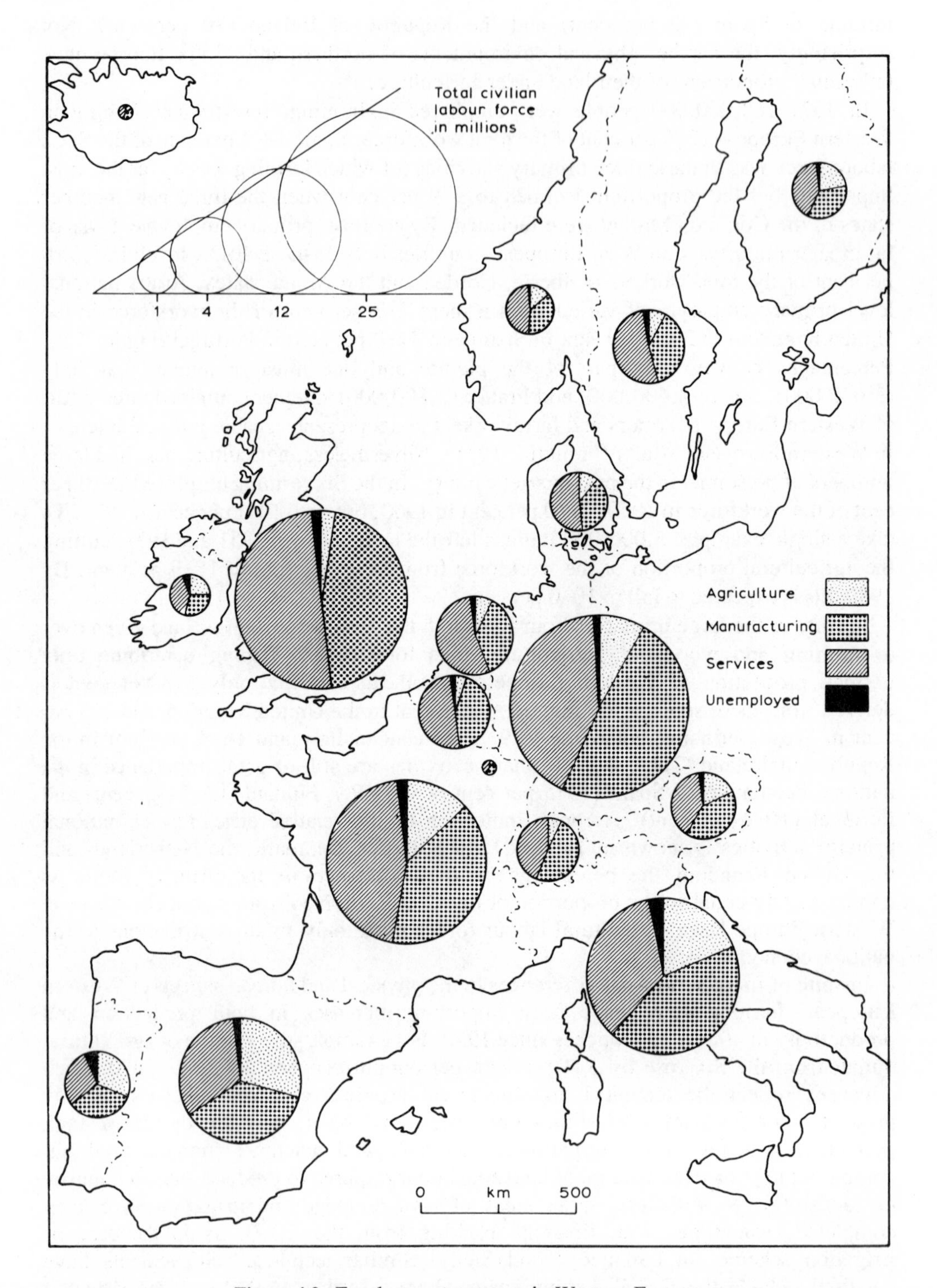

Figure 4.2. Employment patterns in Western Europe.

Table 4.1. Agriculture, Forestry and Fishing as
(a) a Proportion of Labour Force and
(b) a Source of Gross Domestic Product, 1971 (%)

	(a)	(b)		(a)	(b)		(a)	(b)
Belgium	4·4	4·2	Denmark	10·9	7·5	Austria	17·9	6·0
France	13·2	6·3	Republic of Ireland	24·9	16·4	Finland	22·3	13·7
Italy	18·9	9·8	United Kingdom	2·7	2·9	Norway	13·8	5·3
Luxembourg	10·1	4·4				Portugal	31·1	16·2
Netherlands	6·9	5·7	The Nine	9·9	5·3	Spain	29·1	13·5
West Germany	8·3	3·5				Sweden	8·0	N/A
						Switzerland	6·7	N/A
The Six	12·1	5·7						

drainage is allowing large-scale production of cereals, fodder crops and vegetables to gradually replace rice growing in low-lying areas such as the Ebro delta.

Government Intervention in Agriculture

National governments have intervened in West European agriculture in various ways in order to try to achieve a mixture of social, economic and political objectives, which include raising the standard of living of the agricultural population, increasing farming efficiency and, ultimately, trying to capture or to retain the agricultural vote. Four types of intervention have been tried; first, those that directly support farm incomes; second, those that improve the efficiency of existing farms; third, those that alter the shape and size of farms and promote cession of uneconomic holdings; and fourth, those that aid lagging regions through renovating farming and other sectors of the economy and by introducing non-agricultural employment. Intervention has been highly complicated, with individual countries pursuing their own policies but a certain measure of coordination has now been achieved in the E.E.C. under the umbrella heading of the Common Agricultural Policy which will be examined later.

Price supports and import restrictions form the principal means among the first group of measures and have been highly significant in the C.A.P. But, generally speaking, the results of price support have been of limited value. Social objectives are not fully achieved since efficient farmers benefit the most and small farmers receive only slight rises in income through increased production. Land and labour resources continue to be used inefficiently. Economic objectives are not satisfied since marginally increased incomes may induce the already excessive farm population to remain on the land. Discontent and frustration in agricultural circles has provoked clear and widespread political expression. The second form of intervention includes grants and loans for farm improvements, formation of cooperatives for purchasing and marketing, and operating advisory schemes. Although undoubtedly better than straight price fixing this type of action aids larger, better-informed farmers who know how to take advantage of it. Increased efficiency may enable farmers to enjoy a higher living standard, provided that agricultural prices rise sufficiently to keep up with inflation. At the present time of continuing surpluses of some farm products it may be argued that raising the efficiency of every farmer in Western Europe who applies for assistance is no longer desirable.

Reorganizing farm structures forms the third type of action and, if sufficiently radical, may offer a more satisfactory solution. Two distinct problems of farm fragmentation are encountered: division of land into small property units and pulverization of these holdings into tiny strips that may be scattered over a wide area ahd be time-consuming and uneconomic to work with machinery. Fragmentation of holdings originated in several ways, including fossilized plots dating from earlier systems of communal farming, piecemeal land reclamation, and inheritance laws requiring equal division between heirs. Under the technical conditions of the mid-1950s at least half of Western Europe's farmland was considered to be in need of consolidation, ranging from 5 per cent in Denmark and Sweden to 50 per cent in West Germany and Spain and 60 per cent in Portugal. Considerable progress was achieved in the following two decades but thresholds for the desirability of consolidation had also changed and substantial areas still required this structural improvement in 1970 (Table 4.2). Schemes exist in all West European countries but they vary tremendously, from very simple exchanges of parcels for creating larger plots, to ambitious programmes in which consolidation is included with controlling soil erosion, building new roads, irrigation, drainage, removal of rural slums, construction of new houses and even the introduction of industrial employment.

Table 4.2. Land Consolidation Schemes in Five West European Countries, 1969–70

	Total agricultural surface (000 ha)	To be consolidated (000 ha)	%	Consolidated (000 ha)	%	In progress (000 ha)	%
Austria	1,540	850	(55)	472	(31)	60	(4)
France	30,830	15,000	(49)	6,900	(22)	1,800	(6)
Netherlands	2,250	1,140	(51)	360	(16)	600	(27)
Switzerland	1,230	360	(29)	190	(15)	160	(13)
West Germany	20,750	9,000	(43)	5,029	(24)	3,800	(18)

Farm enlargement is another serious problem. In 1970 the average size of farm in the Six was only 12·7 ha and 14·6 ha in the Nine. National averages ranged from 7·7 ha in Italy to 40·2 ha in the United Kingdom and also varied enormously in regional terms within national boundaries. Some 16·6 per cent of farms in the Nine were under 10 ha a further 18·3 per cent between 10 and 20 ha. Within the E.E.C., Italy contained the greatest proportion of small farms, with 38 per cent of her holdings being smaller than 10 ha. The proportion was even higher in Iberia. Such farms are not capable of producing a reasonable income for a family unless they are used for intensive activities, such as factory farming or market-gardening. This is largely the case in Belgium and the Netherlands where agricultural production is highly efficient but over half of the farms are less than 20 ha in size. Unfortunately it is not easy to determine what the desirable size of a farm might be for the future, because of variations in land use, labour input, soil quality, degree of mechanization and a host of other factors.

Legislation in France, West Germany, the Netherlands, Norway and Sweden ensured that when farmland fell vacant it should be used to enlarge neighbouring holdings. But natural changes like this are slow. Attempts to speed up the process have been introduced in Austria, West Germany, France and the Netherlands with annuities and compensation being paid to elderly farmers who retire from agriculture and allow their land to be used for enlarging neighbouring farms. Other schemes give financial assistance to younger farmers who agree to retrain for another job and allow their land to be used for restructuring.

In addition to plot consolidation and farm enlargement, many areas require settlement patterns to be remodelled so that farm buildings can be sited on their farmland rather than in overcrowded villages. Policies of this kind have been implemented in West Germany and the Netherlands, but resettlement is an expensive process. As well as constructing new farm buildings, public utilities must be provided, including hard-surfaced roads between newly dispersed farmsteads and the village. Resettlement in small groups is now preferred to isolated farmhouses. Such a policy reduces the cost of providing utilities and cuts down the social isolation experienced by families who have been moved away from nucleated villages. Land reform has also operated in parts of Western Europe, notably in Italy where objectives were stimulated by social and political conditions in order to provide farms for landless agricultural workers and enlarge existing smallholdings through the division of large estates. The end result has been just the opposite of farm enlargement and will create serious problems for future generations of farmers and agricultural planners.

The final form of government intervention is covered by regional management schemes in which agricultural problems are considered in the context of regional, national and international economies. An integrated approach to rural management has been employed in various parts of France and Spain, in southern Italy and the islands, and the Scottish Highlands and Islands. If such schemes are to be effective, serious appraisals of land potential have to be made for future planning of land use. This kind of operation is undertaken in Sweden and grants and loans for farm rationalization are available only for land that is zoned for remaining in agricultural use. Poor farmlands may be suitable for future afforestation and special grants are available for encouraging this change. In the 1970s it is becoming increasingly clear that radical measures like this are necessary to replace protectionist policies that have shielded inefficient, over-populous and overproductive agricultural systems in most parts of continental Western Europe since the Second World War.

A Common Agricultural Policy

Formulation and implementation of a C.A.P. has formed a focus for harsh political debate in the E.E.C., being of particular concern to France with her large farming population and very extensive area devoted to food production. Nevertheless, in spite of bitter disagreements between member states, the C.A.P. has been viewed as 'the most potent force in the creation of the Common Market' (Charnley, 1973, p. 299) and represents something of a success story when compared with the failure to implement other forms of common policy.

Until the late 1960s each member state operated its own agricultural policy to ensure a food supply base in response to past experience of 'the twin threats of war and scarcity'' which were the principal generators of agricultural policy (Raup, 1971). The result was a series of domestic policies which generated a siege mentality. The emphasis was placed firmly on production and cost considerations were of secondary importance. With the notable exception of the British, Danes and Dutch, all the countries of Western Europe entered the second half of the twentieth century with trade policies designed to provide high levels of output for farm producers, a fixation on expanding agricultural output and no experience of domestic food surpluses. Differing price and support levels ensured protection against imports in each country but there was no way of getting farm produce to flow freely between members of the Six. A Community policy for prices, production levels and marketing was essential.

Progress in establishing such a policy was slow, since governments continued to stress national interests over Community objectives. However, important steps were taken in the 1960s, including the establishment of the Agricultural Guidance and Guarantee Fund (F.E.O.G.A.) to finance the developing C.A.P. and thereby tackle problems of modernization as well as price supports and subsidies when market prices fell below accepted intervention prices. Common price systems came into operation in 1968 for all the main temperate agricultural products and, by that time, the Community had dissolved national systems of support, replaced them by Community support systems and swept away protection between member states.

The net result was to provide relatively high financial returns to farmers in the Six (efficient and inefficient alike) and to continue to increase output. In the late 1960s, payments for the guarantee section of F.E.O.G.A. were accounting for 70–80 per cent of the total budget of the Community, with France as the leading agricultural producer of the Six receiving the greatest share to compensate for financial losses on exported farm goods which had to be disposed of on international markets at prices below protected Common Market levels. The guidance section of the F.E.O.G.A. was much smaller, with Italy being the leading beneficiary and receiving 34 per cent of the total assistance paid up to the end of 1970 (Figure 4.3). The Sud region of the Mezzogiorno alone received 9 per cent of aid to help with schemes for irrigation, marketing and other forms of agricultural advance.

Clearly the F.E.O.G.A. contains a serious internal contradiction. The guidance section helps to improve agriculture and, to a minor extent, the overall economic structure of agricultural areas. But it is undermined in its effect by the price support action of the guarantee section. High guaranteed prices have encouraged even greater output of unsaleable goods (such as the butter mountain of the late 1960s and the beef mountain of 1974) which have had to be stored, exported at subsidized prices, or actually destroyed (in the case of perishable fruit and vegetables). Enormous burdens were placed on the Community's Agricultural Fund and on national exchequers contributing to it, which essentially meant a flow of finance from industrial West Germany to more emphatically agricultural countries, especially France which has received over £600,000,000 more out of the Farm Fund than it contributed to it since 1962.

A savage cut in commodity prices might reduce agricultural production but there is evidence that tampering with prices (which is politically tolerable) will not have this effect. The basic problems may be summarized as too many farmers and agricultural workers in many parts of Western Europe and too much land in agricultural use. It is true that the farm labour force in the Six declined by 3 per cent per annum during the 1960s but productivity per worker increased annually by 7 per cent. The cultivated area of the Six was 5 per cent less in 1970 than ten years earlier and the dairy herd was not significantly larger, but yields of wheat and barley were up by 20–25 per cent, maize

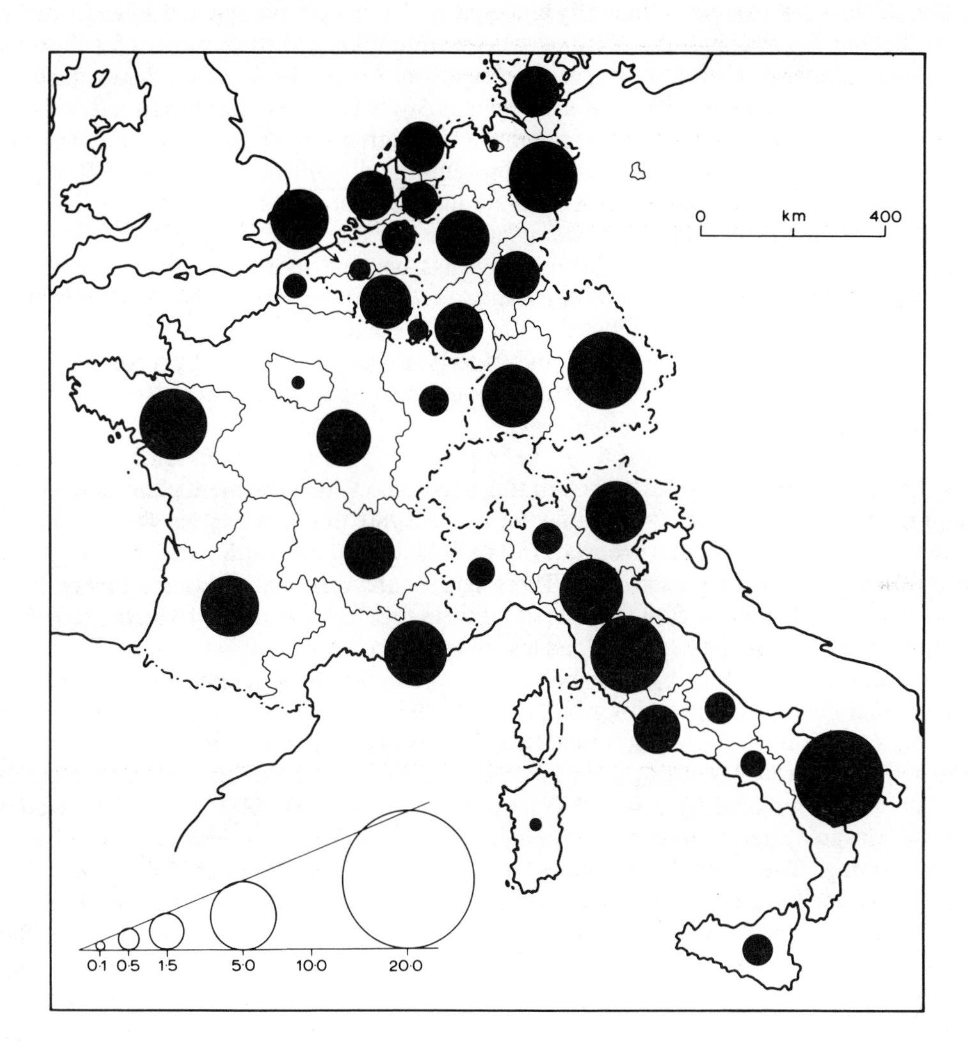

Figure 4.3. F.E.O.G.A. guidance section investment: percentage distribution to 31.12.70 (by region).

by 30–40 per cent, sugar beet by 20 per cent, and yields from dairy cows by 15 per cent.

The situation was reviewed critically in 1968 by Dr. Sicco Mansholt who suggested measures to rescue farming from being an antiquated tradition-bound way of life and make it into a modern business. Admittedly the farming population of the Six was being reduced by 500,000 each year during the 1950s and 1960s but even this was unsatisfactory for reasons of quantity and quality. Young people flocked away from the land leaving the old and unambitious who were unlikely to modernize their farming activities. Mansholt therefore proposed three objectives for 1980: to accelerate the drift from the land, to radically change farm sizes, and to balance out the supply and demand for farm products. Mansholt envisaged that a total agricultural population of 5,000,000 in the Six in 1980 would be desirable, representing only one quarter of the 1950 figure of 20,000,000. In the past, reduction in the number of farmers and agricultural labourers resulted from natural decrease and the fact that better wages were offered in industry. But the task of reducing numbers has become more difficult in the 1970s since the hard core of farmers and small landowners now remains. Every effort should be made to divert children of farming families to take other jobs. A second form of action would involve encouraging the elderly to leave farming by offering more effective annuities and pensions. Efforts should also be made to attract some younger people out of farming and into other jobs.

Small peasant holdings need to be replaced by larger agricultural enterprises practising modern techniques and operating according to development plans that would require official approval. A suitable farm for 1980 might comprise 80–120 ha of cereals, or raise 40–80 dairy cows, 150–200 head of beef cattle, 450–600 pigs, or 100,000 head of poultry each year. These targets are very ambitious and this point is emphasized by the fact that two-thirds of farms in the Six were under 10 ha in the 1960s and two-thirds of all dairy farmers had less than five cows apiece.

The third line of attack by Mansholt involved conditions of supply and demand. He argued that guaranteed prices of grossly overproduced commodities, such as sugar beet and milk, should be slashed to reduce production and hopefully dispose of surpluses. Marketing would need radical reorganization and by 1980 the cultivated surface of the Six should be reduced by 5,000,000 ha out of the total 70,500,000 ha. This would involve an area greater than the farmland of the Benelux countries being withdrawn from farming. The areas that would be affected were not spelled out in the report, but underdeveloped rural regions with limited physical potential for efficient farming would probably bear the brunt (Figure 4.4). They would include southern France, the Massif Central, Corsica, southern Italy and the islands. Mansholt suggested that one-fifth of the liberated surface should be used for national parks and recreation space and the remainder be devoted to afforestation. Financial assistance would help landowners who were willing to convert farmland to other uses provided that this received approval from Community planners.

The general reaction to these proposals was one of hostility, especially from elderly farmers and from associations defending the traditional family farm. The six governments were also dubious. However, Mansholt's shock treatment underlined the predicament of farming in the Six and promoted rational reflection and analysis, especially from young farmers.

Fifteen months after the initial proposals Mansholt announced a modified memorandum, couched in more delicate phraseology, but the general objectives

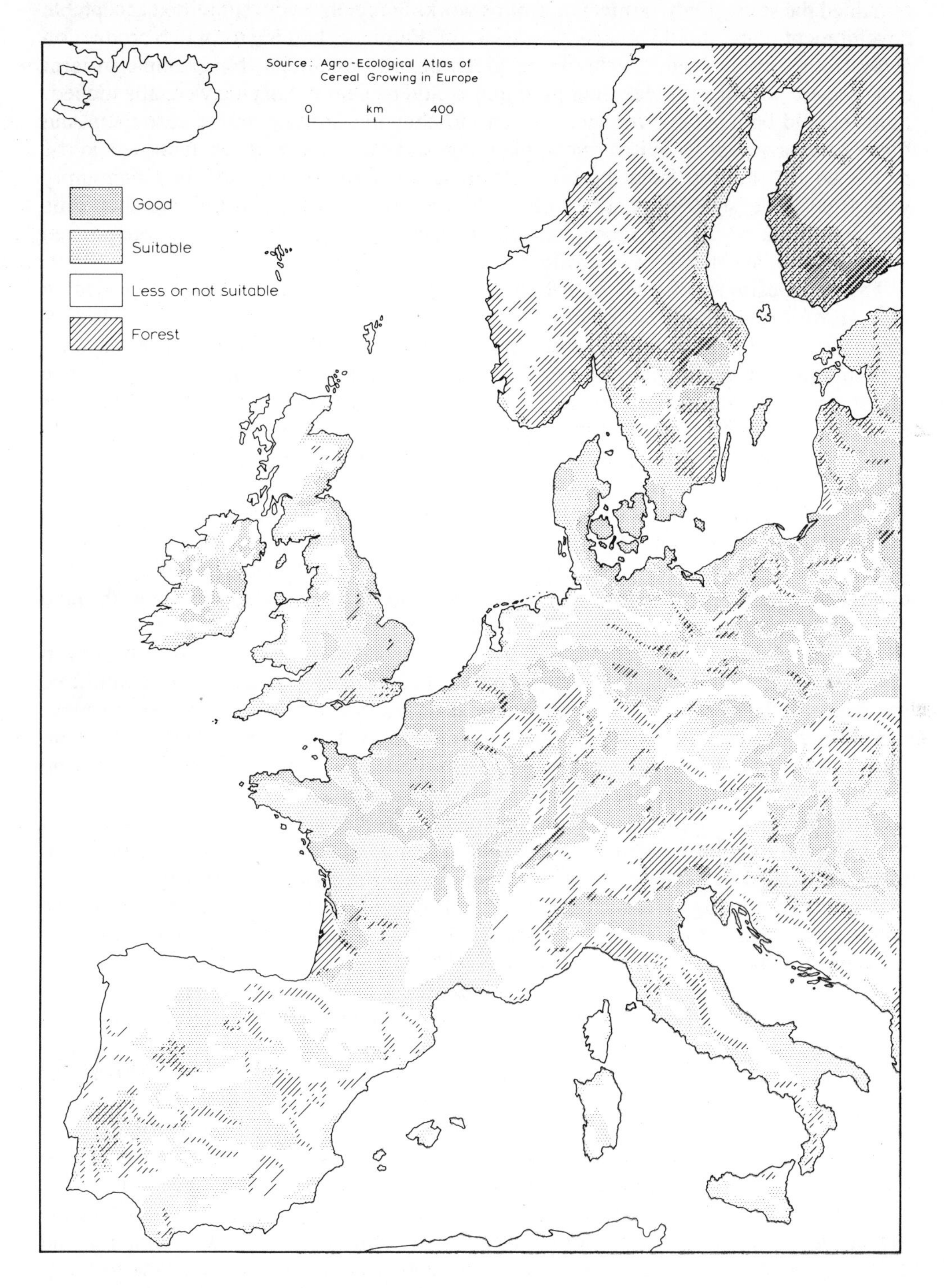

Figure 4.4. Physical potential for arable production in Western Europe. (Reproduced by permission of P.U.D.O.C., Wageningen, The Netherlands, and Elsevier Scientific Publishing Co., Amsterdam.)

remained the same. Only farmers or groups working together who submitted acceptable development plans should receive financial aid. Farmers planning to switch production away from surplus products should be given priority treatment, for example in farm enlargement schemes. A slaughter premium would be paid if dairying were abandoned. Sums should be available for those willing to abandon farming and release their land for whatever purpose suited local planning objectives. No more land would be reclaimed for farming from the sea or from scrub or forest. Instead, the Community would pay at least 80 per cent of afforestation costs. Advice and training should be available more readily for those choosing to remain in the Community's modernized farming sector and also for those who decided to retrain for other jobs.

The implications of the Mansholt proposals are enormous. While providing measures for revitalizing agriculture it is clear that the problems of poverty and lack of alternative employment cannot be solved by programmes that deal only with farming. Agricultural schemes need to be closely integrated with plans for the general economic and social development of the rural residuum. Provision of alternative job opportunities for displaced farmers, agricultural workers and their families requires maintaining a high level of economic activity and employment, and ensuring that suitable training and retraining facilities be made available. Conclusions of this kind are being reached throughout Western Europe since most countries have entered the phase of agricultural overproduction. Sweden, for example, has implemented policies to cut back agricultural output for over a decade. This is a vast new problem which is diametrically opposed to past objectives of increasing Western Europe's farming areas through reclamation and generally raising agricultural output.

A Community marathon in March 1972 approved three directives which moved in the direction that had been proposed by Mansholt. Firstly, interest rebates would be provided for farmers who submit acceptable development plans. However, member countries disagreed as to how national and Community aid should be disbursed to help farm modernization. Secondly, pensions would be available to elderly farmers who quit their land and allowed it to be used for enlarging development farms, for forestry, or recreation space. Farmers need not leave their farmhouses but just give up their land by sale or on long lease. (This measure is of key importance in Italy where no policy of this kind existed beforehand.) Finally, training facilities would be provided for those giving up farming. These three directives are highly significant because they introduce a new balance in the C.A.P., extending the future range of intervention beyond price measures that have dominated policy so far. Their application will not usually involve changing the national legislation of most countries, but they do introduce a common approach to farm reform and offer a cash contribution from the Agricultural Fund which should encourage national governments to speed up their own programmes.

Enlargement of the E.E.C. and application of the C.A.P. to the new members has aroused important new controversies, linked to the fact that Denmark and the Republic of Ireland are important exporters of agricultural produce and the United Kingdom is a major importer. National viewpoints have clashed on many issues, including the formulation of schemes to help the 700,000 or so hillfarmers in the Community who cultivate one-fifth of its farmland. Possible methods of financing schemes and the actual delimitation of hillfarming areas on a combination of demographic and environmental criteria have provoked strong debate from the countries that are most likely to benefit. Thus the Italians, with perhaps 200,000 hillfarmers, wanted to include the whole of the Mezzogiorno. The French, with 150,000, wished to include extensive mountain dairying districts, and the British, with 25,000, wanted to ensure that aid should be available to all areas that benefited from national hillfarm subsidies in the past.

Debates of this kind will continue in the future as each member of the Nine seeks to ensure that it receives a share of whatever financial allocation is being proposed for the support or modernization of farming. However, an important step was taken by the Community in October 1973 when it identified areas qualifying for priority aid from the F.E.O.G.A. at the same time as it issued its list of areas needing regional aid. The criteria used to define priority agricultural zones included: the proportion of the active population involved in farming above the Community average (9·8 per cent); a *per capita* G.D.P. below the Community average (2420 units of account); and a below average proportion of workers in manufacturing industry (43·9 per cent). The resultant pattern was rather similar to that included in the *Thomson Report*, with the exclusion of industrial problem areas (Figure 4.5). Southern and central Italy and the islands, western France, Ireland, and upland Scotland formed the most extensive priority zones, with smaller sections of the other nations also qualifying. The Commission proposed an additional 1,500,000,000 units of account to be spent in those regions in 1974–76 not only to aid agriculture but also to provide a system of incentives to create non-agricultural employment opportunities for those leaving the land. This action is highly important since it represents a shift away from F.E.O.G.A.'s almost exclusive concern with agricultural price supports and introduces a broader and more realistic approach to the problems of the rural residuum.

The Future of the Rural Residuum

The changing position of Western Europe in the world economy, especially with regard to rising energy prices and their implications for living standards, mobility and continued economic growth, would make it unwise to offer precise statements about the relative significance of agriculture and other functions in the rural residuum in the future. Almost certainly, less land will be used for food production and this function will become more concentrated in areas with fertile soils and favourable environmental conditions, such as lowland England and the Paris Basin where pressures from urban expansion are also high. More land will be devoted to timber production and formal recreation space. Already 119,000,000 ha of Western Europe (35·4 per cent of the land surface) are under trees. National variations are great, ranging from 2·8 per cent in the Republic of Ireland to 52·1 per cent in Sweden and 64·5 per cent in Finland (Figure 4.1). Many woodlands form important recreation spaces for urban dwellers and this kind of dual function, combining timber production with outdoor recreation, will become increasingly significant, especially in regions within easy driving distance of major cities. Forestry will gain importance as a user of land and possibly as an employer of labour in the more distant areas, but mechanization of planting and felling means that the number of workers supported directly from timber production will remain small. Even now, it has been estimated that the total timber resources of Sweden could be managed by a workforce of only 50,000. Nevertheless, establishment of wood-processing industries offers valuable possibilities for rural employment once timber has matured.

Many areas of distinctive environment in the mountains, forests, marshes and heathlands of Western Europe have been designated already for nature conservation,

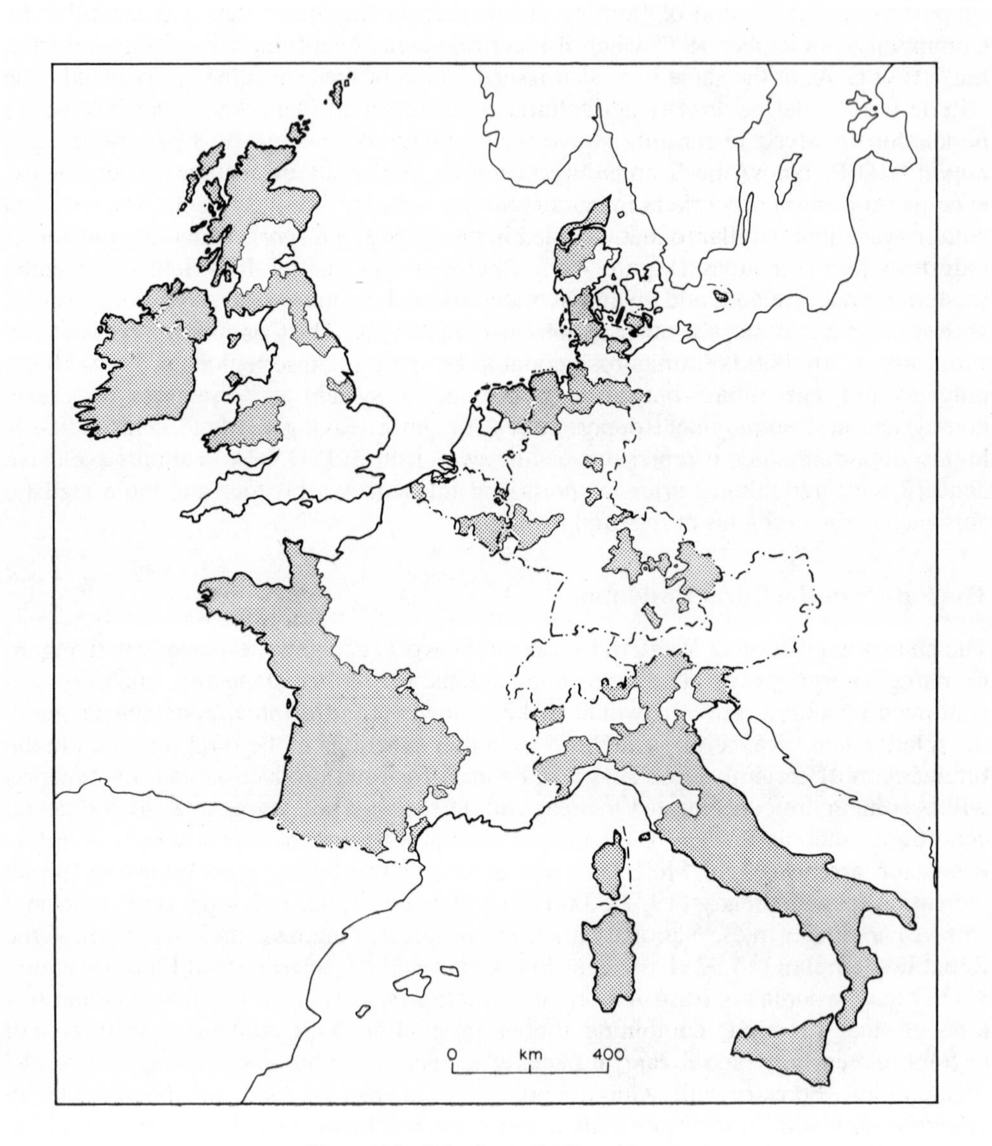

Figure 4.5. Priority agricultural zones.

public recreation, or a combination of the two. Most national and regional parks are post-war creations, varying considerably in organization, size and accessibility. The balance between nature conservation and recreation varies from country to country. At one extreme there are national parks in remote parts of northern Sweden and Finland where the emphasis is on protecting flora and fauna, but in southern parts of Scandinavia some parks are more specifically aimed at conserving traditional cultural landscapes for visitors to enjoy. At the other extreme, national parks in England and Wales come under very heavy recreational use at weekends and in the summer months by virtue of their proximity to major population centres. Similarly, the Dutch Kennermerduinen national park with 1240 ha of coastal duneland has a catchment of 2,000,000 potential visitors within a 30 km radius. Some 700,000 people use the park each year, with 20,000 being an average daily attendance for a fine summer day. By contrast, most national parks in France are distant from large cities and, in any case, are zoned to channel visitors to certain areas and keep others for nature conservation. Most French regional nature parks are quite accessible from large urban centres and cater for short-term recreational demands. Although West Germany has no national parks as such, many of her nature parks offer facilities for recreation as well as nature conservation. The best known of these is the 20,000 ha Lüneburger Heide, 50 km from Hamburg. More national and regional parks will undoubtedly be designated in Western Europe in the future. By virtue of their varied composition of woodland, farmland and heath, they will form multiple land-use zones where farming may be supported for aesthetic rather than for economic reasons in order to maintain attractive, manicured landscapes for visitors to appreciate.

In addition, the rural residuum will fall increasingly under the shadow of permanent and temporary urban uses. Sections of farmland will be abstracted for new housing, industry, roads, airports and many other uses. An interesting commentary on the growing impact of urban land-use demands over the last three decades is provided by the fact that the Dutch Noord-Oost Polder, where reclamation started in 1942, was largely devoted to farming (87 per cent), with only 1 per cent for building, 5 per cent for woodland and recreation, and 7 per cent for canals and communications. By contrast, both the Zuid Flevoland Polder (started in 1968) and the Markerwaard Polder (perhaps to be started in 1975) may well have only half of their areas under agricultural use, with 18 per cent for building, 25 per cent for timber and recreation, and 7 per cent for canals and communications.

Broad zones of land around urban employment centres will retain their essentially 'agricultural' appearance but their settlements will be expanded and will function as residential zones for commuters. Such development depends largely on car-based mobility and may be curtailed by rising oil prices or be channelled along axes with good public transport. Increasing numbers of urban commuters may become hobby farmers but the reverse tendency will also operate, with farmers taking on urban jobs and managing their farms in their spare time or leaving portions of their holdings completely unused. Already 215,000 ha of such abandoned 'social fallow' was in existence in West Germany in 1970, compared with 13,500,000 ha in agricultural use. It has been estimated that the social fallow surface may rise to 700,000–1,000,000 ha by 1980. The rural residuum will come under increased pressure from weekend motoring, provision of other forms of outdoor recreation and the establishment of second homes. Demands will be most intense in areas with good accessibility from urban areas and with pleasant scenery or other attractive features. Once again, the general health of Western Europe's economy and in particular the cost of oil will have an important role in determining the extent of these demands.

Distant areas in the countryside will not escape the impact of tourism which offers important, if only seasonal, support for other forms of employment. As Christaller (1964) noted, tourists are attracted to even the highest mountains, the most lonely woods and the remotest beaches which have a definite appeal for a few weeks each year but have missed out in the race for urban and industrial development and therefore offer limited possibilities for employment for their residents. Similarly in Mediterranean Europe, location, climate and other factors prevented Iberia and southern Italy from participating in the benefits of past industrialization but now offer attractions and provide an influx of wealth to finance new development (Naylon, 1967). In spite of the currency that tourism can bring it is ironic that the more developed a country's tourist industry, the lower the standard of living of its inhabitants (Young, 1973).

Serious environmental disbenefits from tourism have been demonstrated along the shores of the Mediterranean but also affect the rural residuum of Western Europe. Fragile ecosystems may be damaged irreparably as a result of tourist pressure. 'Urban' style accommodation and associated development may completely alter the character of recreational areas. Many of Western Europe's most popular tourist areas have been developed without adequate physical planning and dire environmental problems have resulted. Three examples will suffice to make the point. Every one of the 6000 registered beaches in Italy is dangerously polluted according to standards decreed by the Italian Health Ministry. Further to the west 195 open drains between Cannes and Menton discharge untreated sewage into the tideless Mediterranean; and to the north, the river Rhine is gradually becoming a vast septic sewer stretching from the Alps to the North Sea. As Young (1973) stresses, these consequences are not *inherent* in the development of tourism; they just happen when tourism is developed in a thoughtless way. They do, however, serve as an object lesson to regional planners concerned with the future management of the rural residuum for recreation and other forms of development.

Important though its local impact will be, tourism cannot be the economic mainstay of Western Europe's rural areas. Permanent, rather than seasonal, sources of income are required in the form of light manufacturing or tertiary employment. Installation of such activities in small towns sited away from economic core areas, provincial capitals, or areas with climatic and other environmental attractions has proved to be the hardest test of the efficacy of regional development programmes. Nevertheless it is undeniable that in the long term, the economic health of the rural residuum will depend on neither agriculture nor tourism but on the creation of secondary and tertiary jobs at relatively small 'growth centres' (or 'holding points') where commercial and educational facilities should be grouped to serve broader but still depopulating rural hinterlands. Progress towards rationalization of settlements and service provision has been achieved already in depopulated northern Sweden. This example needs to be emulated, with appropriate modifications for local conditions, in rural regions elsewhere in Western Europe. Without such 'holding points' the future of many of the more remote parts of the countryside will involve little more than a continuing contraction of agricultural population with perhaps a slight financial input from tourists, second-home owners and retired people.

However, this kind of scenario, outlining the possible shape of things to come in the countryside, has been based on the general assumption that living standards in Western

Europe will continue to rise, personal mobility (in the form of car ownership) will increase, and that industrial economies will continue to flourish in order to finance schemes for regional development. Should this assumption prove invalid, in the light of rising energy costs, flows of rural population to urban areas and of townsfolk into neighbouring and even distant parts of the countryside may be curtailed severely. Conservationists may applaud the enforced containment of 'Urban Europe' and the relative stability of its rural areas that may result, but such a condition will not encourage the solution of spatial disequilibria. Rural poverty and urban congestion will continue, and both kinds of environment may be increasingly harmed and polluted. Conservation may be practised under conditions of continuing economic growth but is much less feasible in times of stagnation or recession when it appears as an expensive luxury.

References

Charnley, A. H., 1973, *The E.E.C.: a study in applied economics,* Ginn, London.

Chiffelle, F., 1973, Le remembrement parcellaire au service de l'aménagement régional. Le cas de la Suisse, *Annales de Géographie,* **82**, 28–41.

Christaller, W., 1964, Some considerations of tourism locations in Europe, *Regional Science Association Papers,* **12**, 95–105.

Clout, H. D., 1971, *Agriculture: studies in contemporary Europe,* Macmillan, London.

Clout, H. D., 1972, *Rural Geography,* Pergamon, Oxford.

Clout, H. D., 1973, *The Massif Central,* Oxford University Press.

Desplanques, H., 1973, Une nouvelle utilisation de l'espace rural en Italie: l'agritourisme, *Annales de Géographie,* **82**, 151–164.

Fel, A. and J. Miège, 1972, Transformation et urbanisation des campagnes en Allemagne fédérale, *Annales de Géographie,* **81**, 579–593.

Ford, G. W., 1966, Agricultural policies in the member states of the Common Market, *Agriculture,* **73**, 410–415.

Franklin, S. H., 1969, *The European Peasantry,* Methuen, London.

Franklin, S. H., 1971, *Rural Societies: studies in contemporary Europe,* Macmillan, London.

Gervais, M. and coworkers, 1965, *Une France sans Paysans,* Editions du Seuil, Paris.

Hallett, G., 1968, Agricultural policy in West Germany, *Journal of Agricultural Economics,* **19**, 18–95.

James, P. G., 1971, *Agricultural Policy in Wealthy Countries,* Angus and Robertson, Sydney.

Jansen, A. J., 1974, Explorations into the future of agriculture in Western Europe, *Sociologia Ruralis,* **14**, 45–53.

Kleinpenning, J. G. M., 1969, Geographical stability and change in Ebro delta, *Tijdschrift voor Economische en Sociale Geografie,* **60**, 35–59.

Lambert, A. M., 1963, Farm consolidation in Western Europe, *Geography,* **48**, 31–47.

Lavery, P. (Ed.), 1974, *Recreational Geography,* David and Charles, Newton Abbot.

Mayhew, A., 1970, Structural reform and the future of West German agriculture, *Geographical Review,* **60**, 54–68.

Mayhew, A., 1971, Agrarian reform in West Germany, *Transactions of the Institute of British Geographers,* **52**, 61–76.

Mendras, H., 1970, *The Vanishing Peasant: innovation and change in French agriculture,* Massachusetts Institute of Technology Press.

Mead, W. R., 1973, *The Scandinavian Northlands,* Oxford University Press.

Naylon, J., 1967, Tourism: Spain's most important industry, *Geography,* **52**, 23–40.

Raup, P. M., 1971, Constraints and potentials in agriculture, in R. H. Beck and coworkers, *The Changing Structure of Europe,* University of Minnesota Press, Minneapolis.

Richardson, G. A. and C. Canevet, 1970, The problems of change in a Breton farming community, *Scottish Geographical Magazine,* **86**, 35–40.

Rickard, R. C., 1970, Structural policies for agriculture in the E.E.C., *Journal of Agricultural Economics,* **21**, 407–449.

Rogers, A. W., 1971, Changing land-use patterns in the Dutch polders, *Journal of the Royal Town Planning Institute*, **57**, 274–277.

Simmons, I. G. (Ed.), 1975, *Rural Recreation in the Industrial World*, Edward Arnold, London.

Smith, J. L., forthcoming, *Western Peninsulas of Europe*, Oxford University Press.

Swann, D. and coworkers, 1974, *The European Economic Community: Economics and Agriculture*, Open University Press, Milton Keynes.

Tracy, M., 1964, *Agriculture in Western Europe: crisis and adaptation since 1880*, Cape, London.

Tracy, M., 1972, Structural reform in agriculture, *O.E.C.D. Observer*, **57**, 17–28.

Turnock, D., 1973, *Scotland's Highlands and Islands*, Oxford University Press.

Van Hulten, M. H. M., 1969, Plan and reality in the Ijsselmeerpolders, *Tijdschrift voor Economische en Sociale Geografie*, **60**, 67–76.

Warley, T. K., 1969, Economic integration of European agriculture, in G. R. Denton (Ed.), *Economic Integration in Europe*, George Allen and Unwin, London, 286–306.

Weinschenck, G., 1973, Issues of future agricultural policy in the European Common Market, *European Review of Agricultural Economics*, **1**, 21–46

Wylie, L., 1961, *Village in the Vaucluse*, Harvard University Press.

Wylie, L. (Ed.), 1966, *Chanzeaux: a village in Anjou*, Harvard University Press.

Young, G., 1973, *Tourism: blessing or blight?*, Penguin, Harmondsworth.

II

Regional Development in Practice

5

ITALY

Russell King

Importance and Significance of the Mezzogiorno

Of all the countries in Europe that have to contend with problem regions, Italy presents the most classic model of economic dualism. Italy has the distinction and the challenge of combining one of the most advanced industrial economies of Western Europe with one of the continent's poorest and most depressed areas—the Mezzogiorno, the land of the midday sun. If the problem of underdevelopment within the E.E.C. is to be tackled with any chance of success, the crucial testing ground will be southern Italy, where economic and social backwardness is more pronounced than anywhere else. In fact the problem of the backwardness of the Italian South is of worldwide significance, for nowhere has there yet been devised a satisfactory mechanism for solving the problems of underdevelopment and regional imbalance that affect nearly all countries of the globe. We may think of Italy, including the South, as very much part of modern Europe, as it is of course, but in the recent past the Mezzogiorno exhibited many of the features of underdevelopment characteristic of countries of the Third World. In 1950, the crucial year marking the start of the regional development effort, southern Italy had a mean *per capita* income below the average for Latin America.

The Italian South—considered here as comprising the mainland regions of Abruzzi, Molise, Campania, Apulia, Lucania (also known as Basilicata) and Calabria, together with the island regions of Sicilia and Sardegna (Figure 5.1)—contains two-fifths of Italy's land area and nearly the same proportion of its population (20,000,000 out of 54,000,000). On both counts this is rather larger than the average conception of a depressed region, at least in the European context. The region's population is more than double that of Greece or of Portugal, for instance, and in some respects one is tempted to treat the South as a separate economy altogether, maintaining some sort of loose contact with the North. The ties of association between the E.E.C. and Greece and Turkey, and the application for associate membership by Spain, suggest that additional countries with similar physical and social resource endowments may eventually become part of the Community and thus qualify for the same treatment as southern Italy.

Economic dualism has existed, and continues to exist, in Italy in at least three senses: it is manifested within individual sectors (the factory and the artisan, the modern capitalist farm and the peasant smallholding); it is present in the intersectoral sense (an advanced industry and a backward agriculture); and it has, most importantly in Italy, a geographical expression—the well-known regional dichotomy between the backward, agricultural, peasant South and the more forward-looking, industrialized North. This regional divide is not easily definable on a map—there are some parts of central Italy, in Umbria and Marche, for instance, which are clearly transitional—but

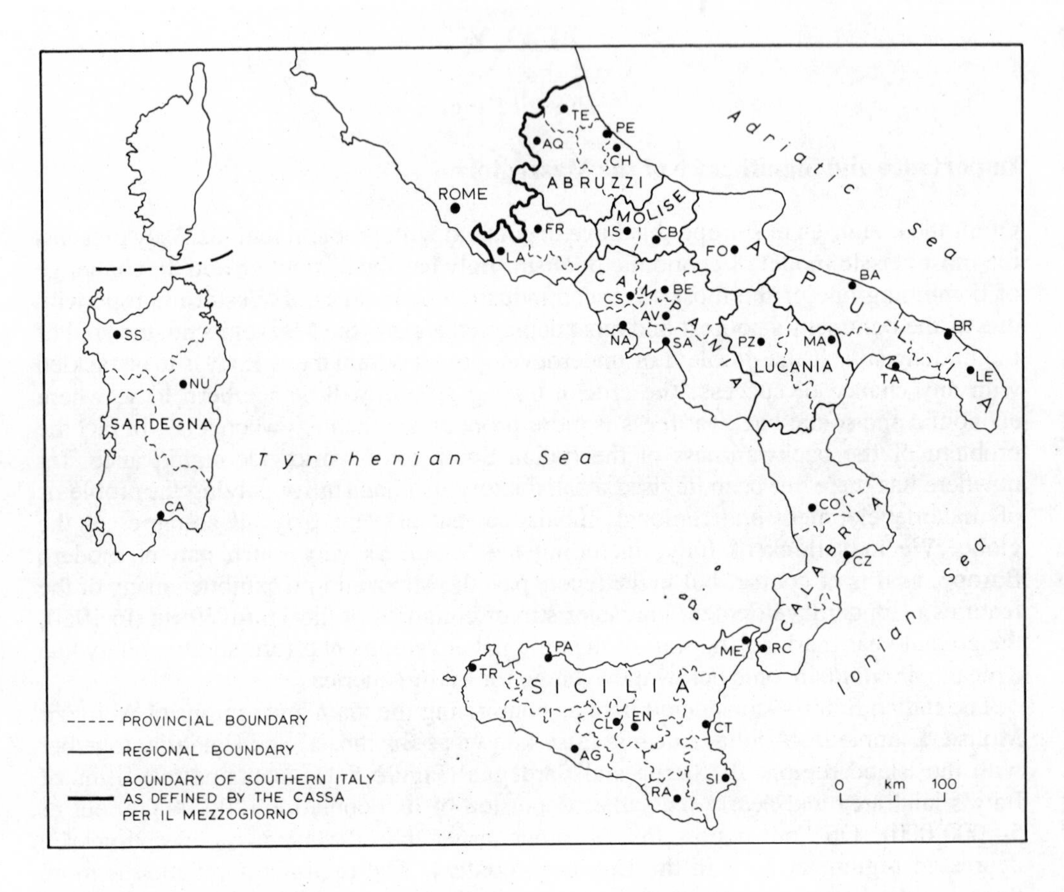

Provincial Capitals:

AG Agrigento
AQ L'Aquila
AV Avellino
BA Bari
BE Benevento
BR Brindisi
CA Cagliari
CB Campobasso
CH Chieti
CL Chieti
CL Caltanissetta
CO Cosenza
CS Caserta
CT Catania
CZ Catanzaro
EN Enna
FO Foggia
FR Frosinone
IS Isernia
LA Latina
LE Lecce
MA Matera
ME Messina
NA Naples
NU Nuoro
PA Palermo
PE Pescara
PZ Potenza
RA Ragusa
RC Reggio Calabria
SA Salerno
SI Siracusa
SS Sassari
TA Taranto
TE Teramo
TR Trapani

Figure 5.1. Administrative areas of the Mezzogiorno.

the two 'halves' of Italy are nevertheless fundamental realities: somewhere south of Rome a change takes place. Even today, the South is part of another world, as if the Spanish feudal heritage lingered on at one end of the peninsula whereas the other part of Italy has almost jumped the Alps into northern Europe. Many of the people of the South still live in poverty and ignorance, with medieval concepts of honour and prestige, of relations between the sexes, between employer and employee and between different social strata.

Perhaps the most practical aspect of the 'southern problem' is that here is an example, rare in the context of the generally unstable political climates of underdeveloped countries, of sustained effort directed towards eradicating regional disequilibria. Only a handful of countries has followed a consistent approach for as long as ten years—generally considered the minimum period necessary for a preliminary assessment of a particular policy—let alone Italy's twenty-five. Partly as a reflection of this worldwide relevance, the case of southern Italy has attracted close attention and has generated a voluminous literature, in turn aided by some of the most detailed official statistics in the world. The region has benefited from the study of a number of leading international economists (Ackley, 1963; Chenery, 1962; Lutz, 1962) and its villages and rural areas have provided a fertile environment for research on the part of sociologists and anthropologists (Banfield, 1958; Galtung, 1971; Lopreato, 1966). It has also been a frequent term of reference for two of the leading theorists of regional development (Myrdal, 1957; Perroux, 1955); the thesis of 'cumulative . disequilibrium' and the notion of *pôles de croissance* have both appeared highly relevant to the Italian experience. For the geographer too, the Mezzogiorno, and the new strategies aimed at developing this problem region, give a new dimension to the study of regional geography, so long straightjacketed in its static, descriptive image.

This chapter consists of two major sections. The first part considers the traditional equilibrium of underdevelopment that existed in southern Italy up to 1950. It is true that many of these features of the traditional South—in particular, geographical factors and some aspects of the social structure—are still present today, but the emphasis is on tracing the evolution of the ecology of backwardness and underdevelopment up to the recent historical past. The second part relates and attempts to evaluate the developmental measures that have been carried out since 1950. The conclusion is that these policies have had partial success, but that geographical changes of at least comparable dimension have been induced by the spontaneous action of the southerners themselves.

Regional Ecology of Southern Italy: the Equilibrium of Underdevelopment

There is no single causal factor in the underdevelopment of the Mezzogiorno; if there were, a solution to the problem of the South would be easy to identify. Rather, a whole complex of factors has been at work interacting variously in time and space to produce a veritable ecology of backwardness. Regional contrasts, in terms of both landscape and prosperity, are very marked. Zones of relative prosperity, in southern Apulia and eastern Sicily for example, make the poorer parts all the more noticeable. Population and settlement patterns range from some of the densest in Europe around Naples and Catania to the wild, untamed mountain tracts of eastern Sardinia. The islands of Sicily and Sardinia in particular can claim to be regional units in their own right; their semi-autonomous status is in part a recognition of their distinctiveness. The Mezzogiorno is not, then, a homogeneous unit. It does, however, possess a number of uniform characteristics, and two groups may be usefully identified at this preliminary juncture: natural factors and human–historical factors. The combination of these historical and geographical influences has produced a number of institutions, such as the large estate or latifundium, the Mafia, the social structure in general, which have acted as a formidable brake on the region's evolution.

Historical Patterns: Colonialism, Unification and Polarization of Industry in the North

The reasons for the South's backwardness have been the subject of vigorous dialectic on the part of Italian scholars for a long time. Historically the *Questione Meridionale* is of special interest because the Mezzogiorno has been a visible and intractable problem for so long. Throughout all Italy's problems runs the theme of the antithesis between the Mediterranean South and the Continental or European North. Ethnically the north–south divide has roots as early as prehistoric tribal migrations and some of the features of the South's relative decline can be traced to the close of the classical era. The South's history is largely one of colonialism. From the Phoenicians onwards the common attitude of successive dominations was to regard the South as a colony to be exploited from coastal footholds for its land and labour. Throughout history the peasants of southern Italy have lived under the dual regime of a remote foreign power and a local oligarchy of feudal lords; nowhere was this pattern more true than in Sicily. Feudalism was vigorously fostered by the Normans but the real foundations of regional dualism and colonialism were laid during the long, decaying centuries of the Middle Ages. During this period a coalition was formed between the local nobles and foreign, usually Catalan, merchants for the purposes of extracting peasant surpluses, chiefly grain, and exporting them, often clandestinely, to foreign markets. With official government remote and incompetent, the fragmentation of power into these 'regional elites' had a lasting impact on the South's social geography (Schneider, 1972). Settlements became orientated to the outside world rather than to each other and the emergence of a functional hierarchy, in which towns existed in economic symbiosis with their surrounding villages, reigning over them administratively and politically, was discouraged.

The *Questione Meridionale* is, however, essentially a national problem and must be looked at primarily in the context of post-1861 political unity. It is generally agreed that the South suffered rather than benefited from union with the rest of Italy. The gap between North and South certainly existed at the time of Unification—Eckaus (1961) estimated that the North's *per capita* income exceeded the South's by about 20 per cent—but it has become wider and more significant during the last hundred years. In the years following Unification, prejudicial taxation drained capital out of the South and the region's embryonic nineteenth century industries, previously shielded by protective tariffs, withered under the blast of competition from the North. Italian industrial development was rapid in the last two decades of the nineteenth century and the early years of the present century, yet this growth was concentrated almost entirely in the North. In fact many factories in the South had closed down by 1890 and Naples was no longer the foremost industrial centre of the peninsula. Agriculture too expanded greatly in the North during the latter part of the nineteenth century, lessening food imports into the region. Southern agriculture, on the other hand, although possessing competitive superiority in one or two crops like citrus fruit and durum wheat, stagnated and failed to provide a basis of savings and consumer demand necessary to support industrial development.

All the geographical factors pointed to the Po Plain as the best location for industry. The North was the seat of the new government; it was nearest to the advanced regions of Europe from which it could draw technological knowledge and capital and in which it could find prosperous markets; it had greater supplies of skilled labour and credit, and better communications. Moreover the North had sources of power: something the South never had in great abundance. The water from the Alpine valleys which had fostered the medieval textile industries was used for electricity generation after the turn of the century and, later, discoveries of oil and gas in the Po Valley further consolidated the region's locational superiority. Under the 'open economy' of the new nation, industrial concentration in the North encouraged the public sector to provide additional social overhead capital, thus creating further external economies for the region, whilst further discouraging capital flows to the South (Schachter, 1967). Because of the overwhelming strength of the polarization effects in the developed region, the market mechanism of the Italian economy over the past hundred years has widened rather than narrowed the regional economic gap; this lends support to Myrdal's contention that the detrimental concentration or 'backwash' effects may be greater than the beneficial 'spread' effects to a depressed region in an open economy.

Geographical Factors

The geography of the Mezzogiorno goes a long way towards explaining the region's fluctuating fortunes. The very factor of location is partly responsible for southern Italy's history of invasion and colonialism. As a natural divide between the eastern and western Mediterranean basins and as a natural bridge between Europe and Africa, southern Italy has been a perpetual meeting-place of cultures and battleground for armies. In contrast to the compact North, with its fertile core of the Po Plain, the very shape of the South, with two peninsulas and two islands, inhibits its spatial integration. Physical geography too plays an important part in the Southern Problem. There is, without a doubt, a kind of 'natural poverty' in the South, though the exact extent of this determinism is the subject of some debate. Mountains and hills occupy 80–85 per cent of the surface area of the Mezzogiorno; steep eroded slopes, deserted formerly malarial plains, stony stream beds, and the general aridity of the climate during the growing season make this a difficult region for agriculture. Throughout the South one is impressed by the strength of the forces of nature—floods, landslides, sheet erosion, gullying, drought and waterlogged plains—and by the inability of the peasant on his own to cope with such forces; indeed, with his simple mattock and wooden plough, and the sheer necessity of cultivating his daily bread, he has brought on much of the environmental destruction himself.

Southern Italy has an unfortunate geological endowment. The High Apennines contain towering limestone blocks reaching nearly 3000 m in the Abruzzi. Intermontane basins are isolated and the highly dissected flanks of the range hinder communications. Especially to the east, the Apennines are bordered by a great belt of rolling clay hills which reappears in south-central Sicily. These laminated clays weather and erode easily. Landslips are particularly frequent in Lucania and Sicily, laying bare whole hillsides. Sheet erosion and gullying are widespread too, and the region is also vulnerable to earthquakes. In the backward South the peasants' traditional view of such phenomena is that they are the work of the devil or the 'evil eye' and that nothing should be done to interfere with such supernatural forces. Ruthless destruction of the original woodland ecology has aggravated the soil erosion problem; the natural water storage capacity of the highlands is reduced, a process which goes hand in hand with the silting up of river valleys and estuaries. The claylands present great problems for peasant agriculture. In winter the soil is heavy and sticky; in summer it is baked hard and equally unworkable. The peasants can only plough lightly for short periods during spring and autumn, and this form of surface ploughing causes the soil layer to become unstable and subject to further erosion. These argillaceous soils thus tend to deteriorate under peasant farming (Dickinson, 1955a).

Older, crystalline rocks form the massifs of Calabria (Sila, Aspromonte), the rugged Peloritani range of northeast Sicily, and much of Sardinia. Apulia is dominated by the long low line of the Murge hills; thin soils and karstic phenomena predominate and the uplands are seamed by canyon-like dry valleys. A similar landscape is found in the limestone tablelands of southeast Sicily. The plain areas of the South, limited to only 15 per cent of total land area, include the Tavoliere, the Plain of Catania and the Sardinian Campidano (Figure 5.2). Although parts of these lowlands were quite richly cultivated in classical times, for the last 1500 years malaria, combined with low rainfall and marshiness, have discouraged exploitation. In the zones of the South which are really favourable for peasant agriculture—the Campania Felix, the richly-weathered *terra rossa* soils of the Bari littoral and the Salentine Peninsula, and the fertile lower slopes of Vesuvius and Etna—population pressure has caused continual fragmentation of farmholdings to the point where many are economically non-viable.

Climate too must be judged as largely a negative factor. With parts of the Mezzogiorno lying further south than Tunis or Algiers, the region has a classically Mediterranean climate. Mean monthly winter temperatures are 6–10 °C; mean summer values around 25 °C, falling with altitude. Absolute maxima, achieved on the plains of the Tavoliere and Catania, can reach 40 °C on midsummer afternoons. Rainfall amounts in southern Italy vary enormously, not only from one location to another (relief being the most important control) but also from year to year. Parts of the southern Apennines receive over 2000 mm annually. At the other end of the scale lowland Apulia, coastal Lucania and southern Sicily receive less than 500 mm. More important than average amounts are yearly differences and seasonal regimes. With individual years producing below 250 mm, drought occasionally becomes absolute; and in most lowland areas a rainless period of 4–5 months is normal. For months on end the brassy sun fires down from a cobalt blue sky. When rain does fall, it often comes in the form of short downpours, the intensity of which limits their usefulness and increases their erosive power. Winter is surprisingly bleak in the mountains, with cold raw winds and snow cover for up to four months of the year. Below freezing temperatures are rare along the coast but Potenza (820 m), a provincial capital which can be considered representative of the hundreds of hill-villages of the South, averages 40 days of frost per year.

Such, then, are the principal features of the physical environment within which the southern peasant has to exist. Without irrigation, crop yields are poor and subject to wide fluctuation. These harsh natural conditions, combined with remoteness from markets, poor communication facilities, backward farming techniques and increasing population pressure, have placed southern agriculture in a generally inferior position

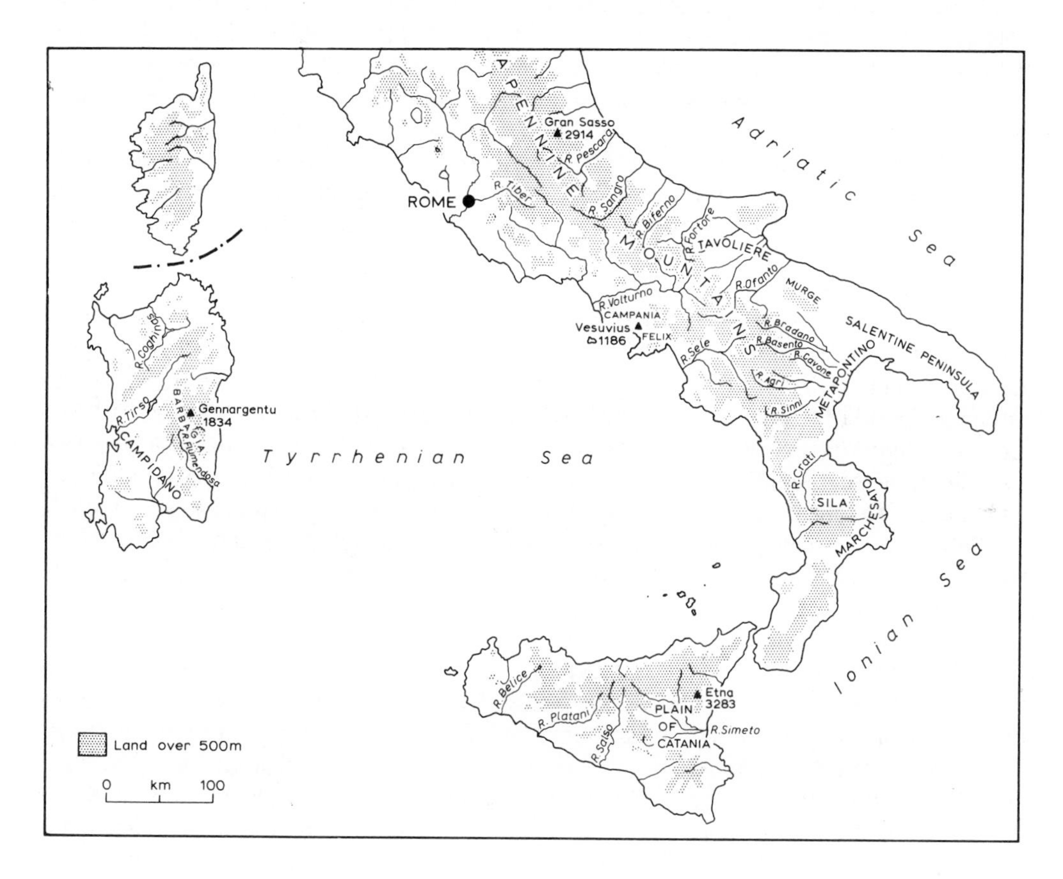

Figure 5.2. Physical environment of the Mezzogiorno.

to that of the North; yet the southern economy has always been more dependent upon the products of the rural sector.

The Ecology of Latifundism

The most important aspect of the historical geography of the rural South is the latifundium; the persistence of this partly feudal, partly capitalistic institution has proved to be the greatest hindrance to the development of southern agriculture. Latifundia were, and are, large landed estates of hundreds, or even thousands, of hectares, mostly orientated towards the production of cereals. Many of the large estates were very ancient in origin, going back to Roman times, but it was the Normans and the Spanish who consolidated the latifundium into such a static, persistent institution. Even when feudalism was abolished in 1806 the rising rural bourgeoisie grabbed most of the former fief land for themselves so that the peasants were, if anything, more destitute than before; although freed from social servitude, the loss, without compensation, of their basic 'survival rights' of pasturage and gleaning plunged them into economic bankruptcy.

Two basic types of latifundium have existed in the South. In its original form—*latifondo capitalistico*—the estate was managed centrally and run with a number of labourers hired on a daily basis each morning at dawn in the local piazza. Unemployment was strongly linked to the latifundian economy for these estate labourers normally found work for less than 100 days per year (Dickinson, 1955b). *Latifondo capitalistico* was commonest in the plain areas of the South: the Apulian Tavoliere, the Metapontino, the Calabrian Marchesato and some of the coastal plains of Campania. In these areas over half the agricultural population were landless day-labourers (*braccianti*), of whom there were 2,000,000 in the South in 1950. Since the nineteenth century most of the estates, however, have been split up into leasehold blocks held by intermediaries who in turn split up the leases into small parcels for renting out to individual peasants: this is the second type of latifundium, called *latifondo contadino* or 'peasant latifundium'. Economically these intermediaries were parasitic, but they wielded tremendous power. As land controllers they were able, in the face of increasing population pressure and land hunger, to continually bid up rents. Virgin soil, often on steep hillsides that should have been left well alone, was ploughed up and overcropped with wheat. Short leases diminished peasant incentive to invest in the land; irrigation, cattle stalls, long maturing crops like vines and fruit trees, and even the application of fertilizer were unknown in this system. The few thousand large landowners (*latifondisti*) who between them owned two-thirds of the land of the Mezzogiorno, were absentee landlords. Apart from occasional seasonal hunting visits, they lived most of the time in the cities of Italy, thereby channelling capital resources out of the rural South into urban areas.

The latifundium, based on a triennial wheat–pasture–fallow regime, was closely influenced by natural conditions. Aridity kept wheat yields to below 10 q/ha—they are not much higher today—with drastic reductions in drought years. From an ecological viewpoint, the land-use regime did give the latifundian system some degree of stability, at least from the standpoint of the landowners and their agents. Its double-sided nature enabled it to switch from cereals to livestock and vice versa as occasion demanded. The exodus of a substantial portion of the rural labour force through emigration during

1880–1914 was countered by expanding the pasture area at the expense of cereals, whereas during the inter-war period Mussolini's 'Battle for Wheat' policy reversed this trend. The cyclical pattern of land use also reinforced the pattern of short-term tenancy contracts. The peasant, denied permanent access or the chance to actually settle on the land, was constrained to be an economic nomad constantly in search of work as labourer, tenant or sharecropper. Rotation and short-term contracts were thus functionally related to another important aspect of latifundism: the settlement pattern. The extremely seasonal nature of labour requirements for cereal cultivation encouraged peasants to be absent from the land for much of the year and the interest of the land controller was in preventing stable settlement on the land on the part of the peasants; hence the existence and persistence of 'agro-towns' which are a clear expression of peasant underemployment and alienation from the land. Peasant agitation against the latifundian system started quite early in the present century in some parts of the South, in the Tavoliere for instance where a strong tradition of peasant communism and unionization developed; in western Sicily on the other hand the strong identification of the Mafia with the land-controlling intermediary class preserved the many quasi-feudal characteristics of latifundism up until surprisingly recently (Blok, 1969a).

Social Structure

Strongly linked to agricultural organization and land-tenure patterns was the social structure of the South. Land ownership was the basic component of social class differentiation, and still is, to a reduced extent, for the medieval structure of lord, vassal and serf survives today as the hierarchy of landowner, tenant and labourer. With land coveted not so much for what it could economically produce, but for its prestige value, with high status attached to leisure and with the very lowest status attached to manually working the land, a vital stimulus to development was lacking. The fundamental distinction in southern rural society has always been between those who work the land manually and those who do not. As absentee landlords the large estate owners were not a familiar sight in the village scene although their castles and palaces often dominated the villagescape. Far more dominant in village life were the *borghesia.* These people, often the agents of the absentee landlords, frequently combined ownership and control of land with professional and administrative posts (doctor, lawyer, clerk etc.). In Sicily this group formed the nexus of the rural Mafia. In contrast to more advanced parts of Europe the southern middle class has exhibited marked failure to evolve from a traditional landed bourgeoisie to a new role in industry and trade. As a class the southern *borghesia* has been reviled by many writers on the *Questione Meridionale.* Its members were essentially *rentiers* obtaining their livelihood from high rents, oppressive tenure contracts and professional sinecures secured by patronage; they supplied few of the functions of true entrepreneurship and were to the social life of the South what malaria was to the physical life (Franklin, 1969; Rossi-Doria, 1958). Such parasitism was an accepted way of life. The *borghesia* is the most strategic class in the explanation of backwardness in the Mezzogiorno at the village scale. Below the *borghesia,* and in part integrated into it, came the artisans, a class which is now rapidly passing out of existence, and below this group came in turn the land owning, renting and sharecropping peasants, and the landless labourers. Other important social characteristics of the southern population, which were partly a cause of underdevelopment and partly a reflection of the general condition of social backwardness, include the extreme rigidity of the family as an institution, capable of excluding all other forms of social cooperation, and the subservient status role accorded to women. A further complication in the social structure of southern Italy has been the existence of regional elites. Geographically these groups are especially important for they are the sociological expression of the various historical–geographic regions in the South and are intimately bound up with regional ecology at different scales. The old absentee *latifondisti*, concentrated in their respective regional capitals of Naples, Bari, Palermo etc., constituted just such a group; an even better example is the regional elite of the Mafia which still holds considerable sway over western Sicily, blocking attempts at socio-economic development. The strength of these regional elites, reflecting the importance of regional fragmentation within the South, has been a considerable obstacle to national economic and political integration as well as to progress at the local scale (Schneider, Schneider and Hansen, 1972).

The persistence of the old social structure for so long is an indication not so much of the system's inherent stability—witness the history of banditry and insurgency in the South—but rather of its preservation by the perpetual exercise of power over the lives of the peasant majority by the rural oligarchy. Banfield (1958) developed the ethos of 'amoral familism' as a partial explanation of this low-level equilibrium. Amoral familism—the inability to concert activity beyond the horizons of the nuclear family, i.e. giving priority to family interests to such a degree that the person or the action become amoral from the point of view of the community at large—is a good description of what has been happening at the local scale in the South; but recent anthropological work (Marselli, 1963; Peabody, 1970; Pizzorno, 1966; Silverman, 1968) has tended to reverse Banfield's causation, maintaining that amoral familism is an expression of a social structure deeply rooted in patterns of land tenure and agricultural ecology rather than a fundamental cause of lack of development. All are agreed, however, that the extreme rigidity of the social structure, the fatalism of the southern peasant and his individualistic, familial nature have prevented the formation of cooperative organizations that could have led to change. The peasants were completely resigned to their inferior position; this point is central to Ciancian's model, which can be considered as an alternative to Banfield's, of explaining backwardness in terms of the peasants' overwhelming sense of inferiority with respect both to the governing classes and to the possibility of improving the environment (Ciancian, 1961). Until recently the Mezzogiorno was in a pre-political state; political institutions were too weak for the effective channelling of pressure for change. Instead, what little peasant protest there was took the primitive form of disorganized brigandage; indeed, banditry still survives in the mountains of Sardinia, home of the most tradition-bound society in Italy, where there is serious poverty too (King, 1973a).

The twin concepts of social structure and population pressure are crucial for understanding the underdevelopment problem in the Mezzogiorno. With so much pressure of population, and with homogeneous categories of peasants all mutually competitive for limited resources of land and employment, any possible cooperative action was thwarted by vertical patron–client chains. Southern society was not composed of coherent interest groups campaigning for their betterment; it would be better to describe it as a gigantic network in which vertical patronage relationships were more dominant than horizontal or lateral dyads (Boissevain, 1966). Clearly if the 'pie' of resources is unalterable and the system is a closed one, as most southern communes traditionally were, then population growth will cause economic insecurity and conflict, so that any individual can improve his position only at the expense of others. This statement explains the remarkable mistrust and factionalism which pervade southern society.

Spatial Structure of Settlement, Land Use and Communications

The South's backwardness is also reflected in the spatial layout of the society and economy. Particularly characteristic of the Mezzogiorno is the large isolated agglomeration sited centrally in each commune. Elongated strip communes aligned from sea level to mountain top, with the settlement traditionally located in the interior, form a sub-type around the coasts of Calabria, Lucania and northeast Sicily. Settlements are usually in the range 2000–8000 inhabitants, though several of the so-called 'peasant cities' of Apulia attain 20,000–40,000. This heavily agglomerated nature of southern rural settlement reflects geographical influences that for the most part no longer obtain (Blok, 1969b). Malaria, piracy and, with occasional exceptions in Sicily and Sardinia, banditry and insecurity no longer afflict the southern population. The 'crisis of the antique settlements', to use a phrase coined by Compagna (1963), represents a powerful obstacle of physical inertia to the socio-economic evolution of the rural sector, for modern agricultural development tends to require a more dispersed settlement pattern with the farmhouse located on, or at least near, the holding. In contrast to northern Italy, where towns, based on reciprocal ties both between themselves and with surrounding rural regions, have been centres of industry and commerce, in the South town–country relationships are immobile. The towns are not really towns at all, but huge dormitories, peasant-cities or 'agro-towns' of rural workers. Monheim's (1969) study of Gangi in Sicily as a typical southern agro-town shows how these settlements, which are dominant over most of the South, are intermediate between town and village. A settlement of 10,000 like Gangi, or even of 50,000 like Cerignola in Apulia, is a village in its dependence on agriculture (usually supporting 50–90 per cent of the population), in its social structure and in its absence of truly urban functions; it is a town only by virtue of its size and its 'piazza life'. Each agro-town is linked to provincial and regional capitals by loose ties of administration and to the outside world by emigration, but there is no real hierarchy of settlement—city, town, village, hamlet, farm—in the *latifondo* zones of the South. The link between settlement type and the *latifondo* ecology has always been strong. Whether the peasant was a landless labourer on the large estates, or whether he rented scattered strips of *latifondo* on a yearly or half-yearly basis, there was no possibility of his actually settling on the land and, for his periods of unemployment, there was more chance of finding occasional scraps of supplementary work by living in a large population agglomeration where social contacts could be maximized. If southern rural settlement was heavily concentrated, originally for reasons of defence against marauders or malaria, or even because of the desire of the feudal baron to control better his subservient peasant population, a strong reason for the preservation of this pattern in more recent decades has been the work-finding and income-saving opportunities fostered by the complex of social interactions inherent in concentrated settlement.

The distances separating one agro-town from its neighbours and the immobility of town–country relationships have exerted a strong influence on land-use patterns. Elaborations of the 1951 census indicated that nucleated settlements were three times as widely spaced in the South as the North (4, as against 12, settlements per 100 km^2) and that each southern settlement was on average three times as big as its northern counterpart (2588 inhabitants as against 842). Despite the fact that many agro-towns are sited on hilltops or mountainsides, which means that land adjacent to the settlement is of generally poor quality, the immediate environs of the settlement—the so-called *corona* or 'halo' about 2–4 km wide—are intensively cultivated whereas the fief land farther out, lacking application of labour and capital because of its distance from the centre of habitation, remains extensively cultivated. This is a classic von Thünen situation and means that total agricultural productivity has been considerably lowered by the settlement pattern. Where parcels of land are particularly widely scattered, peasants often spent up to four hours a day trudging out to their distant fields and back. This pattern, conditioned basically by the factors of distance and primitive transport, survives very much intact, being broken down only slowly by the diffusion of more rapid means of transport.

Communications have always been less developed in the South; and they functioned less efficiently as a means of regional coordination than did northern transport networks. Interior southern Italy is very difficult country for communications. River valleys, generally the principal routeways for communications axes, are mostly short, narrow, tortuous in course and lead nowhere in particular. They are subject to winter flooding and valley-side slumping. Since most southern settlements originated as fortress villages or as colonies founded to valorize the feudal lands, they were not located in valleys or at the intersections of trade routes. Even a provincial capital like Enna has its railway station 300 m in the valley below; only extremely costly engineering would get a railway line up to the 1000 m plateau on which the town sits. Whereas the landowning and land-controlling classes were united on a regional basis by bonds of coalition, the structure of settlement and communications, and the social and spatial isolation of one agro-town from another, greatly hindered the development of regional integration at the level of the peasant masses. In terms of network structure, the roads, even today, tend not to meet at nodal settlements. Junctions occur in open country and settlements are linked to each other in haphazard linear fashion by meandering roads which nevertheless fail to connect some agro-towns with their immediate neighbours (Schneider, 1972). Regional and local isolation has been a dominant feature in the evolution of the South; nowhere is this more true than in the remote island of Sardinia. With each settlement largely self-sufficient there was no incentive for travel or trade between neighbouring villages, for they produced the same limited range of goods and services. This largely explains the extraordinary spatial immobility of the southern villager, his sense of attachment to his home town, and his hostility towards, and his ignorance of, neighbouring settlements. This *campinilismo*—meaning the villager will never go so far from his village as not to be able to see the belfry of his church—is still characteristic of the rural South. Two quite recent anthropological studies (in the Lucanian hills and the Salentine Peninsula) revealed many old village women who have never in their lives been outside their native settlement (Banfield, 1958; Maraspini, 1968).

Mezzogiorno at Mid-Century

In 1950 the situation in the South was critical, if not explosive. It is true that pockets of considerable poverty existed in the North, in the Po Delta for instance, but what was most significant was that the South's was a regressive situation. Real *per capita* income actually fell 7 per cent over the period 1938–51, whereas in the North there was an 8 per cent increase (Tagliacarne, 1955). Bad housing, inadequate schools, poor health and sanitation facilities, malnutrition, high incidence of infectious diseases—almost any indicator of socio-economic development gave the same picture of southern backwardness *vis-à-vis* the North (Table 5.1). Even these data understate the true nature of the problem because of marked contrasts within the South between urban and rural conditions and between rich and poor. The poorest southern province, Agrigento in Sicily, had a *per capita* income of only 20 per cent that of Milan. Abject poverty characterized the majority of the southern population, with incomes, living conditions and a subsistence orientation akin to those of the peoples of the Third World.

Table 5.1. Socio-Economic Indices before the South Italian Development Effort

	North	South
Annual *per capita* income (£), 1951	220	90
Annual *per capita* electricity consumption (kWh), 1950	98	30
Roads (km/1000 km^2), 1951	720	349
Roads (km/1000 population), 1950	432	246
Vehicles per 1000 population, 1949	20	6
Telephones per 1000 population, 1949	24	5
Percentage of active population in agriculture, 1951	35	56
Percentage of agricultural land under irrigation, 1948	11	2
Annual use of fertilizer (kg/ha), 1952	73	32
Annual *per capita* sugar consumption (kg), 1949	12	4
Average natural increase per 1000 population (annual average, 1947–9)	7	17
Percentage illiteracy, 1951	6	24
Percentage of families classed as 'poor' ,1951	4	27
Percentage of persons living in conditions of overcrowding (more than 2 persons per room), 1951	19	53
Percentage of dwellings without any form of sanitation, 1951	16	40

Basic to the problem of the South was the fact that industrial and agricultural production had nowhere near matched the growth of a population that had doubled in the previous 70 years. Population natural increase in the South was 2½ times the rate of the North. By 1951 it was calculated that southern Italy had 38 per cent of the population, 50 per cent of the births and 75 per cent of the natural increase of Italy as a whole (Dickinson, 1955a). Between 1880 and 1930 some 7,000,000 southerners had emigrated, but after 1930 this 'safety valve' was firmly shut by Mussolini. The southern agricultural labour force grew by 17 per cent during 1930–50, from 3,100,000 to 3,600,000. In Apulia the rise was 38 per cent. In North Italy, as indeed in the rest of Europe, the same period saw a considerable drop in numbers employed in agriculture. Most southern provinces had about 70 per cent of their populations engaged in agriculture at the 1951 census, the average figure for the region, 56 per cent, being heavily affected by the 20 per cent figure of the populous Naples province. Orlando (1952) calculated that 45 per cent of labour potential was underemployed—used inactively or sub-marginally. Mounting population pressure meant mounting pressure on the land. In spite of, indeed because of, the dominance of latifundia over large areas of the South, small, fragmented properties were just as characteristic of the

Mezzogiorno. One-fifth of the South's farmland was in peasant holdings composed of at least five fragmented plots.

With land monopoly by *latifondisti* blocking attempts to develop southern agriculture through increased employment and with industrial employment virtually non-existent outside a handful of large towns, the situation in the Mezzogiorno in the immediate post-war period was desperate. Men returning home from the war swelled the ranks of unemployed. These men, having fought in the North and abroad, had seen how much better conditions of life could be; they became a strong element in activist movements such as the illegal squatting on *latifondi* in Sicily, Calabria and Apulia in the late 1940s. There was revolution in the air and, quite apart from economic and social considerations, the post-war government of Alcide De Gasperi was compelled to do something if the country was not to dissolve into violent civil disorder.

Attempts at Regional Development

Italian policy aimed at solving the regional problem of the South can be crudely divided into four phases, the dates of which are only approximate. The first and least important phase covered 1945–50 and was a period of post-war reconstruction during which much of the energy of the nation was directed towards regaining pre-war levels of production in all sectors of the economy; American Marshall Aid was instrumental in this achievement. The other three phases of policy are considered in the rest of this chapter. The second, 'pre-industrialization', phase ran from 1950 to 1957 and was characterized by emphasis on land reform, land reclamation and general infrastructure. In the third phase, 1957–64, attention was switched to stimulating industrial development directly by locating publicly owned industry in the South; the impact of 'growth pole' theory also began to be felt. In the last phase, policy has become more regionally concentrated within the South as a whole and more integrated into national planning.

The Land Reform

On 30 October 1949 at Melissa, a poor mountain village in Calabria, a group of peasants occupying a large estate were fired on by police, resulting in three fatalities and thirteen wounded. A nationwide outcry followed the incident and the government, which had for some time been considering a project for a general agrarian reform, rushed through legislation for the expropriation of land from large landowners and its redistribution to landless and semi-landless peasants. Eight special regions were delimited for the carrying out of the reform, these regions being the principal areas of large estates in the country (Figure 5.3). The reform was not specifically a southern measure, but the fact that six out of these eight regions were in the Mezzogiorno meant that land reform was primarily directed at alleviating poverty in the South. Apart from Calabria (300 ha) and Sicily (200 ha) the private landownership ceiling, above which expropriation took place, was determined by a sliding scale whose aim was to cull most land from extensively cultivated large holdings (King, 1971a). The total area thus expropriated was 767,000 ha, 535,000 ha (70 per cent) of which was in the South. Assignments were in the form of complete farms (usually with specially-built farmhouses attached) of around 4–10 ha (though there was considerable variation outside these limits), or small plots of 1–2 ha designed to supplement income from

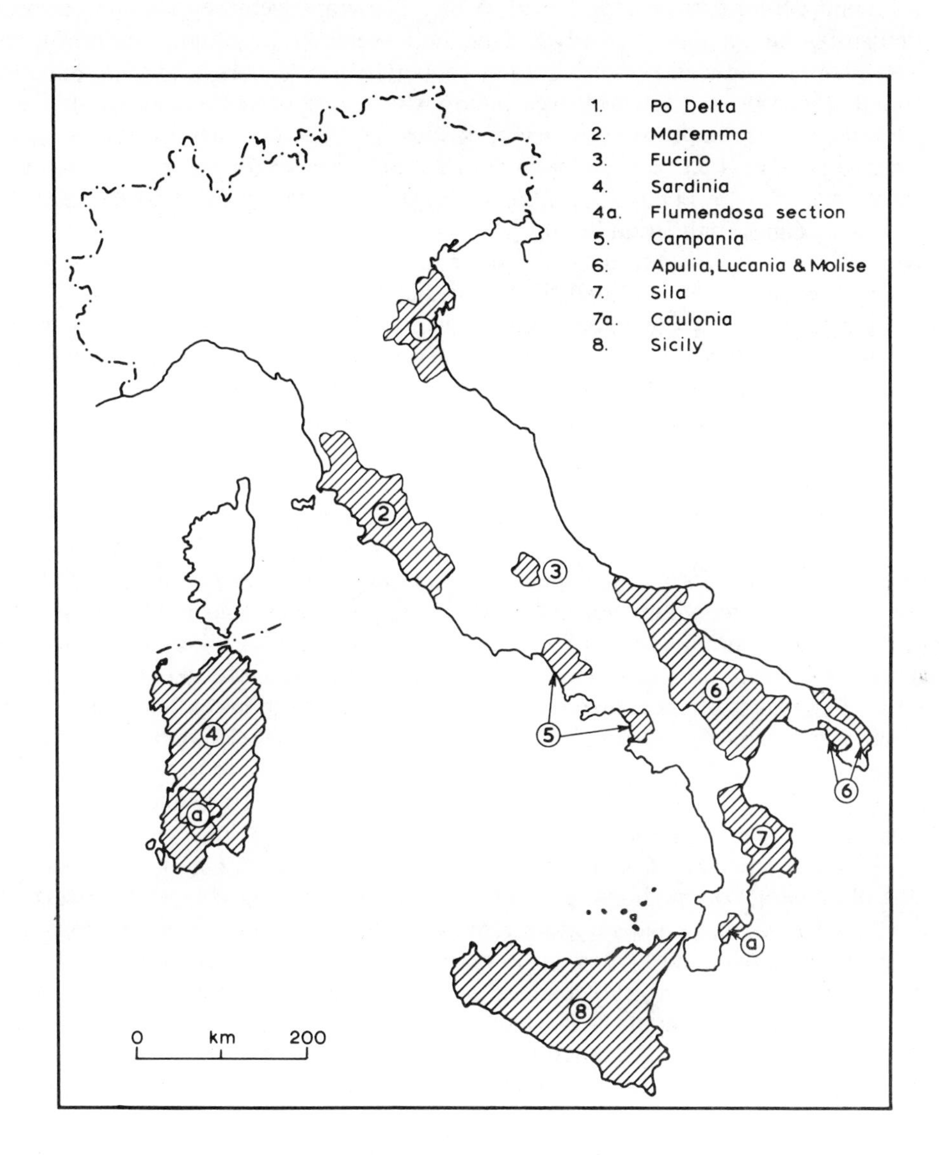

Figure 5.3. Land reform regions in Italy.

other sources. The beneficiaries totalled 113,000 families, of whom 91,000 (81 per cent) were in southern reform territories.

In spite of its hasty origin, the land reform was in many ways a model of detailed planning and administration (McEntire, 1970). The work went beyond expropriation and redistribution, in that it involved land improvement, irrigation, settlement and land-use planning, education, setting up of cooperatives and marketing of produce. As most of the expropriated territory was previously devoid of settlement, several new service villages were built to act as service centres for the scattered reform farmhouses. This conversion of bare *latifondo* land to a landscape of intensive farming with regularly planned farmhouse settlement is perhaps the most visible post-war geographical change in the rural South.

Economically, the land reform scored some notable successes. Land reclamation and agricultural intensification enabled the per hectare output of expropriated land to be raised by a factor of two to three. Similar increases were verified in assignees' *per capita* incomes. Although the cost of land reform proved to be much higher than originally anticipated, a crude cost–benefit analysis showed that the reform had paid for itself, in terms of increased marketed agricultural output, by the mid 1960s (King, 1972). The economic results of the land reform do, however, vary greatly from zone to zone. They are perhaps most spectacular in the Metapontino, the Ionian coastal strip of the provinces of Taranto and Matera. Here irrigation, low demographic pressure and a large expropriated area permitted massive state intervention and the setting up of several thousand new farms. A linear programming model run for these 5 ha farms show that they have a high income and productivity potential (Dean and De Benedictis, 1964). The Metapontino is one of the showpieces of the Mezzogiorno and similar, smaller-scale oases of prosperity have been created by the land reform in Campania and Sardinia. These areas are not, however, typical of land reform in southern Italy. At the other end of the spectrum, reform in central Sicily has been a fiasco, a 'pseudo-reform' sabotaged by the Mafia and by the peasants' unwillingness to move out of their agro-towns to the isolation of the countryside (Blok, 1966). Here and in the clayey Marchesato of Calabria the peasants' refusal to forsake gregarious village life, poor land quality and the small size of the new holdings (3–6 ha), combine to produce the paradox of modern, commodious, but empty houses surrounding villages of hovels. On the Apulian Tavoliere, perhaps a more representative situation, reform holdings have been abandoned over the years as economic development raises the minimum acceptable standard of living and changes the aspirations of the beneficiaries; although this represents a certain amount of wasted investment (about £1500 per farmhouse), the reform has at least fulfilled the function of holding the peasants on the land at a higher standard of living until such time as they can find more remunerative employment elsewhere.

The single most important achievement of the reform was probably the destruction of the social and economic power of the *latifondisti*. The *borghesia* class, upon which the ruling Christian Democrat Party depended for power in the Mezzogiorno, were, however, left untouched by the reform; and nothing was done to help those unlucky peasants who received no land (in some areas assignments were made on a lottery basis). South Italian land reform was, therefore, more notable for what it destroyed than for what it created. Although for regional development purposes the reform was a movement in the right direction, it was a rather small step that did not greatly affect the socio-economic structure of the South as a whole. For every new job provided by the reform, probably seven others were found by emigration by southerners during the 1950s. In Sicily, during the same period that the island's reform agency struggled to redistribute 60,000 ha, the peasants acquired 280,000 ha on the open market. Molise in 1950 had 63,000 small peasants of whom 15,000 were landless *braccianti*: reform beneficiaries in this region numbered just 500. Yet it may be that the most important effects of the reform are those which are not immediately apparent or measurable. To the downtrodden peasantry, who before 1950 lived in a hopeless world of poverty and unemployment, it brought some relief and the hope of better things to come; viewed in the long-term structural growth sense, the reform fulfilled, albeit partially, the important function of stabilizing the peasants on the land to provide a point of departure for future development; even in the short term it probably averted a disorder whose costs could have been immeasurable. It was inevitable, however, that such a one-dimensional programme should have serious limitations as a regional development policy. If the reform had been more sensitive with regard to differing geographical areas, to changing conditions over time and to the all-important human frailties of the South, it would have been more effective and less costly. Because of the boom that Italy, especially the North, was to experience from the late 1950s on, a boom that was totally unforecastable ten years earlier, the context of the land reform changed faster than could the policy itself. Land reform became obsolete—a victim of its own timing. Although finance for completion of the reform was provided well into the 1960s, and although many thousands of viable commercial farms have been set up (as well as an equal number of non-viable units), the reform-based policy orientation was quietly dropped in favour of more realistic measures.

The Cassa per il Mezzogiorno

To a certain extent the geographically and sectorally limited nature of land reform policy was recognized right from the start, for in the same year that the land reform laws were passed the 'Fund for the South' was set up, with the task of providing at least some of the functions that the reform authorities did not. The Cassa, a supra-ministerial executive body, was also designed to remedy the lack of coordination and planning which had characterized the few attempts at southern regional development before 1950. The Cassa was not a replacement of normal ministerial planning; rather the Fund was conceived to deal with 'extra-ordinary' works which went beyond the scope and resources of individual ministries. Territorially the Cassa operates over the seven regions shown on Figure 5.1, plus the two southern provinces (Latina, Frosinone) of the region of Latium, and the small islands off the coast of Tuscany. Initially a ten-year period of operation was envisaged, concentrated on provision of basic infrastructure for the South, and a sum of £600,000,000 was allotted for the decade 1950–60. As early as 1952, however, it became apparent that both the time scale and the amount of finance were too restrictive. The Cassa's life was subsequently extended to 1962, then 1965 and, most recently, to 1980. A great deal of extra money was also made available: nearly £5,000,000,000 was spent by the Cassa over the period 1950–70. In addition to allotments from the Italian government the Cassa was given mandate to apply for foreign loans. The World Bank has been an important source of loan capital, accounting for 30 per cent of the Cassa's investment during the first decade, and the E.I.B. contributed £200,000,000 during the 1960s (Mountjoy, 1973). The scale of Cassa spending has increased considerably in recent years. Although in 1950 it was obvious that a dramatically new approach to the southern problem was required, the severe economic conditions of the time decreed prudence rather than audacity: a policy, in other words, of 'first aid' or assistance rather than of positively promoting increased productivity (Allen and MacLellan, 1970). Emphasis in the early 1950s was put on the rehabilitation of agriculture, the Mezzogiorno's principal economic activity, and on modifying the structural environment. The chief items of intervention were river and erosion control, marsh reclamation, afforestation of uplands, road, aqueduct and sewer construction and, in support of the land reform, the building of farmhouses and service villages. The initial £600,000,000 was assigned as follows: 77 per cent for agriculture, 12 per cent for sewers and aqueducts, 9 per cent for transport and communications, 2 per cent for tourism. No mention was made of industrial development before 1957. In the early 1950s the government was primarily concerned with improving the Mezzogiorno's social environment; particularly important were the objectives of bringing drinking water to every village and increasing the metalled percentage of road length from 13 to 61 per cent. It was argued that the 1950 measures in themselves were never aimed at bringing about autonomous economic development in the South; any upswing in industrial development in the first ten years as a result of infrastructural improvements would have been regarded as a bonus rather than a policy expectation (Watson, 1970). Where criticism does appear justified concerns the geographical dispersion of efforts throughout all zones of the South. Even here the issue is far from clear-cut. This problem was a thorny one for Cassa decision-makers, reflecting strong conflicts between objectives of economic benefit and social–spatial equality. Various factors operating in the past had resulted in a spatial and altitudinal distribution of population and economic activity rather out of character with modern, strictly economic, requirements. In particular, the areas of new irrigation in the lowlands were sparsely settled (though the land reform was rectifying this to some extent), whereas the highland areas, especially of central Sicily, Calabria and Lucania, where there was little potential for real agricultural improvement, had high population concentrations. As with most policies which attempt to steer a middle course, the Cassa's received criticism from both flanks: on the one hand that too little was being done for the still closed world of the southern uplands, on the other that too much money was being wasted in these areas on projects that were to be of little value in economic development terms. Calabria especially became known as the graveyard of public works because of the large number of isolated, and often unfinished, projects scattered throughout its area. The Cassa was also accused of aiming at outwardly visible results which could be exploited politically by the Christian Democrats. Half a million pounds went on building a soccer stadium in Naples and £30,000 was spent on extending a church in a Sicilian village when the commune lacked a sewerage system costing £13,000. La Palombara (1966) concluded that the Cassa's expenditure was determined as much by the ballot box and local political interests as by economic rationality. Large and medium landowners in the South could get up to three-quarters of land improvement costs defrayed by Cassa grants; the conflict with the aims of land reform, which favoured lower-status rural classes, went apparently unnoticed. In fact the publicity surrounding the land reform effectively obscured the massive aid given to the privileged landowning class of the South (Franklin, 1969).

Industrial Policy

The year 1957 saw an important reorientation for the Cassa's policy. Largely because the infrastructural and agricultural policies did not seem to be having any pronounced development effects, the Cassa turned its attention to industry; under the Industrial Areas Law the Cassa was authorized to encourage the establishment of industry in specially favoured parts of the South.

The idea of industrial development in the Mezzogiorno had in fact for some time been brewing up. In 1952–53 the Cassa began to participate in industrial credit through its support of the three regional credit agencies—ISVEIMER for the mainland South, SOFIS for Sicily and CIS for Sardinia. In 1954 Budget Minister Vanoni presented his ten-year scheme for the development of production and employment in Italy. This scheme paid particular attention to the Mezzogiorno and considerable industrial development was implicit in the plan's ambitious aim of raising southern mean *per capita* income from 50 per cent of that of the North (1954) to 75 per cent within ten years. According to Jucker (1960), the Vanoni Plan was a 'pious resolution'; few believed that age-old interregional differences could be eradicated in so short a time; yet although Vanoni died in 1956 and his scheme never became operational, its basic premises became the watchword of government planning for many years to come.

The various incentives conceded by the 1957 law to industry locating in the South included exemption from profits tax for industrialists' capital reinvested in the South, direct subsidies on machinery and plant, customs exemption, and loans at specially low interest rates. These measures could reduce capital costs of firms locating in the South by up to 30 per cent and a 'normal' profit rate of 6–9 per cent could be increased by half (Ackley and Dini, 1959). A further very effective law compelled state and semi-state agencies such as ENI (the state hydrocarbons agency), Italsider (iron and steel) and IRI (the important state holding company for industrial reconstruction) to locate at least 60 per cent of their new investments in the South. More industrial development initiatives were taken by the Cassa in the early 1960s. In this decade, with the Italian political climate more conducive to large-scale public participation in southern development, the approach to planning gathered momentum. In 1961 a new law set up IASM (the Institute for Assistance to Development in the Mezzogiorno). Composed of the Cassa and the three credit organizations, IASM was designed to assist the establishment and modernization of southern industry by preparing technical and feasibility studies. In 1963 a new public shareholding company, IN-SUD, was created, with capital from the Cassa, to enable the government to participate more directly in financing southern industrial development; a parallel organization, FINSARDA, operates under the aegis of the regional government in Sardinia. This increasing concern of the Cassa with promoting industrial development is reflected in the sectoral apportionment of its spending. Agriculture accounted for 77 per cent in 1950, 56 per cent in 1957, falling to 25 per cent during 1966–69. Industry, although Cassa-stimulated industrial development did not really get under way until 1959–60, rose steadily from zero in 1950 to 60 per cent in 1962.

Since the late 1950s, in the light of 'unbalanced growth' and 'growth pole' theory and owing to the initiative of Saraceno (1955) in adapting these concepts to the South Italian situation, the Cassa has devoted much attention to the creation and management of selected industrial zones in the Mezzogiorno. The famous 'opening to the Left' in 1960 (the coalition between the Christian Democrats and the Socialists) provided the necessary political environment to put a Saraceno-type growth-centre policy into effect.

Two types of industrial growth have been envisaged: areas of industrialization and industrialization nuclei. Decisions as to the locations of these growth centres were not imposed by Rome but were left to the initiative of local 'consortia' of administrators, industrialists and traders. The consortia had to prepare plans for their respective zones and submit them for approval by the Cassa. 'Areas' had to demonstrate an industrial vocation already in existence and normally consisted of cities or urban regions possessing a population of at least 200,000. Abundance of flat land and sufficient power supplies were further requirements. Less stringent conditions were applied to localities qualifying for nucleus status; most of these places have populations of less than 75,000 and the development envisaged is of small and medium firms. In this industrial growth pole strategy, parallels with Spanish regional development strategy can be suggested, and in fact Italian planners were commissioned by the Spanish government to advise on setting up of growth poles in Andalusia in the late 1960s.

So far over forty localities have presented themselves for consideration for industrial development zone status at either area or nucleus level. Their locations are given on Figure 5.4. The question immediately arises as to whether the areas and nuclei are too numerous and too scattered to fulfil the original conception of spatial concentration of development. In fact the 42 areas and nuclei considered by 1970 comprise 29 per cent of the South's area and 45 per cent of its population. The large number of scheduled localities results largely from southern provincialism and local political influence. Many of the smaller nuclei, such as some of those in Calabria, have only limited development prospects and seem to have been selected more to give an impression of balance of effort throughout the regions of the South. Regional and provincial factionalism is still very strong in the Mezzogiorno. In any economic considerations of industrial development Calabria, the poorest, most rural and most mountainous region of Italy, is bound to suffer. The recent riots in Reggio against another province (Catanzaro) being selected as regional capital indicate the depth of feeling and the importance of the role that government policy can play in relieving backwardness and unemployment.

Other criticisms can also be made, indicating that the growth pole policy has not contributed as much to southern development as it might have done. No distinction was made between the aids available to the two categories of area and nucleus, so that in practice the distinction between the 'intensive areas' and the 'diffused nuclei' became blurred. No overall assessment was made of the relative suitability of different locations for certain types of industry. No real coordination exists between the various areas and nuclei (except for the Bari–Brindisi–Taranto 'triangle'); the consortia have responsibility only for their own zones and there is no organization which effectively takes the important wider regional view. No interrelation exists between town planning and industrial development. Even more vitally, no new instruments to promote the location of northern private industry in the South were adopted in order to make the policy more effective; in fact the reinvested profit tax was retained in industrial zones.

Considerable developments have, nevertheless, resulted. As early as 1959 the Bari–Brindisi–Taranto trio of towns in Apulia had been identified as particularly suited to industrial expansion, and this triangle was designated a European pole of regional development by the E.E.C., with careful studies set in motion of the zone's planned development. Italconsult, a Rome industrial consultancy concern, drew up a comprehensive input–output table of the region and its possible industrial developments. From an analysis of this table new industries that would find the zone's location

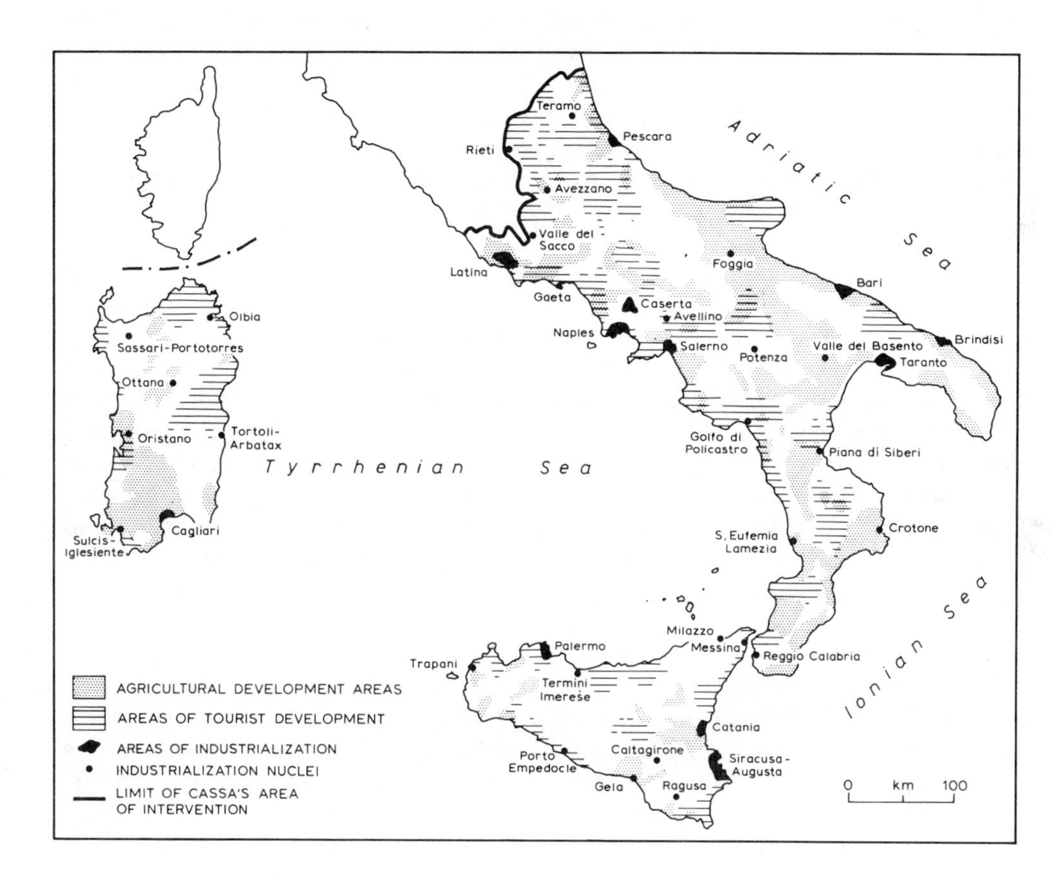

Figure 5.4. Areas for the development of agriculture, industry and tourism in the Mezzogiorno.

particularly advantageous were determined. A list of over 200 'candidate' industries was sifted through in this way. Italconsult's proposals, made public in 1965, involved a concentration on mechanical engineering industries at Bari and Taranto and on petrochemicals at Brindisi (Mountjoy, 1966). In 1960 had begun the construction of the great Italsider iron and steel plant at Taranto, the largest in Italy, incorporating an initial investment of £200,000,000. This plant alone, with annual output of 10,000,000 tons targeted for 1975, employs over 5000 operatives and provides a considerable basis for future industrial development; in French terminology it is identified as the principal *activité motrice* of southeast Italy. Italconsult's analysis proposed the installation of a single wave of about 30 firms, composed of nine larger factories making a range of major engineering products, using as basic raw material the output of the steel plant, and some 20 smaller firms providing specialized, auxiliary inputs. The basic ideology behind the scheme as an operational concept is that whilst an individual or small number of firms would clearly not regard prospects for setting up in business here as very good, when there is provision for their proceeding in conjunction in a group amounting to a real industrial complex, the outlook will appear far more favourable (Watson, 1970).

Similar feasibility studies have been made to plan the industrial expansion of the Siracusa–Augusta node in eastern Sicily. Here expansion of oil-based industries is seen as the major trend, with diversification into processes involving petroleum byproducts (plastics, detergents, paints etc.) and into non-ferrous metal refining (Mountjoy, 1970). Already the industries of the Siracusa–Augusta axis employ 15,000 persons, a third of whom work in chemicals. Other important enterprises which have recently been established in the South include the Bagnoli iron and steel plant and the Alfa-Sud car factory, both in the Naples region, ENI's petrochemicals complex at Gela, Fiat's assembly plant near Palermo, SIR's petrochemicals plant at Porto Torres and, most recently of all, the oil-based synthetic textiles complex at Ottana in central Sardinia. This last concern is of special interest as it is motivated not so much by strict economic criteria but by socio-political considerations of bringing steady work to the notoriously delinquent shepherds of the Sardinian mountains.

These examples are sufficient to indicate the character of industrialization in the South in the 1960s. Most of the development was due to public rather than private initiative and the majority of plants were capital-intensive rather than labour-intensive. In chemicals and metallurgical industries, where much of the Cassa's industrial spending went, each job created cost over £100,000 (Watson, 1970). The majority of higher status managerial and technical workers were northerners who had to be persuaded to shift south. Experience at Brindisi, Taranto and Gela revealed a pattern of sharp increase in the number of low status construction workers in the initial phase, followed by layoffs, and ultimate replacement at a reduced level by skilled personnel and engineers of predominantly non-local origin. The construction workers, many of whom had shifted from former agricultural pursuits, either joined the unemployed or found marginal, low-wage tertiary employment (Rodgers, 1970). Apart from a concentration of firms around the *Autostrada del Sole* in Latina province (located just south of the Cassa boundary in order to take advantage of incentives, yet situated as close as possible to the North and to the Rome market), private entrepreneurs have been wary of the South. The employment problem has not yet been overcome since the 1,000,000 or so jobs that resulted, directly or indirectly, from Cassa policy during the period 1950–70 failed to balance an out-migration from the region of nearly 2,000,000 workers. Despite the Cassa's investment, only 30 per cent of Italian industry was located in the South in 1970; and for large firms (employing over 500 employees) the figure was only 8 per cent.

Regional Planning and the Problem of Coordination

The problem of coordinating the 'extraordinary' intervention of the Cassa with the 'ordinary' activities of the standard administrative machinery has always been recognized; but it has never been satisfactorily resolved. Indeed, because of the Cassa's special autonomy, relations with government ministries were bound to be limited (although they need not have been so few and casual as they were) and this in turn led to lack of understanding and to conflict. The problem does not, however, concern just coordination and possible duplication of initiative, but also the fact that the very existence of the Cassa tended to produce a lessening of commitment to the South on the part of the ordinary ministries who saw the Cassa either as a substitute for their own activities or as a dumping ground for problems they found difficult to cope with themselves.

Reforms introduced in the 1965 law which extended the Cassa's life up to 1980 attempted to confront these deficiencies. Coordination of intervention at national, regional and sectoral levels, was aimed at by the establishment of medium-term five-year national plans (the first of which ran from 1966 to 1970) and by the recasting of the Cassa's administration on a more streamlined basis. The object was to reform the Cassa into a more specialized executive agency concentrating on relatively fewer but more important strategic planning projects.

The indications are that this new planning system is bringing better results, at least at the global level. For one thing, the spending of the 'ordinary administration' on the South has, after many years of decline, started to increase in terms of its proportion of total expenditure (Allen and MacLellan, 1970). Rather more important, and quite distinct from the greater coordination which the insertion of Cassa work within the overall framework of the national plans has produced, is the fact that the Cassa's planning now has a more concentrated geographical basis. In addition to the 42 industrial areas and nuclei, some 82 agricultural and 29 tourist development areas have been laid out, and present and future spending will be almost totally confined to these zones. Agricultural expenditure for example is mostly limited to irrigation development in the plains. Responsibility for parts of the South falling outside these zones of high potential reverts to the Ministry of Agriculture. In fact, for reasons which will appear obvious to the geographer familiar with southern Italy, many of the three types of area—agricultural, industrial, touristic—coincide, for they are for the most part coastal plains suited to all three types of economic development. There are some half dozen parts of the South where this overlapping is particularly marked (Latina, Naples, Pescara, Bari–Brindisi–Taranto, Catania–Augusta–Siracusa, Cagliari) and these, due to receive the lion's share of Cassa attention over the next few years, are tentatively designated as major growth poles.

It is too early yet to evaluate the progress of this new line of more coordinated regional development policy, but the general patterns of intervention, sectoral and geographical, are clear. While agriculture suffers a relative, and indeed a small absolute, decline in terms of overall expenditure, industry and tourism increase in importance. Spending on industry during the five years 1966–70 was over ten times the total amount spent during the fifteen years 1950–65. In all sectors there has been a marked shift away from policy of a 'public works' character: this trend was already apparent by 1960 but is now much clearer. The key words now are concentration and growth (Allen and MacLellan, 1970). It is likely that this geographical and sectoral concentration of Cassa activity will leave considerable social problems in its wake—the conflict between industry and tourism already apparent in the Naples and Siracusa areas, for instance. There is, moreover, no guarantee that the official doctrine of 'concentration of intervention and diffusion of effects' will reach the interior mountain regions which cover a quarter of the Mezzogiorno and contain a tenth of its population. The neglect of these 2,000,000 southerners who live outside any designated area and who receive only 2 per cent of total Cassa funds is, to say the least, unfortunate; but this is the policy which will ensure the fastest economic growth for the Mezzogiorno as a region.

Tourism

Unlike before 1950 when hardly any visitors ventured further south than Naples, Amalfi and the Paestum temples, tourism is now regarded as an important sector which can contribute a great deal to the economic and cultural evolution of the Mezzogiorno. The *Autostrada del Sole*, now complete to the very toe of Calabria, with extensions in Sicily and down the Adriatic coast, has broken down the isolation of the South and the stage seems set for rapid tourist expansion. With its ample reserves of sunshine, historic monuments, sandy beaches, inland scenery and native friendliness, the Mezzogiorno is well endowed with the basic ingredients for a sizeable tourist activity. Calabria, Sicily and Sardinia in particular are rich in natural beauty and landscape contrast. Lack of local initiative probably accounts for the fact that tourism is still little developed outside a few well-known localities (Naples, Sorrento, Amalfi, Capri, Taormina, Alghero, the Costa Smeralda), and there is a certain lack of concern for environmental preservation: planning controls are minimal and lack of civic pride causes much of the South to have a rubbish-dump appearance.

Especially since 1965, the financial incentives offered by the Cassa for tourist development are generous. Previously, Cassa funds were limited in this field to restoration of ancient monuments and the building of museums, but now grants and loans are available for building, enlarging and modernizing hotels and tourist villages. As with industry and agricultural policy, the intention of the Cassa is to concentrate on selected tourist areas. Loans can encompass up to 70 per cent of admissible expenditure, to be repaid at 3 per cent interest over 20 years. Grants are also available but are lower than those for industry, being set at a maximum of 15 per cent of total capital outlay (Allen and MacLellan, 1970). Three types of tourist area exist with different policies for each: zones as yet undeveloped but with potential for rapid valorization, areas already undergoing some development, and 'mature' areas which could, if due care and control were not exercised, become spoilt and saturated. There is no doubt, particularly when one considers the tourist developments which have transformed the Mediterranean coasts of Spain and France, as well as parts of northern Italy, and the increasing preference of North Europeans for Mediterranean holidays, that the Mezzogiorno deserves to be much better known than it is at present. Tourism should, therefore, contribute in no small measure to the development of southern Italy. However, its geographical restriction to the coastal strips (plus one or two exceptional upland areas such as Mount Etna) makes it unworkable as the basis of a general development policy.

Emigration

In spite of the positive achievements in the fields of agricultural and industrial development on the part of the official agencies, many authors (e.g. Lopreato (1966) and Peabody (1970)) maintain that the prime impetus for many of the drastic socio-economic changes that have occurred in the South since 1950 has been emigration. More than any other single cause, geographical mobility has enabled the southern peasants to break the vicious circle of poverty and despair which formerly bound them to their villages, their landlords and their tiny scraps of land. As is well known, emigration is not new to the Mezzogiorno, but in contrast to the earlier 'New World' phase of migration, the main direction of post-1950 current is to northern Italy and other E.E.C. countries, especially West Germany. Between 1950 and 1970 nearly 3,000,000 people left the South. The scale of this movement surprised Italian geographers and demographers and the process received sudden and widespread documentation (Barberis, 1960; Cafiero, 1964; Fontani, 1966; Galasso, 1959). The extent to which emigration can aid an underdeveloped area like the Mezzogiorno is a line of research insufficiently explored. Demographic and migratory trends are parameters vital to southern development planning and it seems that sufficient recognition of their importance has not always been made. The idea of emigration as a positive contributing factor to the South's economic development was first widely canvassed by Lutz (1961). Although this view initially caused some consternation in Italy, in retrospect it seems to have been substantially correct; emigration has been more important in solving the South's unemployment problem than land reform, irrigation and industrialization combined. The issue is not, however, as simple as this. Emigration solves the problem of unemployment and poverty at the individual level but not necessarily at the regional scale. The selective nature of emigration—it being mostly young, vigorous, single males who leave—causes a human impoverishment of the communities left behind; and the disordered nature of the movement has not always resulted in higher incomes for those who remain. Much of the emigration is the result of a declining agricultural workforce, itself consequent upon the progressive rationalization and mechanization of southern agriculture. Whereas in 1950 the policy was to absorb as many people in the rural sector as possible—hence the labour intensive land reform—the assumption of the 1968–80 period is for continuing rapid decline in the farm workforce. During 1950–65 the southern male rural labour force declined by over 1,000,000 (from 2,500,000 to 1,500,000) and recent estimates forecast a further drop of 930,000 between 1965 and 1980. The decade 1970–80 will see a slowing down of the rate of off-farm movement and migration (66,000 per year as opposed to 74,000 in 1965–70), if only because the human well is about to run dry. It is difficult to estimate what proportion will actually leave the South, for Italian migration figures are compounded by rural–urban migration within the South, by returning migrants, and by seasonal and short-term movements. Whatever the actual number, emigration will continue to be a feature of the South for at least a decade yet and Cassa regional policy is at last firmly latched on to this fact.

Recent detailed studies at the regional scale by Rodgers (1970) and at the local and individual scales by Galtung (1971) reveal much information about the geography and sociology of southern Italian emigration. Taking the period 1952–68 Rodgers (1970) showed how, on the macro-scale, annual migration from the South correlated strongly with the pattern of investment and expansion in northern Italian industry, indicating this to be the major 'pull' factor. Periods of boom throughout the late 1950s, to reach a peak in 1962, recession during 1963–66, and resurgence 1967–68, were exactly reflected in the yearly total of migrants leaving the South. The intensity of migration is not uniform within the South; on the contrary, some very marked spatial variations occur. Provinces with high out-migration (20–30 per cent of the 1960 population having emigrated during 1952–68) are those where government-generated industrial growth has been slow. These provinces are the mountain Apennine provinces of L'Aquila, Campobasso, Avellino, Benevento and Potenza, together with the three Calabria provinces and the central Sicilian provinces of Enna and Caltanissetta. On the other hand, provinces with low out-migration ratios (0–10 per cent of the 1960 population having migrated during 1952–68) are precisely those where government-fostered industrialization has had most impact; these are Latina, Naples, Brindisi, Taranto, Catania and Siracusa. Rodgers (1970) demonstrates how industrialization in the Mezzogiorno is the most important single element accounting for spatial variations in southern emigration ratios, explaining 39 per cent of the variance. If a 'wider' variable of 'socio-economic health' is used, the explanation level rises to 65 per cent. The fact that these levels are not higher is partly due to the data aggregation at the province level, whereby the very considerable rural to urban migrations within southern provinces are concealed. Positive residuals from the industrialization–out-migration correlation analysis (indicating higher than expected out-migration) such as L'Aquila, Teramo, Salerno, Campobasso, Enna and Caltanissetta can be partly explained by these provinces' nearness to other provinces which are industrializing rapidly (Naples, Pescara, Catania, Siracusa) and which attract labour which might not otherwise have moved. The evidence suggests that certain key growth poles such as Pescara and Catania have succeeded in stimulating migratory movements within the South; it also suggests that the major embarkation points for emigration (Naples and Palermo) have succeeded in intercepting some erstwhile movement out of the South by acting as stopover points in a two-stage migration process (Rodgers, 1970). Negative residuals (of which Lecce and the three Sardinian provinces are prominent examples) are less easily accounted for, except perhaps in the vague terms of isolation, both from northern Italy and recent developments in the South, and the population's stronger cultural attachment to these provinces. Sardinia in particular has never participated as fully as other southern regions in emigration; this is very much a reflection of isolation, low population density and the strength of the island's regional culture.

Galtung (1971), who compared migration characteristics in three Sicilian villages (one located coastally, one in the hill zone and one in the mountainous interior), points out another interesting geographical aspect of emigration: the closer one comes to the coast, the stronger the orientation becomes towards the New World as opposed to other parts of Italy. The mental maps of aspiring migrants are not simple geographical maps; they are based on sociological or psychological proximity. As one moves from the interior to the coast the link with the local region weakens; the link to Italy strengthens then weakens; and finally the link to overseas strengthens. To migrate from the interior is to move to a job in Turin and Milan; to be audacious on the coast is to dream of other coasts far away, linked not by bonds of common nationality but by form of communication—the sea.

The effects of emigration are widely known but variously interpreted. The outflows have been termed the 'haemorrhage of the Mezzogiorno' by the southern press and there have been cries that, quite apart from social costs, 'raising men for export' is hardly a profitable economic activity. Yet it can also be pointed out that the 'bleeding' might have been far freer had the development programme never been implemented; and where would northern Italian, indeed much German, industry be without southern Italian labour? For those that remain behind, emigration has two positive aspects: the emigrants send back capital to their villages, thus giving life to stagnant economies; and their absence reduces the number of competitors for scarce resources.

Modernization rather than Development

To evaluate what has happened in the Mezzogiorno over the past 25 years, to decide whether results have been worthy of the effort put in, are not easy tasks. The progress of the South, with its contradictions, sorrows and victories, unfolds along a complicated path. Impressionistic opinions are, it is true, fairly common and, of course, have their value. Observers who knew the South just after the war are surprised to find the region's poverty and backwardness have disappeared, shrunk to much smaller levels or confined to isolated pockets such as Calabria and western Sicily. Large-scale irrigation and industrial projects are radically changing the geography of the South. Towns like Taranto, Siracusa and Oristano grow fast and now have substantial middle class and relatively prosperous factory-based working class populations: groups which were conspicuously absent in the past. Employment in agriculture has been halved, living standards have risen and considerable social progress has been made. Unemployment, though still higher than in the North, has fallen. Infrastructure, roads in particular, has improved remarkably. Illiteracy is lower and overcrowding in housing has fallen 20 per cent. In 1950 agriculture employed 57 per cent of the working population, industry 20 per cent and the tertiary sector 23 per cent. In 1970 the respective figures were 33, 32 and 35 per cent. Since 1950, in short, the Mezzogiorno has progressed further than during the previous 90 years since Unification. Geographically, there has been a movement of population and settlement downwards from the hilltops and mountainsides to the plains, coasts and valleys. The land reform provoked a dispersion of settlement which has been paralleled in the private sector, although in marginal upland areas there is a retreat from isolated farms. Perhaps the most profound revolution that has occurred in the South since the war is a psychological one; the peasants have discovered themselves and the intellectuals preach a new *meridionalismo*, marked less by rhetoric and emotion and more by concrete and objective argument. The southern policies do, of course, have their faults, as should have been made apparent. The early stress on infrastructure to the exclusion of industry has been questioned; many of the allocations under the land reform consisted of plots which were too small to be economically viable; there has been some mis-spending of funds, lack of coordination and poor quality sectoral and spatial planning. As is perhaps inevitable in any programme of public largesse, some of the riches lined unintended pockets; bureaucrats and civil servants proliferated to take away some of the resources intended for peasants and workers. All these criticisms ring true, but mistakes are inevitable in any such intervention; southern policy has probably had no more than its fair share, and such mistakes should not lead to a wholesale condemnation of the effort.

But there are difficulties in being more precise about the effects of the development measures. In particular, it is not possible to separate out the effects of the official intervention from autonomous development which would have happened anyway. Furthermore, much of the evidence is of a contradictory nature, which enables both critics and protagonists to state their case with force. In terms of the overall development of the Mezzogiorno *vis-à-vis* the North, individual indices are often more misleading than useful. For the majority of the common measures of development, such as those set out in Table 5.1, the important thing to note is that the absolute gap between North and South has continued to increase, but at a decreasing rate. Because of the much lower base levels for the South, recent levels expressed as percentages over 1950 are higher than those of the North, even though the absolute gap continues to widen. Thus, for example, *per capita* income within the South increased 338 per cent over the period 1950–66, compared to 284 per cent in the North; yet the absolute gap widened from £120 to £250. If, then, the gap is being narrowed, it is closing very slowly; and, in fact, the greater part of it remains, 25 years after the development effort spearheaded by the Cassa got under way.

The question begs itself, however: against what objective should southern development policy be evaluated? In fact, objectives have rarely been spelled out. It is true that the Vanoni Plan utopistically aimed at North–South parity, but other policy statements have never been so explicit. It was soon realized that, so fundamentally different from each other were the 'two Italies', it was unrealistic to have expected the South to catch up with the North within a short period. By the early 1960s the talk was of propelling the South from a static to a dynamic economy, to a level of 'autonomous development' or 'self-sustaining growth'. As Allen and MacLellan (1970) point out, it is difficult to gauge to what extent the autonomous growth objective has so far been realized. It is true that all the trends have been accelerating, but the point of 'take-off' has yet to be reached. The next decade may well be critical in this respect. For the time being, however, the fact that the Cassa is still operating indicates that the South is not yet ready to be left to its own devices.

Another important question which arises is whether it is fair to measure the South's progress against the North, which has had one of the fastest average rates of growth in the world since 1950. Perhaps a better yardstick would be the growth of other southern European countries such as Greece, Turkey, Portugal and Spain. In this comparison the Mezzogiorno emerges with the fastest growth rate of all, faster in fact than many northern European countries. Indeed, perhaps it is surprising that the South managed to keep pace with the burgeoning economy of northern Italy. On the other hand, a large proportion of southern growth is the result of outside help. Net imports average 27 per cent of southern income and 45 per cent of investment in southern industry since 1957 is accounted for by state firms (Allen and MacLellan, 1970). None of the other Mediterranean countries listed could have really tolerated what is tantamount to a permanent balance of payments deficit of 20–30 per cent of gross national income. Looking at the amount of external aid engenders disappointment that the region has not performed better and doubt as to whether the prosperous countries of Europe could sustain a similar level of support to those relatively less developed countries of the Mediterranean which may become further associated with the E.E.C. in the future.

Because of the capital-intensive nature of much of the industrial investment in the South, a dual industrial structure has been created comprising the modern, semi-public, northern or foreign-financed giants working mostly in petrochemical and metallurgical products, and the small artisan traders producing furniture, clothing, foodstuffs etc. Locally-financed small-scale and medium-scale industries have shown disappointing development. This size of firm is geographically rather immobile and its widespread emergence must await the further development of industrial and entrepreneurial talent within the South. According to one observer the entrepreneurial activity and investment that have gone into the South from the North have resulted in the development of particular segments of the southern economy which can, in a sense, be regarded as southern outposts of the basic economic framework of the North (Singh, 1967). The increasing congestion of northern cities like Milan, Turin and Genoa may, in the absence of any positive deterrent policy like the British Industrial Development Certificate, provoke some movement southwards of industry; but recent extension of industrial incentives to some of the less developed northern regions (Umbria, Marche, Friuli–Venezia–Giulia etc., all of which, incidentally, are more prosperous than the richest southern region) may well dilute the South's relative attraction. What is certain is that the funds available to the Cassa are insufficient to deal with the real magnitude of the problem or the objectives specified. The current annual allocations to the Cassa are actually lower, as a proportion of total Italian national income, than they were in the early 1950s. Not only are present funds out of line with what Italy could afford and with what is required by the objectives of the National Plan, they are also well below current U.K. expenditure on regional policy. Continuing spatial differences in prosperity levels within the South reflect the differential impact of the Cassa's intervention. The main contrast is between the poorest regions of Calabria, Lucania and Molise and the more prosperous ones of Campania and Apulia. Calabria's mean *per capita* income is 70 per cent that of Apulia and only one-third that of Lombardy.

For several observers, regional policy to develop the Mezzogiorno has been a case of 'too little, too late'. The Mezzogiorno is an 'old face under a new mask'; the mask constitutes a veneer of modernization, but the age-old features still poke through in many places. The geography of the South is still characterized by certain contradictions—poverty and riches, population growth and unemployment, skyscrapers and slums, modern industry and street-hawkers—indicating that the economy is still in an unbalanced, dualistic, transitional state. One analysis describes this transitional nature as 'modernization without development' (Schneider, Schneider and Hansen, 1972). This analysis points out that the rise in incomes, the decline in agricultural population, the wide availability of consumer products, the access, particularly of young people, to information and education, and the penetration of mass media to all parts of the South are the consequences of closer contact with northern urban and industrial centres rather than the result of significant productive expansion in the South itself. Most of what is consumed in the South, from furniture to cosmetics and convenience foods, is produced elsewhere. Even when new productive facilities are located in the Mezzogiorno, they are not the result of local initiative but are transplanted from the established industrial centres in the North. Capital, technology, managerial personnel and raw materials often all come from without and the parent establishments remain in control of the profits. Because Fiat assembles cars in Sicily does not mean that Sicily is capable of developing a large-scale car industry; it does mean that Fiat are willing to take advantage of cheap land, government subsidies, tax concessions, abundant labour and new consumer markets. The fact that such traditional coalitions as the Mafia in western Sicily, and more nebulous patron–client relations elsewhere, still control the new political and economic structures means that social development too is at a crossroads; the use of terms such as 'post-peasants' (Weingrod and Morin, 1971), 'peasants no more' (Lopreato, 1966) and 'members of two worlds' (Galtung, 1971) illustrates this present cultural ambivalence. It is true that new status groups such as government agency employees and returning emigrants have altered the social structure of the traditional South, but these new groups, in particular the civil servants, still function according to the old rules of clientelism and patronage; it is just that the old landlord–tenant subordination is replaced by *clientelismo della burocrazia.* Bureaucracy, as is well known, flourishes in Italy; and this has been especially true of the South over the last twenty years. Naples, for example, saw its municipal employees increase by over 400 per cent to 15,000 during the period 1953–67. In many ways, the least successful aspect of the development of the South has been the failure to negotiate much-needed political and administrative reform. The irony of regional planning in Italy lies in the fact that the goal of coordinating development efforts has led to the proliferation of would-be coordinators whose very multitude inhibits effective intervention: the particularism which is the enemy of regional planning has been reproduced in the ranks of the regional planners themselves—indeed the allegation has been made that the whole planning effort is geared as much to the interests of the bureaucrats themselves, in expanding civil servant job opportunities, as to benefiting directly the poor people of the South (Fried, 1968).

Four major factors resolve the apparent enigma of where the money to underwrite the new consumer economy of the South comes from. Franklin (1969) characterizes what has been happening in the South over the past two decades as 'welfarism writ large'; price subsidies, loans, grants, land redistribution, unemployment benefits, pensions, medical improvements and the expansion of bureaucratic employment all have the effect of improving the standard of living of southerners, but rarely do these inputs radically alter the productive capacity of the Mezzogiorno. A second important factor in resolving the paradox is the enormous expansion of credit and hire purchase; according to Weingrod and Morin (1971) in Sardinia this has been the major factor facilitating the transition of an otherwise still predominantly rural economy to a consumer economy. The third major source of new purchasing power in the Mezzogiorno is emigrant remittances. In contrast to the earlier, more permanent migrations to the New World, the post-war pattern is for the export of labour rather than persons, a temporary movement by men who leave their families behind and return often, eventually to reinvest their savings in a better life back home. At about £400,000,000 per year, this source of ready cash has an important impact on the southern economy, especially at the village level. Settlements in Calabria or the interior of Sicily receive up to half their income in this way, the money being mostly channelled into house improvement and construction. The final source of modernization is tourism. Like emigration, tourism has the function of attracting foreign currency and employing peasants forced off the land by mechanization and the decline of the rural economy. So far, tourism has had much less impact on southern Italy than it has in Spain or Greece, but the process can be foreseen. Fundamentally, welfaristic hand-outs, pensions, credit, emigration, tourism, even state-controlled industrialization, are ways of strengthening the dependence of the underdeveloped South on the metropolitan, industrialized North without a radical reordering of the economy and society of either region. All of these measures create a redistribution of wealth, from North to South and back again, which does not threaten, indeed it reinforces, the basic patterns of production and trade in both places. In spite of the undeniable achievements in the South, and in spite of great transformations in the landscape of the Mezzogiorno, the essential difference between North and South remains indelible. The South will long continue to be a dependent province of the North, at least until a fundamental reordering of social and economic values occurs. Italy is still a country of great regional diversity and disparity, and will for some time be regarded as one of the best laboratories in the world for the study of regional dualism and regional policy.

References

Ackley, G., 1963, *Un Modello Economico dello Sviluppo Italiano nel Dopoguerra,* SVIMEZ.

Ackley, G. and L. Dini, 1959, Tax and credit aids to industrial development in southern Italy, *Banca Nazionale del Lavoro Quarterly Review,* **51**, 339–369.

Allen, K. and M. C. MacLellan, 1970, Regional Problems and Policies in Italy and France, Allen & Unwin, London.

Banfield, E. C., 1958, *The Moral Basis of a Backward Society,* Free Press.

Barberis, C., 1960, *Le Migrazioni Rurali in Italia,* Feltrinelli.

Barzanti, S., 1965, *The Underdeveloped Areas within the Common Market,* Princeton University Press.

Blok, A., 1966, Land reform in a west Sicilian latifondo village: the persistence of a feudal structure, *Anthropological Quarterly,* **39,** 1–16.

Blok, A., 1969a, Mafia and peasant rebellion as contrasting factors in Sicilian latifundism, *Archives Européennes de Sociologie,* **10**, 95–116.

Blok, A., 1969b, South Italian agro-towns, *Comparative Studies in Society and History,* **11**, 121–135.

Boissevain, J., 1966, Patronage in Sicily, *Man,* **1**, 18–33.

Cafiero, S., 1964, *Le Migrazioni Meridionali,* SVIMEZ.

Carlyle, M., 1962, *The Awakening of Southern Italy,* Oxford University Press.

Chenery, H. B., 1962, Development policies for southern Italy, *Quarterly Journal of Economics,* **76**, 515–547.

Ciancian, F., 1961, The south Italian peasant: world view and political behaviour, *Anthropological Quarterly,* **34**, 1–18.

Compagna, F., 1963, *La Questione Meridionale,* Garzanti.

Dean, G. and M. De Benedictis, 1964, A model of economic development for peasant farms in southern Italy, *Journal of Farm Economics,* **46**, 295–312.

Dickinson, R. E., 1955a, *The Population Problem of Southern Italy: An Essay in Social Geography,* Syracuse University Press.

Dickinson, R. E., 1955b, Geographic aspects of unemployment in southern Italy, *Tijdschrift voor Economische en Sociale Geografie,* **46**, 86–97.

Eckaus, R. S., 1961, The North–South differential in Italian economic development, *Journal of Economic History,* **20**, 285–317.

Fontani, A., 1966, *La Grande Migrazione,* Riuniti.

Franklin, S. H., 1961, Social structure and land reform in southern Italy, *Sociological Review,* **9**, 323–349.

Franklin, S. H., 1969, *The European Peasantry: The Final Phase,* Methuen, London.

Fried, R. C., 1968, Administrative pluralism and Italian regional planning, *Public Administration,* **46**, 375–392.

Galasso, G., 1959, Geografia delle emigrazioni dell'Italia meridionale, *Nord e Sud,* **50**, 75–106; **51**, 60–87; **52**, 48–60.

Galtung, J., 1971, *Members of Two Worlds: A Development Study of Three Villages in Western Sicily,* Columbia University Press.

Holland, S. K., 1971, Regional underdevelopment in a developed economy: the Italian case, *Regional Studies,* **5**, 71–90.

Jucker, N., 1960, The Italian state and the South, *Political Quarterly,* **31**, 163–173.

King, R., 1971a, Italian land reform: critique, effects, evaluation, *Tijdschrift voor Economische en Sociale Geografie,* **62**, 368–382.

King, R., 1971b, The Questione Meridionale in Southern Italy, *University of Durham Department of Geography Research Paper,* **11**,

King, R., 1972, Cost perspectives on the Italian land reform, *FAO Land Reform, Land Settlement and Co-operatives*, **1**, 36–42.

King, R., 1973a, Poverty and banditry: pastoral anarchy in Sardinia, *Geographical Magazine*, **46**, 127–132.

King, R., 1973b, *Land Reform: the Italian Experience*, Butterworth, London.

La Marca, N., 1970, Industrialization of southern Italy, *Banco di Roma Review of the Economic Conditions of Italy*, **24**, 33–45, 145–162, 212–227.

La Palombara, J., 1966, *Italy: The Politics of Planning*, Syracuse University Press.

Lopreato, J., 1966, *Peasants No more*, Chandler.

Lutz, V., 1961, Some Structural aspects of the southern problem: the complementarity of emigration and industrialization, *Banca Nazionale del Lavoro Quarterly Review*, **59**, 367–402.

Lutz, V., 1962, *Italy: A Study in Development*, Royal Institute of International Affairs, London.

Maraspini, A. L., 1968, *The Study of an Italian Village*, Mouton, The Hague.

Marselli, G. A., 1963, American sociologists and Italian peasant society—with reference to the work of Banfield, *Sociologia Ruralis*, **3**, 319–338.

McEntire, D., 1970, Land policy in Italy, in D. McEntire and D. Agostini (Eds.), *Towards Modern Land Policies*, University of Padova Institute of Agricultural Economics and Policy, 171–221.

Molinari, A., 1953, Aspetti e misure della depressione del Mezzogiorno, in *Problemi dell'Agricoltura Meridionale, Cassa per il Mezzogiorno*, 97–136.

Monheim, R., 1969, Die Agrostadt im Siedlungsgefüge Mittelsiziliens: Untersucht am Beispiel Gangi, *Bonner Geographische Abhandlungen*, **41**.

Mountjoy, A. B., 1966, Industrial development in Apulia, *Geography*, **51**, 369–373.

Mountjoy, A. B., 1970, Industrial development in eastern Sicily, *Geography*, **55**, 441–444.

Mountjoy, A. B., 1973, *The Mezzogiorno*, Oxford University Press.

Myrdal, G., 1957, *Economic Theory and Underdeveloped Regions*, Duckworth, London.

Orlando, G., 1952, Metodi di accertamento della disoccupazione agricola Italiana, *Rivista di Economia Agraria*, **7**, 18–59.

Peabody, N. S., III, 1970, Toward an understanding of backwardness and change: a critique of the Banfield hypothesis, *Journal of Developing Areas*, **4**, 375–386.

Perroux, F., 1955, La notion de pôle de croissance, *Economie Appliquée*, **8**, 307–320.

Pizzorno, A., 1966, Amoral familism and historical marginality, *International Review of Community Development*, **15**, 3–15.

Rodgers, A. B., 1970, Migration and industrial development: the south Italian experience, *Economic Geography*, **46**, 111–135.

Rossi-Doria, M., 1958, The land tenure system and class in southern Italy, *American Historical Review*, **64**, 46–53.

Saraceno, P., 1955, *The Development of Southern Italy*, Giuffrè.

Saville, L., 1967, *Regional Development in Italy*, Edinburgh University Press.

Schachter, G., 1967, Regional development in the Italian dual economy, *Economic Development and Cultural Change*, **15**, 398–407.

Schneider, P., 1972, Coalition formation and colonialism in Western Sicily, *Archives Européennes de Sociologie*, **13**, 255–267.

Schneider, P., J. Schneider and E. Hansen, 1972, Modernization and development: the role of regional elites and non-corporate groups in the European Mediterranean, *Comparative Studies in Society and History*, **14**, 328–350.

Silverman, S., 1968, Agricultural organization, social structure and values in Italy: amoral familism reconsidered, *American Anthropologist*, **70**, 1–20.

Singh, B., 1967, Italian experience in regional economic development and lessons for other countries, *Economic Development and Cultural Change*, **15**, 315–322.

Tagliacarne, G., 1955, Italy's net product by regions, *Banca Nazionale del Lavoro Quarterly Review*, **35**, 215–231.

Watson, M. M., 1970, *Regional Development and Administration in Italy*, Longmans, London.

Weingrod, A., and E. Morin, 1971, 'Post-Peasants'; the character of contemporary Sardinian society, *Comparative Studies in Society and History*, **13**, 301–324.

6

France

Hugh D. Clout

Geographical Context of Post-War France

By contrast with stagnation in the 1930s France has experienced striking growth since 1945. Her planned economy has forged ahead and her population has risen from 40,500,000 (1946) to 52,000,000 (1973). Between 1935 and 1945 the crude birth rate had averaged only 15 per thousand and the crude death rate 16 per thousand, but after the Second World War these values changed dramatically. The birth rate was inflated to an average of 20 per thousand between 1946 and 1954, and then gradually stabilized to 16·5 per thousand in 1970, by which time the death rate had fallen to a mere 10·6 per thousand. The French 'baby boom' lasted longer than in other West European countries, but immigration also played a vital complementary role in what has been popularly called the French 'demographic miracle' (Clarke, 1963). Decolonization, and especially the loss of Algeria in 1962, led to repatriation of 1,260,000 French nationals during the last quarter century (McDonald, 1965, 1969). In addition, France attracted foreign workers in large numbers to plug gaps in her indigenous labour force resulting from an excess of deaths over births during both world wars and the depressed years of the 1930s. The general objective of maintaining the labour force has been achieved, indeed it actually grew from 18,850,000 to 19,960,000 between 1954 and 1968. Urban employment expanded rapidly, with jobs in transport and communications rising from 24·2 per cent to 29·6 per cent of the national total, those in banking and administration from 13·8 per cent to 16·6 per cent, and those in building from 6·7 per cent to 9·6 per cent. By contrast, employment in manufacturing and mining rose only from 28·7 per cent to 29·2 per cent of the enlarged workforce, and jobs in agriculture, forestry and fishing declined from 26·6 per cent to 15·0 per cent.

Population growth and changing emphasis in employment have been linked to rapid urbanization, stimulated by high rates of natural increase among existing city dwellers and large flows of migrants from the countryside, so that in 1968 70 per cent of the nation were recorded as residents of urban areas. However, France is a large and relatively empty country. The national density of population is low (92 persons per square kilometre) and France stands out as a reserve of open space for the rest of Western Europe. The bulk of her territory is countryside, served by villages and small market towns. Areas of dense settlement are relatively few and far between, including greater Paris, major communications axes in the Rhine, Rhône and Seine valleys, the industrial provinces of Nord and Lorraine, ports and coastal resorts in Provence, and free-standing cities such as Toulouse, Bordeaux and Nantes (Brunet, 1973) (Figure 6.1). In brief, over half of the French live on one tenth of the national territory. But spatial diversity in France is related to many other groups of geographical and historical characteristics of which perhaps six merit attention at this stage.

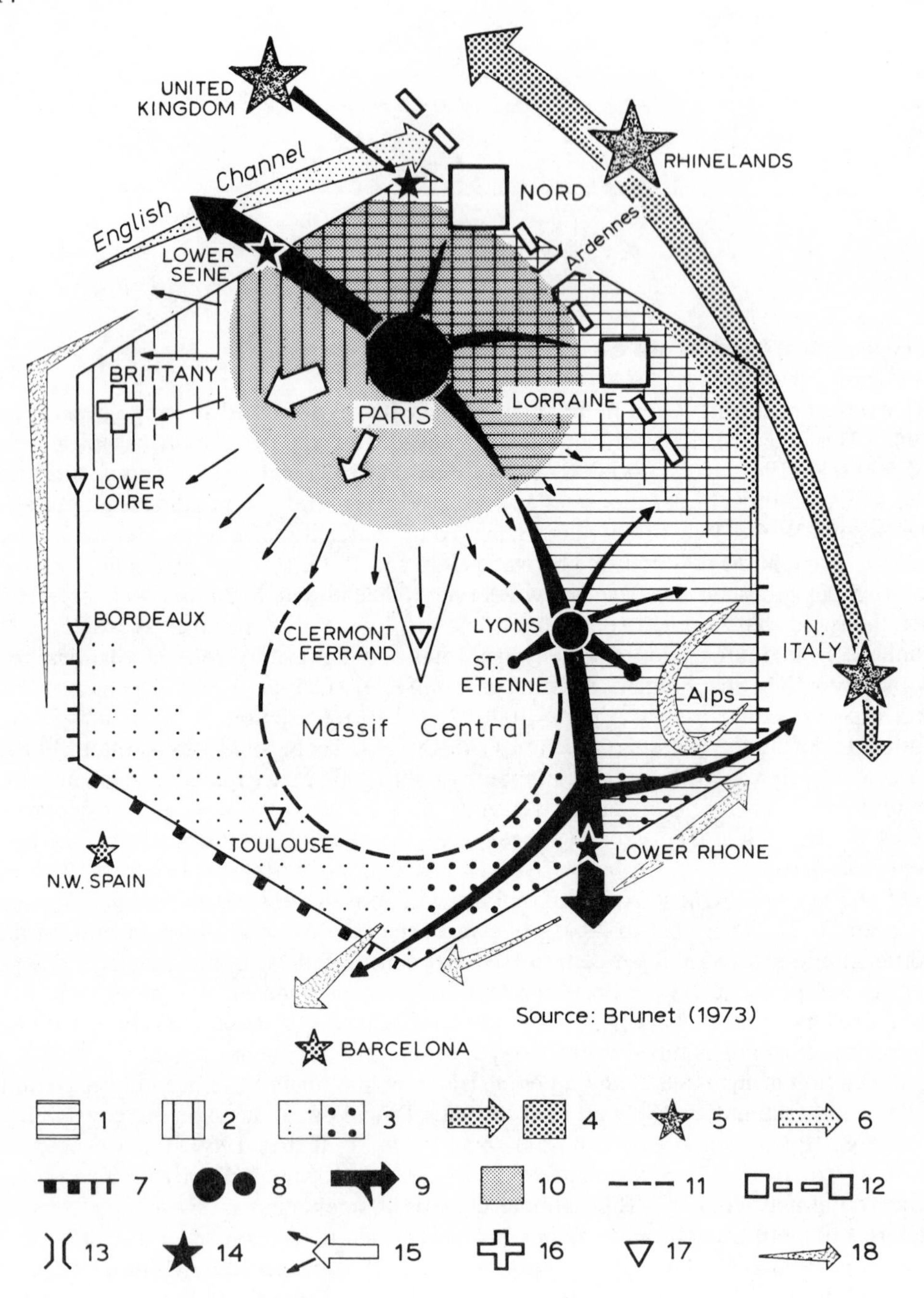

Figure 6.1. Spatial structure of France. 1 Industrial and urbanized areas. 2 High rates of natural increase of population. 3 'Southern France'. 4 Rhinelands axis and its 'overspill' into France. 5 Major industrial growth centres in neighbouring countries. 6 Main maritime routes. 7 Mountainous international frontiers. 8 Major industrial focus. 9 Main axes of communication. 10 Major national growth area. 11 'Area of repulsion'. 12 Old industrial regions in need of renovation. 13 Upland core, dividing an axis of nineteenth-century industrialization. 14 Major port complex. 15 Outward expansion of economic activities from the inner Paris Basin. 16 Rural labour surplus. 17 Large provincial urban centres. 18 Major areas of tourism. (Reproduced by permission of R. Brunet and *L'Espace Géographique*.)

First, the land of France contains an abundance of differing physical resources which are undergoing reappraisal for human use. By way of example, rich soils and uncomplicated terrain are at a premium in this age of agricultural mechanization, which emphasizes more strongly than ever before the great and still only partially realized potential of the Paris Basin as the focus of cereal growing. By contrast, technological change has combined with poor soils, heavy rainfall and harsh relief to make farming in upland areas emphatically 'marginal'. Yet some mountain zones, especially parts of the Alps, are blessed with climatic and scenic conditions which augur an even more prosperous future through tourism than farming ever provided. Less well-endowed uplands, such as the Massif Central, are increasingly beyond the fringe for farming and are struggling to sell their modest but relatively unspoiled charms to tourists.

Second, the 'hexagon' of French territory has access to numerous routes by land and sea to European neighbours and to more distant parts of the world. The significance of each of these contacts has changed many times in the past and will continue to do so. For example, the small harbours of western France were vital in the history of trade and colonial development in the sixteenth and seventeenth centuries, but then gave way to larger ports such as Bordeaux and Nantes which in turn have been massively eclipsed by Dunkirk and Le Havre since the 1950s. Hard-won land frontiers in northern and northeastern France were 'vulnerable zones' in the nineteenth century from which all but basic industries were steered away; but since 1957 these same areas have become contact zones with neighbours in the E.E.C. offering great prospects for trade and international cooperation.

Third, France is furrowed by major valleys which form natural routeways of crucial significance for communications systems from Roman roads to railways and motorways. However, routes in the basins of Aquitaine, Paris and the Rhône, avoiding the highlands of the Massif Central, are of unequal significance. Those through the Paris Basin are of crucial importance, linking the national capital to its nearest ocean ports and to surrounding regions. Similarly, routes along the Rhône afford contact between the political and economic core of Paris and the trading ports of the Mediterranean South. By contrast, the valleys of Aquitaine are orientated perpendicular to the main land routes fanning out between Paris and Iberia, and the valleys of northern France do not bind frontier areas more closely to the Paris Basin but rather facilitate links across man-made political boundaries.

Fourth, important cultural differences survive. Even though southward invasion continues, the customs and dialects of the South (*langue d'oc*) remain different from those of the North (*langue d'oïl*). At a regional scale, Basques, Bretons, Corsicans and bilingual Alsatians are strongly aware of their cultural distinctiveness and may even provoke minor explosions to prove it (Lebesque, 1970; Sérant, 1965). At a yet more local level, Auvergnats know that they are different from men from the Limousin or the Bourgogne. Such examples could be multiplied throughout the country.

Fifth, the component parts of France are irregularly endowed with the necessary natural resources for industrial development. It is true that small quantities of metallic ore and timber were found in many areas in the past and fed primitive manufacturing activities, but adequate supplies of coal and other minerals for industrialization in the nineteenth century were restricted to sites of limited accessibility bordering the Massif Central and to vulnerable frontier zones such as Nord and Lorraine, where coal and iron ore were shared with rival nations.

Last, but certainly not least, the administrative organization of France has long been moulded by powerful centralization, thereby adding the contrast between capital and provinces to the numerous types of interprovincial difference evoked so far.

It comes as no surprise to learn that modern France is a country with striking regional contrasts and consequently striking regional problems. But whilst a system of national planning was formulated as early as 1946 to stimulate and steer the whole economy, action for tackling regional issues developed at a more leisurely pace and in a more piecemeal fashion. Such relative tardiness reflects the immediate post-war priority of restarting the national economy, the necessity of accurately identifying regional problems, and the slow acceptance of the 'region' as a suitable unit for planning purposes. Nevertheless, France managed to anticipate many of her neighbours in devising machinery for regional management and has probably been more successful than them in tackling regional problems.

Identifying Regional Problems

Economic and social conditions in provincial France were investigated in the 1940s and early 1950s by unofficial regional expansion committees, composed of academics and representatives of local agricultural, commercial and industrial interests. Inventories of resources were prepared for many parts of the country, but these reports were of no significance for policy-making nor did they relate to official regions, since none had yet been defined. At the same time, economists and geographers were describing spatial variations in the national economy. The earliest and perhaps most influential study was undertaken by J. F. Gravier (1947) whose book, *Paris et le désert français*, depicted a core/periphery model which showed that Paris had absorbed a grossly inflated share of the nation's population, economic activities, decision making, cultural facilities and higher education. To echo a phrase used by nineteenth-century observers, 'Paris was devouring the provinces'.

Elevation of the capital to such a position was partly due to political centralization, with provincial parliaments being abolished at the Revolution and replaced by over eighty *départements* controlled from Paris. Schemes for rebuilding parts of the capital during the Second Empire drew large numbers of provincials to the city, but the really important phases of Parisian economic growth came later at the end of the nineteenth and early in the twentieth century. Dynamic engineering industries were established in the capital, at the focus of national road and rail networks for collecting coal and raw materials and for dispatching finished goods at home and abroad. Industrialists were able to enjoy the life style of the capital and to choose workers from the largest pool of labour in the country, which was also its largest domestic market. At this stage at least, Paris was considered to be sufficiently distant from France's frontiers to escape immediate threats of land attack and many finishing industries in Paris had links with basic industries in the North and North-East. By contrast, provincial France became a cultural and economic 'desert', sucked dry by the urban monster of Paris. It was clear to Gravier that further expansion of the core of France would only serve to weaken the periphery more. If the problems of the provinces were to be relieved a policy was needed to decentralize some of the strength of Paris.

More detailed models of spatial diversity were proposed by other researchers. Thus the economist J. R. Boudeville (1966) recognized three subdivisions of national territory after analysing employment, incomes, functional links between town and country, and other factors. A rectangular area from Lyons to the Mediterranean Sea contained attractive climatic conditions and valuable supplies of hydro-electric energy and was deemed to have great potential for employment growth. The Paris Basin and northeastern France had inherited legacies of nineteenth-century urban growth and coal-based industrialization, and needed to renovate city environments and economic life to meet future demands. Finally, central and western France were characterized by low incomes, lack of urbanization and non-agricultural employment, and had limited potential for modern economic growth, which would only be achieved through powerful assistance from outside.

Other workers contrasted eastern and western France, separating the two along an imaginary line from Lower Normandy to the Rhône delta, and then isolated Paris from the rest of the country. The West, covering 56 per cent of France, contained 48 per cent of her population in 1846 but only 37 per cent in 1946, having experienced a net loss of 2,000,000 through out-migration to Paris and to employment centres in the East during the intervening hundred years.

A more detailed regional typology has been offered by American economist N. Hansen (1968) who recognized 'congested', 'lagging', and 'intermediate' zones. Paris formed the classic case of congestion, with respect to population, commerce, economic activities, traffic and poor housing. A public opinion poll in 1965 showed that only 42 per cent of Parisians wanted to stay in the capital. Lagging zones offered few, if any, attractions for economic innovation and were typified by much of western France. Intermediate zones contained supplies of industrial raw materials, plus qualified labour and cheap power. These conditions were fulfilled in northern and eastern France, but important local differences were recognized between Nord or Lorraine with their old industrial heritage, on one hand, and the Grand Delta of the lower Rhône, on the other.

In addition to these academic summaries, programmes for regional action were drawn up after 1955, by which time official planning regions had been defined. Great debate had occurred over many decades with a view to grouping *départements* into larger units for more efficient administration (Gravier, 1970). As early as 1911 geographer Vidal de la Blache had examined spheres of influence of major towns and thereby defined functional regions; and *départements* had been linked together for economic administration and policing under the Vichy regime of the early 1940s. *Ad hoc* groupings operated later in that decade but it was necessary to tackle the problem of regionalization in a more rigorous way, through analysing urban hinterlands, university catchment areas, agricultural variations, areas covered by administrative services, and the survival of provincial loyalties. The net result was 22 'regions', which were reduced to 21 and later raised to 22 when Corsica was separated from Provence (Figure 6.2). These regions have been criticized because old boundaries remained unchanged and *départements* were simply pieced together. The particular groupings were open to debate, with, for example, the official Paris region covering only 2 per cent of France and in no way corresponding to the capital's hinterland. The number of regions in France might be considered excessive when compared with larger regions in Federal Germany and Italy. In fact, various informal grouping procedures have been required to associate regions in recent years. Thus, the 22 regions were linked into eight macro-regions for regionalizing the Sixth National Plan and special arrangements were devised for integrating regional management in the Paris Basin (Figure 6.2).

The programmes for regional action for each of the official regions contained important collations of factual information but they lacked imagination as planning documents. Standardized data describing conditions in each part of France were not

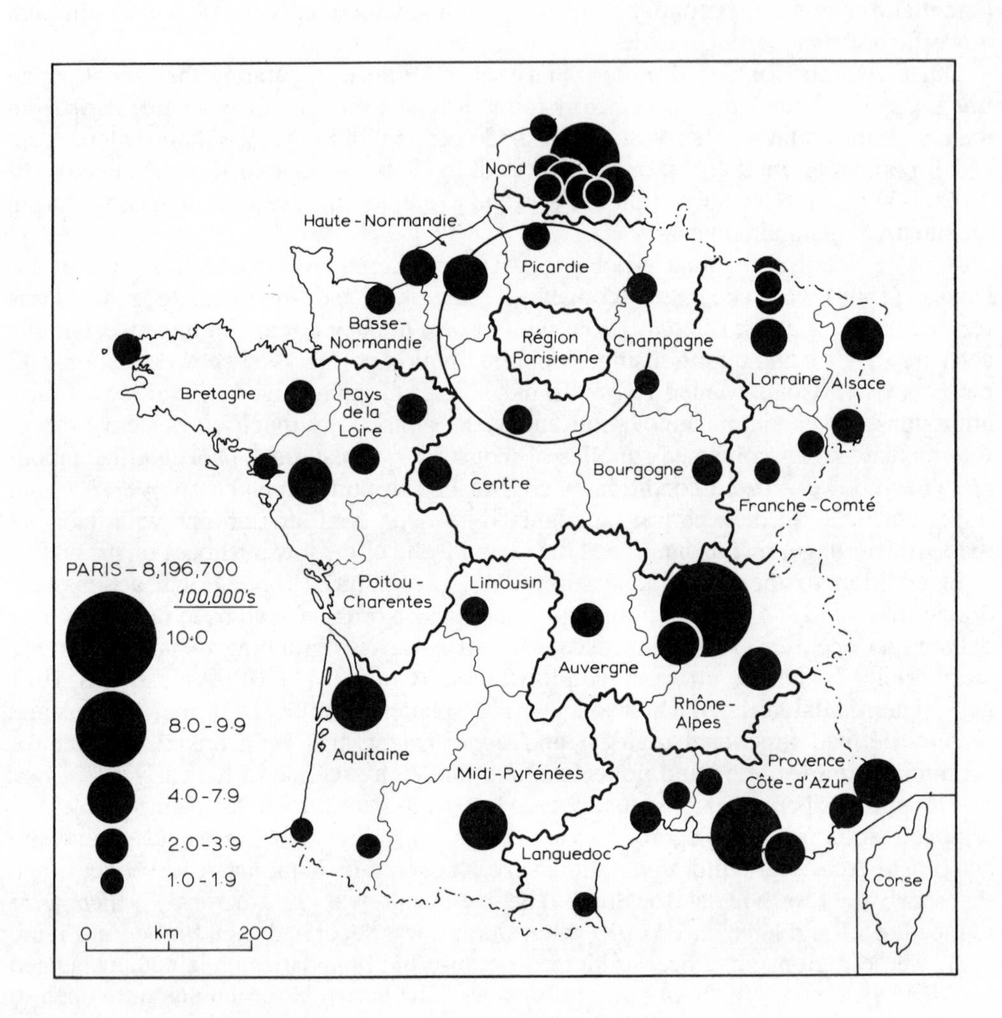

Figure 6.2. France: planning regions and major cities.

available until the mid-1960s, by which time regional issues were being introduced into national economic plans. Thus the Fourth National Plan (1962–65) drew attention to the under-industrialized West and the Fifth Plan (1966–70) emphasized the need to modernize farming and to provide more manufacturing and tertiary jobs in that half of France. In addition, it described employment problems in old industrial and mining areas in the North and East, the need to replan Paris, to strengthen services and employment in major provincial cities, and to introduce replacement jobs in declining market towns. Each of these objectives was incorporated in the Sixth Plan (1971–75).

At first, planning regions were used simply for translating central decisions into the provincial context, but their function was elaborated in 1964 when an advisory committee for economic development was set up in each region. As well as drawing attention to provincial problems the Fifth Plan introduced the idea of regionalizing national plans and their budgets. Discussions were held between advisory committees and regional prefects to assist in this task and a report was drawn up for each region outlining projects that were anticipated in forthcoming years. On the basis of this evidence, decisions were taken in Paris for financial allocation between regions. From the beginning of 1974 two advisory councils have operated in each region with limited financial powers being transferred to regional authorities through collection of dues paid on driving and vehicle licences and on property transactions. These embryonic regional revenues are being used to implement local development schemes and promote appropriate research.

However policies for the 22 planning regions represent only one element in the complicated array of strategies developed over the past 20 years for modifying patterns of economic and social life in France. Since 1963 such spatial strategies for planning urban growth, relocating manufacturing and tertiary activities beyond Paris, and reorganizing rural areas, have been directed by the *Délégation à l'Aménagement du Territoire* (D.A.T.A.R.), which is an elite strike-force of fewer than a hundred technocrats.

Planning Urban Growth

As Gravier had suggested in the 1940s and others demonstrated in a more scientific fashion, Paris stood out as the focus of economic and cultural life for the whole of France. Below this level, eight cities, or more precisely *groups* of cities, supplied broad ranges of services to populous provincial hinterlands. In the northern industrial area, Lille plus Roubaix and Tourcoing served a densely populated area truncated by the Belgian border (Figure 6.3). Comparable functions in northeastern France were provided by Strasbourg in Alsace and the rival cities of Metz and Nancy in Lorraine. To the south, Lyons controlled a broad hinterland comprising the Rhône valley and the eastern edge of the Massif Central, with Grenoble and Saint-Etienne performing important sub-regional functions. Finally, southern France was serviced by Bordeaux, Marseilles and Toulouse, with Nantes emerging as the leading urban focus for the North-West. Remaining parts of France in the Massif Central and the outer Paris Basin looked to relatively small provincial towns for normal services and then directly to Paris for higher-order activities.

Recognition of the functional importance of Paris and the provincial cities provided a context within which the aspatial 'growth pole' (*pôle de croissance*) concept formulated by economist F. Perroux might be translated into geographical reality.

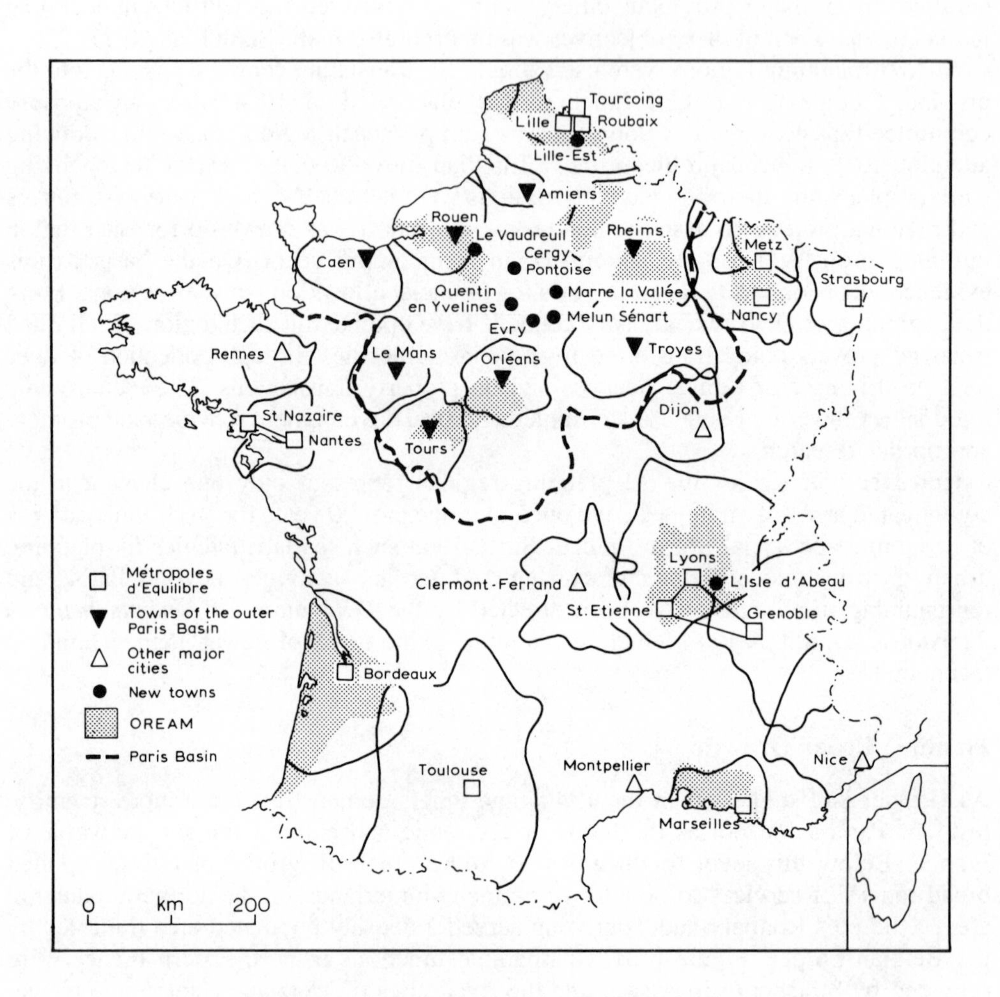

Figure 6.3. France: major cities and hinterlands.

Additional employment opportunties and cultural facilities needed to be installed in selected provincial centres so that they might serve their hinterlands more effectively and operate as *métropoles d'équilibre* to counterbalance Paris and attract migrants who would otherwise have moved to the capital.

Opinions varied on the number of *métropoles* that would be desirable. Some observers believed that only Lille, Lyons and Marseilles, with roughly one million inhabitants apiece, were sufficiently large and well-equipped to have any chance of counterbalancing Paris. Others favoured a score of cities of varying sizes, distributed more evenly throughout the country. In practice, eight *métropoles* were designated by the D.A.T.A.R. in 1964 as part of the Fourth National Plan (Figure 6.3). They varied considerably in terms of services provided, geographical structure and population size from the rather illogically linked *métropole* of Lyons/Saint Etienne/Grenoble with 1,520,859 inhabitants to the city of Strasbourg, with only 302,772 (Table 6.1).

Table 6.1. Population in Métropoles D'Equilibre 1962–68

	1962	1968	Average annual growth 1962–68
			%
Bordeaux	498,429	555,152	+1·8
Lille/Roubaix/Tourcoing	821,228	881,439	+1·2
Lyons/St. Etienne/Grenoble	1,520,859	1,738,660	+2·3
Marseilles/Aix/Fos	909,911	1,056,847	+2·5
Nancy/Metz/Thionville	508,783	560,657	+1·6
Nantes/St. Nazaire	453,632	504,628	+1·8
Strasbourg	302,772	334,668	+1·7
Toulouse	365,482	439,764	+3·1
Paris region	8,469,863	9,250,674	+1·5

This selection of cities was criticized on several grounds. The largest cities were chosen to be *métropoles* but they were not distributed regularly throughout France. Large parts of France lacked *métropoles* of their own since Clermont-Ferrand and Dijon, serving the Massif Central and Burgundy respectively, were not designated; and the exclusion of Rennes from the list in favour of Nantes brought forceful protest from the Bretons. Sometimes groups of neighbouring cities were chosen as *métropoles* and under such circumstances it would be vital to integrate schemes for town planning and public transport if these centres were to hang together as meaningful growth poles. In any case it might be argued that in planning for the future it was not necessarily appropriate to choose urban sites that had functioned successfully under varying past conditions.

In spite of these and other reservations, the *métropoles* have been strengthened to varying degrees in recent years as part of associated policies operated by the D.A.T.A.R. for relocating manufacturing and tertiary jobs. Certainly the population of each *métropole*, save Lille/Roubaix/Tourcoing, grew more rapidly than the population of Paris in the last decade (Table 6.1). Nevertheless, there is still great variation in size, service provision and environmental attractiveness between these *métropoles* with, for example, the Alpine backcloth of Grenoble contrasting with the legacy of nineteenth-century industrialization in Lille and Metz or with the reputation of Nantes as a cultural backwater. In fact, the objective of building up provincial growth centres was extended beyond the *métropoles* in 1967 when Besançon, Clermont-Ferrand, Dijon, Limoges, Montpellier, Poitiers and Rennes were selected to receive financial aid for attracting tertiary employment.

At the same time as strategies were being prepared for strengthening provincial cities, plans were being produced for guiding the future of Paris. The Plan d'Aménagement et d'Organisation Générale (P.A.D.O.G.) was finalized in 1960 (Hall, 1966, 1967; Thompson, 1973). This worked from the premise, perhaps more appropriate to the 1930s than the 1960s, that migration to the capital might be controlled by steering new employment opportunities to cities in the outer Paris Basin. The P.A.D.O.G. also emphasized that environmental conditions in the capital needed improvement and plans were outlined for constructing ring and radial motorways, creating an express *métro*, implementing urban renewal programmes, and constructing large new service centres in the outer suburbs. However, the volume of natural increase taking place in Paris made the P.A.D.O.G. target population of 9,000,000 quite unrealistic. In addition, from a nationalist viewpoint it was considered undesirable to limit the growth of Paris, which performed important supranational functions. Accordingly, a new strategy, the Schéma Directeur d'Aménagement et d'Urbanisme, was presented in 1965 which envisaged the population of the Paris Region as reaching 11,600,000 in 1985 and 14,000,000 by A.D. 2000. Many projects outlined in the P.A.D.O.G. were retained in the new strategy and two important additions were made. The first was the idea of devising growth axes parallel to the middle Seine and the second involved accommodating population increase in new cities detached from the main body of Paris and located between 15 km and 35 km from Notre-Dame (Merlin, 1971). Creation of these new urban foci would hopefully avoid problems of suburban sprawl, provide local employment and services, and enable new styles of urban environment to be designed. At first eight new cities were proposed but these were scaled down to five following later schemes for the integrated management of the Paris Basin.

The proposal to build new cities was an innovation in two respects. The idea of small and fairly self-contained new towns had been rejected at earlier stages by French planners. Detached housing estates had been constructed in the countryside near the natural gas deposits of the South-West and in the steelmaking and coalmining areas of the North-East; but these estates had not been planned as complete 'towns'. Post-war urban growth was accommodated more typically in estates of apartment blocks tacked on to existing towns. The second innovation was the size of the Parisian new cities, with populations of *c.* 500,000 to be housed by the end of the century.

The Schéma Directeur did not deal directly with surrounding areas of the Paris Basin but had obvious implications for them. The Lower Seine valley, for example, had the two growth axes directed into it and would be particularly vulnerable to pressure from economic and demographic growth. Other parts of the Paris Basin would be affected in less urgent but still important ways since they would have to accommodate population and jobs steered by the D.A.T.A.R. away from the capital and to provide land for new communication links to the city and recreation space for its residents. Hence new measures were instituted to allow the regions of the Paris Basin to be managed more effectively and in harmony with the capital city. An inter-ministerial committee was set up in 1967 to coordinate planning schemes throughout the Paris Basin, which was redefined in extended form in 1970 to include not only the regions of Paris, Haute-Normandie, Basse-Normandie, Centre, Champagne and Picardie, but also parts of Bourgogne and the Pays de la Loire (Figure 6.2). A white paper for the Paris Basin was drawn up in 1969 and proposed that inter-regional policies should be integrated for communications, education, leisure and cultural facilities, and that activities needing to be near but not actually *in* the capital should be deflected into towns and cities in the outer Paris Basin. As a result of this proposal to boost outer parts of the Basin the number of Parisian new cities was reduced from eight to five.

In the Lower Seine valley an experimental planning team was set up to produce a strategy for the area to accommodate 2,500,000 inhabitants by A.D. 2000, compared with 1,100,000 in 1962. The resultant plan proposed that the bulk of future population growth be accommodated downstream from Rouen. Continuous urbanization of the Lower Seine valley was to be avoided by conserving green wedges, and satellite towns were to be expanded around Le Havre and Rouen, with a complete new town being planned at Le Vaudreuil to the southeast of Rouen to accommodate 40–70,000 inhabitants by 1985.

Similar organizations were established to prepare strategies for other parts of France where large-scale urbanization was envisaged in the remainder of the present century. These include the Oise Valley, northern Champagne, the middle Loire Valley, and areas surrounding *métropoles d'équilibre* in the Nord and around Bordeaux, Lyons/Saint-Etienne, Marseilles/Aix/Fos, Nancy/Metz, and Nantes/Saint-Nazaire (Figure 6.3). Each of these strategies prepared by Organisations d'Etudes et d'Aménagement d'Aires Métropolitaines (O.R.E.A.M.) deals with urban renewal and growth, highway construction and industrial location. For example, the O.R.E.A.M. plan for Lyons/Saint-Etienne embraces the Part-Dieu urban renewal scheme, the city's first *métro* line, new motorway links, Satolas airport and the new town of Isle d'Abeau which will house 135,000 in 1985. Similarly, renovation schemes and new housing estates around cities in the Nord will be complemented by the new town of Lille-Est (Villeneuve d'Ascq) to accommodate 120,000 residents by 1985 plus 40,000 students on a university campus site. New town status has not been granted to urbanization schemes around the major port and industrial site of the Gulf of Fos near Marseilles, but the number of residents living immediately east of the Etang de Berre is likely to grow from 53,000 in 1972 to 150,000 by 1985, and those to the west from 80,000 to 350,000.

French urban planning has largely been concerned with schemes for *métropoles d'équilibre*, greater Paris and the new towns. As a result, criticisms have been raised that a series of provincial versions of 'Paris' have been fostered, each with their surrounding 'deserts', and that the problems of lower-order settlements have been ignored. Undoubtedly there has been much truth in such allegations and, as a result, two important changes in policy have been made. First, in 1967 regional capitals qualified for financial assistance for relocation of tertiary employment. Second, more attention has been paid in the 1970s to improving housing, transport, job provision and cultural facilities in medium-sized towns of 100,000 or less. Each of these schemes for urban development cannot operate in isolation and has been planned in association with policies for relocating employment opportunities which form the backbone of the concept of *l'aménagement du territoire*.

Relocating Employment in Manufacturing Industries

Policies for relocating branches of French manufacturing industry have had various spatial objectives since 1955 (Clout, 1970; Durand, 1972). The idea of transferring key industries from Paris to the provinces had been implemented in a limited way in the 1930s when 62 aircraft workshops and 15,000 workers plus their families were moved to towns in the outer Paris Basin and in southern and western France. This operation was purely strategic in nature, since the capital was considered vulnerable to bombardment should war occur. Towns receiving these factories were often ill-prepared for additional population and its social and commercial needs. Much has been achieved under the heading of 'industrial relocation', following the writings of Gravier and others and the introduction of special financial measures. The concept has broadened since the 1950s from straightforward concern for transferring jobs from Paris to the provinces, so that three objectives are now involved.

Decentralization remains the first of these objectives, whereby the overwhelming dominance of Paris in French economic life is to be curtailed to the benefit of the whole country. Certainly the capital grew more rapidly than other parts of France after 1800, with the number of residents in the Paris region increasing from 1,500,000 to 8,600,000 in 1962 while the national total rose by only 70 per cent from 27,300,000 to 46,500,000. However, rates of new job provision and population growth in the capital have been irregular in the present century. For example, the number of Parisians rose by 1,370,000 between 1911 and 1931, then remained stable at about 6,700,000 during the depression years and the Second World War, only to grow even more rapidly after 1945. Numbers were swollen by 700,000 between 1946 and 1954 to reach 7,400,000; and in the following six years the Paris region grew by a further 1,200,000. By comparison, it had taken two decades for the city to grow by that amount at the beginning of the century. Less than half of the capital's population growth in the 1950s was produced by natural increase, the remainder stemming from in-migration to the city's attractive job opportunities, involving over 60 per cent of employment in the nation's car industry, 60 per cent in aeronautical, electrical and mechanical engineering and pharmaceuticals, 50 per cent in printing, administration, commerce and banking, and 75 per cent in research work. A case for decentralizing some of that employment to other parts of France could be made both from the viewpoint of the densely-packed and expensive capital and from that of the deprived provinces.

The second objective of industrial relocation complements the first and involves enlarging employment possibilities in the provinces to reduce the volume of migration to Paris. As the various studies of spatial variations in France suggest, important differences in employment and living standards exist not only between the capital and the provinces and between eastern and western France, but also between regions and between town and country. A flexible system of regional aid is therefore required to attract new jobs to a wide range of problem areas.

Finally, labour redeployment and industrial renovation is required in areas with contracting sources of employment, such as coalmining, textile manufacture, ironmining, and steelmaking. Taking France as a whole, employment in the coal industry was reduced from 330,000 to 218,300 between 1948 and 1960 in order to rationalize and raise production from 45,000,000 tons to 60,000,000 tons. Thereafter, further reductions in labour were accompanied by planned cutbacks in output, with 132,500 workers producing 40,000,000 tons in 1970. Similarly, the cotton textile industry reduced its workforce from 138,000 in 1955 to 72,500 in 1971 and the woollen industry cut back labour from 100,700 to 59,800. Mechanization, rising productivity and changing demand reduced workers in iron-mining from 25,000 to 10,500 between 1962 and 1970, and employment in steelmaking dropped slightly from 199,000 (1953) to 180,000 (1968). Very often a number of labour shedding activities co-exist in the same regions, notably the Nord, Lorraine and old industrialized parts of the Massif Central. Provision of replacement jobs is just one of a series of major social, economic and environmental problems in such areas.

Local authorities offer a wide and variable range of financial, fiscal and material encouragements for factories to be installed or expanded in their areas. In addition, a system of measures has been devised by the central government, on the one hand applying controls on industrial growth in overcrowded areas and on the other offering incentives for relocation in the provinces. Controls on manufacturing growth in the Paris region have operated since 1955, whereby firms needed to obtain permission if additional premises in excess of 500 m^2 of floorspace were to be occupied. Withholding planning permission and levying special dues on permitted industrial expansion provided useful measures for steering manufacturing jobs beyond the capital. But, in practice, these controls have been weakened several times with the threshold being raised first to 1000 m^2 and then to 1500 m^2 in 1972. Similar attempts to control industrial growth around Lyons were enforced between 1966 and 1970 but were then abandoned.

Complementary to these restrictions are measures for encouraging industrial relocation in provincial France. Finance has been available since 1950 for setting up factory estates and special corporations have been established to equip them, combining assistance from State, private and local authority sources. Over 500 estates have been laid out, varying in size and quality, from small factory estates on the edges of market towns to large and well-equipped sites near Dunkirk, Le Havre and Marseilles/Fos which have been designated as the leading growth points for port-based heavy industrialization in France for the remainder of the century.

Financial schemes for encouraging provincial industrialization were started in 1955 and have been modified several times since then. At first grants were offered to firms opening or extending factories in a score of *zones critiques* which included coalfields, textile-producing towns and other parts of the country where ceramics, cutlery, shoemaking and other old industries were shedding labour. Four years later grants were extended to larger *zones spéciales de conversion*, involving cotton manufacturing valleys in the Vosges, coalmining areas around Béthune, and the shipbuilding area of Nantes and Saint-Nazaire. In 1960 grants were made available more widely provided local employment conditions were deemed to justify them and, as a result, aid was extended to areas of underemployment in the countryside as well as to old manufacturing areas with high rates of unemployment. But this system had the disadvantage of not allowing industrialists to know in advance the degree of financial assistance they might expect to receive, since local employment conditions had to be scrutinized. This 'case by case' approach was duly replaced in 1964 by a new regime which provided 'adaptation grants' in mining and old industrial areas and 'development grants' in predominantly rural western France. Highest rates of assistance were available in problem areas such as Bordeaux, Brest, Limoges and the Lower Loire, with a median rate operating in northwestern France and a lower rate available in the South-West, the Massif Central and Corsica. In 1972 the system was simplified yet

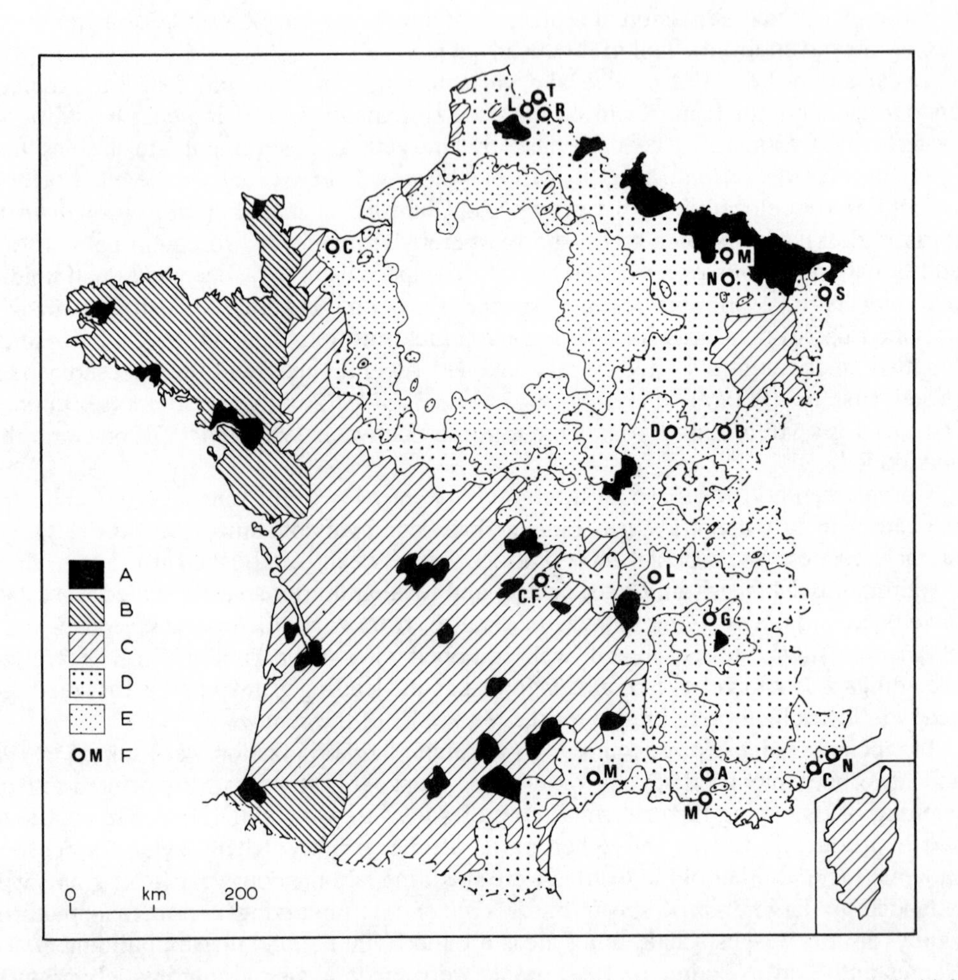

Figure 6.4. France: graded assistance for regional industrialization. A Maximum rates of aid for industrial and tertiary development. B Medium rates. C Low rates. D Fiscal exemptions for regional industrial development. E Fiscal exemptions for industrial decentralization from Paris and Lyons. F Urban centres in receipt of aid for tertiary development.

again and 'regional development grants' for creating or extending factories were offered at the highest rate in a scatter of manufacturing and mining areas and in border zones in northeastern France (Figure 6.4). The North-West and the *département* of Pyrénées-Atlantiques received a median rate, with a lower one available in remaining parts of the West and in Corsica.

Other forms of government aid include cheap loans for factory construction and indemnities to cover part of the costs of dismantling, transporting and reassembling machinery taken from Parisian factories to the provinces. Land prices are pegged at special low levels on estates where factory jobs are needed urgently and finance is available to help cover costs of training or retraining workers. In addition, new factories in western France and in some sections of the East may be eligible for exoneration from taxation for a number of years (Figure 6.4). Over and above these inducements, the Société Sidérurgique de Participation et de Dévéloppement Economique (S.I.D.E.C.O.) and the Société Financière pour Favoriser l'Industrialisation des Régions Minières (S.O.F.I.R.E.M.) provide finance for creating new jobs in steelmaking and coalmining areas. Finally, 'European' sources of assistance may be available from the E.C.S.C. and the E.I.B.

The number, value and distribution of grants for industrial development have undergone important changes since 1955. In the early years, relatively few grants were awarded but, from 1961 to 1966, over 200 were made each year, worth collectively between 50,000,000 and 100,000,000 francs per annum. Schemes in the Aquitaine, Bretagne and Midi–Pyrénées regions were particularly numerous. Fewer grants were awarded in 1967 and 1968 but thereafter their number and value increased to reach a peak of 446 in 1970 at a total value of 426,000,000 francs. Nevertheless, the annual amount of government investment for regional aid in France during the early 1970s was less than one-tenth of the sum being used for the same objective in Great Britain. Regions other than the three already mentioned came to the forefront for French government aid in the late 1960s and early 1970s, with large sums being contributed to old industrial and mining areas of Lorraine and Nord and to the rural regions of Poitou–Charentes and Pays de la Loire. Attention to other parts of western France declined. Investment in mining and old industrial areas was also boosted by the S.O.F.I.R.E.M., which financed schemes to provide 9000 jobs, mainly in the coal basins of the Massif Central, and by the S.I.D.E.C.O. which is providing 1500 new jobs in Lorraine. The number of factories transferred from Paris plus provincial extensions of Paris-based firms reached an annual peak in 1961 when 289 plants were involved. More than 250 were recorded in 1962 and in 1963 but since then annual figures have declined, standing at little more than 100 after 1969. In total, almost 3000 schemes were negotiated between 1955 and 1971 to provide about 500,000 new industrial jobs in the provinces. More than four-fifths of these jobs are already in existence and the remainder will become available when new factories reach peak production.

These results are extremely important but it is undeniable that the pattern of job creation is almost the inverse of what had been anticipated. Regions recognized as having the most serious employment problems, and therefore qualifying for the highest rates of assistance, have received relatively few jobs (Figure 6.5). Bretagne, for example, received only 3·5 per cent of new factories and 5·4 per cent of the new jobs created, and results are even less impressive in Aquitaine, Limousin and other western regions (Table 6.2). By contrast, regions in the Paris Basin have done extraordinarily

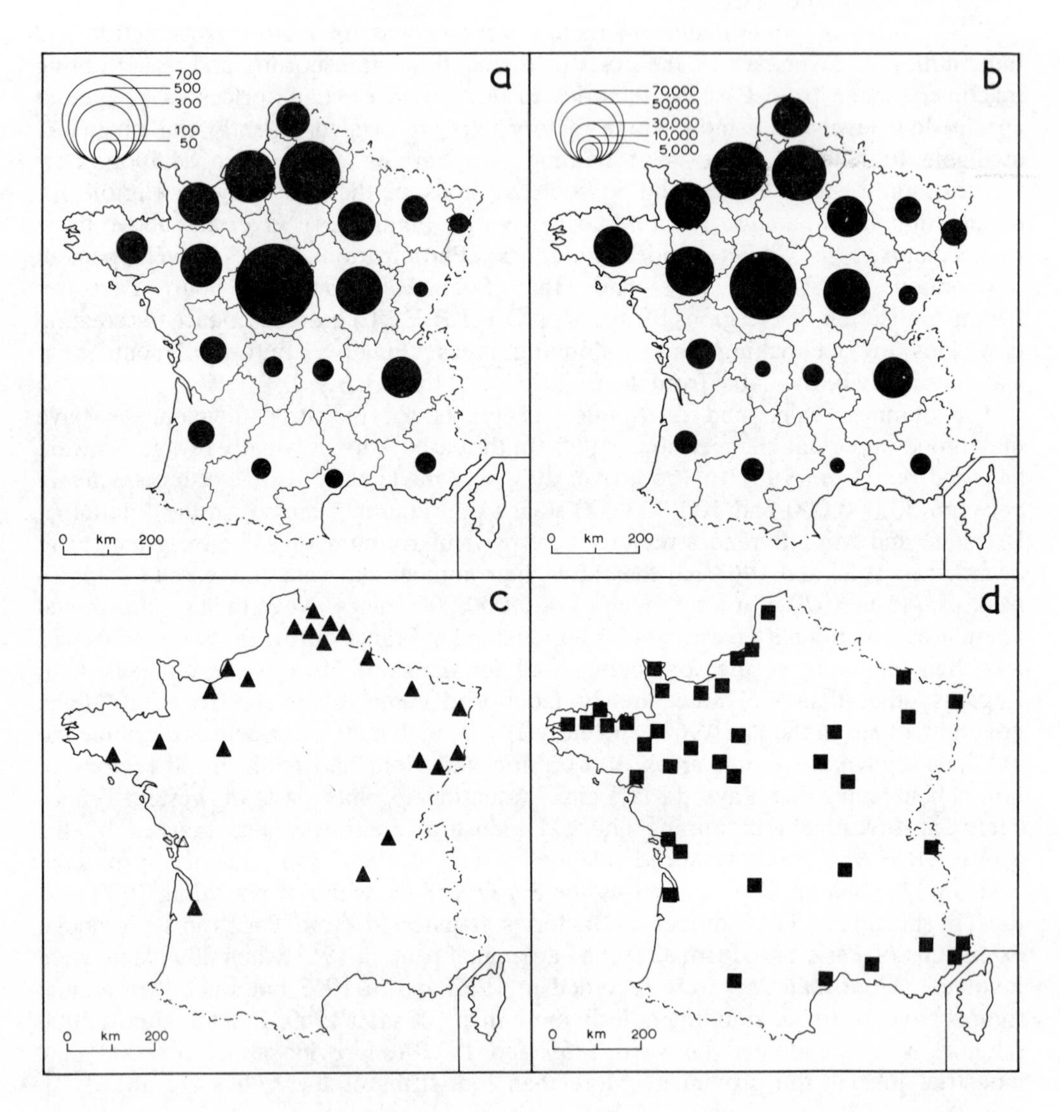

Figure 6.5. France: regional industrial development, 1955–71. (a) Jobs still to be created when schemes completed. (b) New industrial jobs. (c) Decentralized car assembly and components works with more than 1000 workers apiece. (d) New provincial electronics factories.

Table 6.2.

	Percentage relocated factories 1955–71	Percentage new industrial jobs 1955–71	Employment in manufacturing 1968	Percentage in manufacturing employment 1968	Change in manufacturing employment 1954–68	Percentage change in manufacturing 1954–68
Alsace	1·4	2·4	199,980	45·3	+ 7,580	+ 3·9
Aquitaine	2·6	2·0	208,140	31·9	+16,740	+ 8·6
Auvergne	2·0	1·6	139,900	36·5	+11,720	+ 9·2
Bourgogne	7·9	7·7	163,850	38·3	+25,450	+18·3
Bretagne	3·5	5·4	151,300	27·2	+20,730	+15·9
Centre	22·1	15·0	212,880	38·8	+54,670	+34·6
Champagne	5·2	5·6	158,960	43·9	+10,950	+ 7·3
Franche-Comté	0·7	1·3	166,920	50·4	+31,070	+22·9
Languedoc	1·1	0·8	99,900	29·9	+ 3,110	+ 3·2
Limousin	1·3	0·8	65,820	30·6	+ 6,920	+11·7
Lorraine	2·3	2·5	342,680	50·4	+ 5,280	+ 1·6
Midi-Pyrénées	1·3	1·7	173,680	32·4	+ 5,930	+ 3·5
Nord	4·0	5·5	575,480	51·0	−64,950	−10·0
Normandie (Basse)	6·2	7·0	105,960	30·0	+34,270	+47·8
Normandie (Haute)	9·2	11·2	201,180	42·6	+45,010	+28·8
Pays de la Loire	6·6	8·3	213,500	34·7	+52,320	+32·5
Picardie	13·0	10·3	245,820	45·2	+43,510	+21·5
Poitou–Charentes	2·3	2·9	110,820	31·0	+19,020	+20·7
Provence	1·6	1·7	242,980	34·4	+19,630	+ 8·8
Rhône–Alpes	5·6	5·5	668,140	47·2	+59,870	+ 9·8
Région Parisienne			1,379,980	41·0	+ 8,450	+ 0·6

well, even though development grants were not generally available (Figure 6.4). Thus the Centre, Basse-Normandie, Haute-Normandie and Picardie regions together accounted for 50·5 per cent of industrial relocation operations and 43·5 per cent of new jobs created. In total, half of the relocation schemes arranged between 1955 and 1971 involved reception areas where no government grants were provided (Bastié, 1973).

Taking France as a whole, 45 per cent of relocated jobs were installed in towns of more than 50,000 inhabitants, 35 per cent in towns of 5–50,000 population, and the remainder in smaller settlements. A great ring of towns within 200 km of the capital were the major beneficiaries, including Amiens, Beauvais, Blois, Caen, Orléans, Rheims and Tours. One-fifth of relocation operations involved the departure of firms' headquarters from Paris; but in only one-sixth of cases were all workshops and offices removed from the capital. Ten major industrial corporations provided one quarter of the jobs produced by the relocation programme, with Citroën, Renault and Saviem (cars and heavy vehicles), Thomson-CSF and Radiotéchnique (electronics), and SNIAS (aeronautics) occupying leading positions (Parry, 1963).

Changes in the location of vehicle manufacturing have been of critical importance. In 1955 the industry had been concentrated in the Paris region, Peugeot and Berliet being the only provincial producers and small branches of Parisian firms operating at Annonay and Le Mans. Over the last twenty years three dozen car-assembly and component plants have been established in eastern, northern and western parts of the country, with those employing more than 1000 workers apiece in 1972 being identified in Figure 6.5(c). The D.A.T.A.R. not only encouraged French car firms to relocate in the provinces but also attracted major international corporations to invest there. Thus the old industrial and mining region of Nord has welcomed factories operated by Chrysler and Volvo, as well as by Peugeot and Renault. These have been joined by works run by international tyre companies, including Dunlop, Firestone and Goodyear. A total of 150,000 provincial jobs has been created in the vehicle industry alone, a figure identical to the number of car workers recorded in the Paris region in 1955. Employment in the capital's vehicle industry has risen to 183,000, compared with over 230,000 in the provinces.

Other activities associated with car manufacture have also undergone important relocation. The tyre manufacturing industry had been formerly concentrated in Paris, Clermont-Ferrand and Montluçon, but is now distributed much more widely, with 20,000 new jobs being installed in western France, the outer Paris Basin and old industrial areas since 1955. Following limited pre-war dispersal the aeronautical industry has also experienced important provincial growth so that by 1970 54,000 of the workforce of 106,000 were employed outside the capital, compared with only 30,000 of the 80,000 workers in 1955. Development of the aerospace educational complex at Toulouse has had much to do with this. Finally, the electronics industry, once concentrated in Paris and Grenoble, has seen remarkable growth elsewhere in the provinces, particularly in northwestern France (Figure 6.5(d)).

It is difficult to appraise the full impact of industrial relocation in the broad framework of *l'aménagement du territoire*. At a general level, it is striking that half of the schemes were accomplished *without* State aid and that the pattern of job creation is almost the inverse of the map of financial assistance. Nevertheless, thousands of manufacturing jobs were steered away from the capital. In spite of this, the population of the Paris region continued to increase by 780,000 between 1962 and 1968, involving a greater volume of growth than any other region. However, this represented only a 9·1 per cent rise, with higher rates being recorded in Provence (17·1 per cent), Rhône–Alpes (10·4 per cent), and Languedoc–Roussillon (9·8 per cent), and in each of the *métropoles d'équilibre* except Lille (Table 6.1). Also it must be remembered that over half of recent population growth in the capital was due to natural increase. As one might expect in a major world city, the overall growth of employment registered in Paris in the last twenty years was in response to expansion in the tertiary rather than the secondary sector. This fact suggests both the relative success of industrial relocation schemes and the need for employment policies to pay much more attention to relocating tertiary jobs.

It is undeniable that relocated and expanded industries have brought substantial new employment opportunities to some parts of the provinces. Local incomes and spending power have increased and social and economic advances have been achieved in towns where 'industrial decentralization appeared to be the miracle of the second half of the twentieth century' (Rochefort and coworkers 1970, p. 88). Taking the provinces as a whole, manufacturing employment rose by 10·2 per cent from 4,041,000 in 1954 to 4,454,000 in 1968, with relocated activities making a substantial contribution in that upward trend. Growth in industrial employment in the Paris region was negligible during those years, rising by 0·6 per cent from 1,371,530 to 1,379,980. But percentage increases were most pronounced in neighbouring regions within the Paris Basin, with a record 47·8 per cent growth in industrial jobs being recorded in Basse-Normandie from 1954 to 1968 and advances in excess of 25·0 per cent occurring in the Centre, Haute-Normandie and Pays de la Loire regions (Table 6.2). It is well to remember that these figures relate to net increases and that the absolute volume of job creation was larger, since old provincial industries were shedding labour. Yet in spite of twenty years of effort, western regions are still under-industrialized (Table 6.2) and the Sixth Plan (1971–75) rightly stressed the need to accelerate factory construction in that half of France.

Relocating Employment in Tertiary Activities

Experience in the 1960s showed that relocation of employment in offices, higher education, and research activities needed to be included in schemes for regional development if these were to keep in line with major trends in French economic life. This conclusion is undoubtedly logical but it raises serious questions relating to the prestige of Paris. Paris, as Professor Pierre George (1973, p. 195) explains, is not just the capital city of France, 'it is one of the great business centres, political meeting places, and cultural capitals of the world. As national capital, Paris provides the French counterbalance to the concentrations of interests and activities which challenge France's authority in competition with Europe's major economic regions, especially the Rhinelands'. One might therefore argue that any reduction of the tertiary power of Paris would weaken the international standing of France. It would not be unjust to suggest that this dilemma is reflected in the rather ineffectual way that recent policies for tertiary relocation have been implemented.

Tertiary employment in the Paris region grew rapidly during the 1960s with 320,000 jobs being added in six years to the 2,200,000 in this sector in Paris in 1962. The western part of inner Paris and the western suburbs were particularly attractive to prestige office developments in the private sector. By contrast, manufacturing jobs fell by over 100,000 from the 1,500,000 in existence in 1962. But more recently the role of the capital in accommodating new tertiary jobs has declined, with the Parisian proportion falling from 70 per cent of those created in 1969 to 50 per cent in 1971. This reduction is in response to policies for tertiary relocation initiated in 1967. However the system appears to have been implemented too loosely to be really effective. For example, only 28 applications for relocating tertiary concerns beyond Paris were lodged from 1968 to 1971, by comparison with 2100 requests for relocating manufacturing. But at the same time the annual volume of office construction that received planning permission in the Paris region doubled from 767,000 m^2 in 1969 to 1,435,000 m^2 in 1971 and during the same three years the volume of office extension authorized in Paris was 60–70,000 m^2. Demand was growing much more rapidly than these figures suggest, since the projected tertiary floorspace that failed to obtain planning permission rose from a mere 17,000 m^2 in 1969 to 656,000 m^2 in 1971. It is clear that most large firms continue to view Paris and its new commercial quarters, such as La Défense, as a prestige focus for opening or maintaining head offices and research facilities. Official reactions to this fact are mixed, since more tertiary activity in and around Paris will enhance the international status of the city—and of France for that matter—and will be vital to the capital's new cities and suburban service centres; but it will also contribute to further congestion in the Paris region.

New measures for relocating tertiary growth have not been implemented strictly enough to have much influence on the private sector. However, the State has taken the initiative in relocating three branches of tertiary activity. First, new provincial universities have been opened to reduce pressure on the enormous and inefficient University of Paris (by constructing colleges at Amiens, Orléans, Rheims, Rouen and Tours) and to build up higher education in more distant cities, such as Brest, Limoges, Nantes and Saint-Etienne, that previously lacked universities. Second, important branches of the civil service and government research units have been transferred to provincial cities. Thus Toulouse has become a centre of international status for research and education in aerospace and important progress has been made in the field of electronics research in Rennes and other western towns. But ample scope still remains for relocating more State-directed tertiary activities. Third, a start has been made on decentralizing cultural facilities from Paris. The best-known examples are the Maisons de la Culture, of which nine are already open in the provinces, three more are planned and a further three will be established in the outer suburbs of Paris. Relocated activities of this kind combine to strengthen the functions of *métropoles d'équilibre* and other provincial towns and make them more attractive as reception centres for further manufacturing and tertiary employment.

In order to increase efficiency, policies for relocating tertiary jobs were reorganized in 1971. Controls on office development in central Paris are being enforced more stringently and emphasis is being placed on steering offices to new cities and to outer suburban service centres, such as Creteil, Nanterre and Saint-Denis. Special dues are being levied on new office developments authorized in the capital, with high rates of payment required in western inner quarters, medium rates in eastern Paris, and lower rates in suburban service centres. In addition, Parisian employers have to make regular contributions toward meeting the cost of providing public transport in the capital. Financial aid for tertiary development has been extended to all areas that qualify for industrial aid, as well as the *métropoles d'équilibre*, plus Besançon, Caen, Cannes, Clermont-Ferrand, Dijon, Montpellier and Nice (Figure 6.3). Finally, the State is entering into new agreements with banking and insurance firms for further relocation schemes and is also intensifying its own programme for transferring administrative and research activities. It remains to be seen whether these measures will prove more effective than those operating in the late 1960s.

Planning Rural Areas

So far attention has been devoted to planning urban areas and predominantly town-based activities, but important measures have also been implemented for managing sections of the rural residuum and accommodating changing pressures exerted upon them. These schemes are in addition to programmes of farm enlargement and plot consolidation that operate throughout France. Legislation in 1960 contained proposals for identifying rural problem areas and channelling aid to them to speed agricultural modernization and also to establish factory employment, improve public transport, install equipment for tourism and provide better facilities for vocational training and retraining. Not until 1967 were zones for rural renovation defined, covering northwestern France, the Massif Central and dispersed mountain areas elsewhere (Figure 6.6). Over a quarter of France and a third of all her farms were involved. Structural problems and physical conditions varied greatly within and between renovation zones, but each was characterized by low average incomes from agriculture, high average age of farmers, poor communications and shortage of off-farm employment. Special finance was made available for five years but was then extended to 1975. In the North-West funds have been devoted to streamlining facilities for processing agricultural products and improving the notorious road network, thereby reducing the isolation of Brittany. Dual carriageways are being constructed parallel to the northern and southern coasts, with a new highway through the centre of the peninsula. Schemes have been started in other renovation zones for improving roads, rationalizing stock rearing, providing piped water supplies and installing facilities for tourism. The Sixth Plan stressed that such action was merely a beginning and argued that stronger measures should be implemented for tackling rural problems on an integrated and comprehensive basis.

This challenge has also been faced by seven mixed-economy planning corporations, which combine public and private finance (Figure 6.6). The corporations vary in their objectives but each is attempting to modernize and diversify the local economy. In northeastern France, clearance, deep ploughing and fertilization of parts of the 400,000 ha of heathland, has been started to produce good grazing land. New farms have been laid out and the corporation's work has encouraged local farmers and landowners to start their own improvements. Land clearance has also been undertaken in southwestern France to tackle broader problems in the Landes forest, where fires damaged or destroyed almost 500,000 ha of timber during the 1940s and 1950s. Plans for protecting the forest and diversifying the local economy were produced involving cutting firebreaks and establishing 200 new farms, averaging 40–50 ha apiece. Reclamation and fertilizing techniques were criticized by the settlers who sued the corporation on the grounds that low fertility and small farm size rendered viable agriculture impossible. In 1968 a replacement corporation was set up to cover the whole of rural Aquitaine and, as far as the Landes are concerned, has concentrated on improving stock farming and developing coastal tourism.

Irrigation has been the chief means of improving the hillslopes of Gascony using

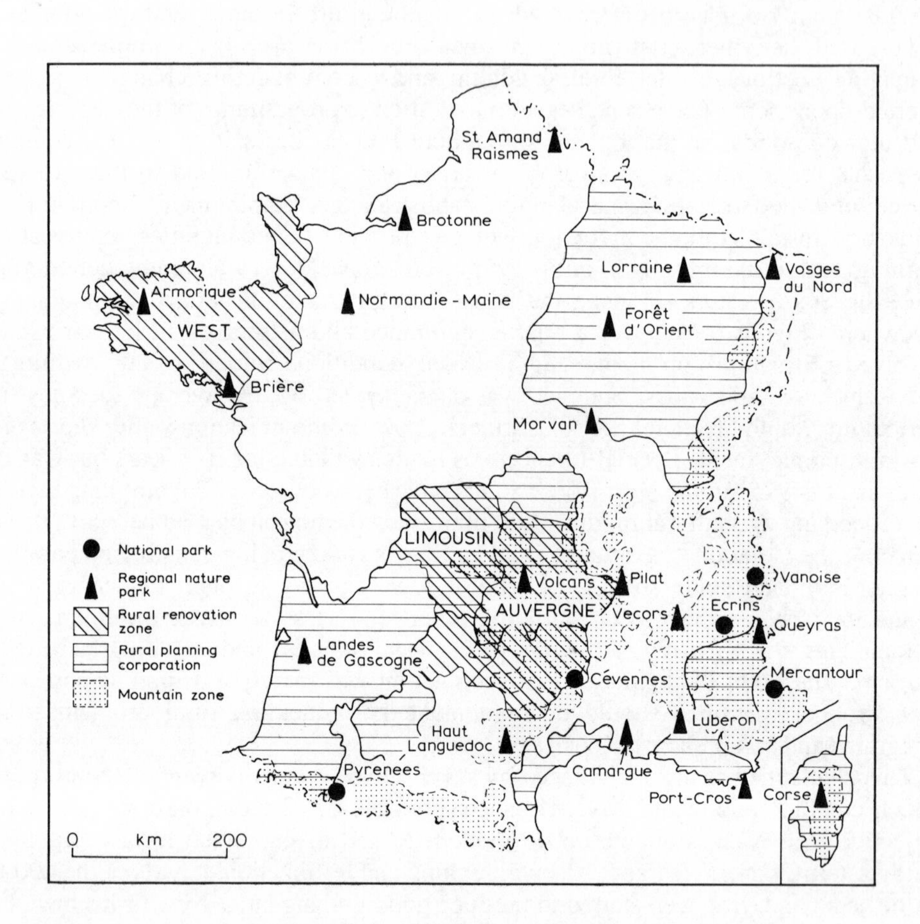

Figure 6.6. France: rural management.

water from the river Garonne and its headstreams and from a series of small reservoirs. Crop production is being improved and more attention is being paid to modernizing marketing. Water supply is of prime importance in other rural management schemes in the South. For example, the Bas-Rhône–Languedoc corporation operates in a 250,000 ha concession area and has five main objectives: to improve irrigation; provide facilities for storing and processing farm produce; to offer technical guidance to farmers; to encourage development of local industry and tourism; and to participate in any other way in schemes for regional development. The first objective is the key to other aspects of rural change and irrigation is being tackled by means of a canal running westwards from the Rhône to Agde and by a series of channels taking water from local headstreams to the lowlands of western Languedoc. Important, but patchy, progress has been achieved to encourage farmers to change from vine mono-culture to producing fruit and vegetables and raising livestock. East of the Rhône delta the Provence corporation was set up to make use of the water resources of the Durance valley, improve irrigation, and guarantee water supplies to Marseilles, other Provençal cities and the area of industrial development on the Gulf of Fos. In Corsica an intensive effort is being made to reclaim land on the eastern coastal plain, free it from malaria and install new farms. Finally a broad-based programme of rural management has been introduced in the Massif Central. Agricultural improvements include schemes for making rational use of high summer pastures, improving livestock breeds and installing irrigation in the Allier valley. Conversion of unused moorland to timber production has also made progress and numbers of new camp sites and holiday villages are functioning successfully.

Growing demands on rural and coastal resources for purposes of recreation are being resolved in two main ways by government planners. Following legislation in 1960 national parks have been designated with the seemingly contradictory objectives of conserving the environment and meeting recreational demands (Clout, 1975). These aims are reconciled by defining zones with differing degrees of access. Nature reserves occupy innermost areas of national parks where access is restricted to researchers, partly by virtue of isolation. Surrounding zones are subject to conservation measures which defend flora, fauna and traditional ways of life, and prevent introduction of activities that would degrade the environment. Facilities for tourists and for providing local employment may be installed in outer zones. Six national parks have been designated, of which five cover extensive areas of mountain land (Cévennes, Ecrins, Pyrénées-Occidentales, Vanoise, Mercantour); the sixth involves Mediterranean islands plus surrounding water areas at Port-Cros. More national parks are planned in Corsica and in the Alps, but at the end of the Sixth Plan only 2 per cent of France will have national park status.

On a more restricted scale, regional nature parks have been established in areas with distinctive landscapes, such as marshland (Brière, Camargue), woodland (Landes, Fôret d'Orient) and upland territory (Volcans, Vosges du Nord). The nature park formula is fluid, varying according to local conditions and likely intensity of visitor pressure. Thus Saint-Amand, close to industrial cities in the Nord, contains sports equipment as well as wildlife reserves. In more remote parks, such as the Volcans, the emphasis is more on conserving landscapes than providing recreational facilities. The nine parks in existence vary in size from 10,300 ha at Saint-Amand to 206,000 ha in the Landes. Ten additional park sites are being studied and suggestions have been made for three more.

Unlike these relatively modest parks, the State has participated in major schemes for developing tourism that have provoked striking environmental changes. The most advanced of these is in Languedoc, involving reclamation of marshy coastland and installing accommodation. In fact, the project has experienced problems linked to rising costs and a lack of moderately-priced accommodation. The scale of the operation has now been reduced from six new resorts to three development areas at La Grande Motte, Cap d'Agde, and Leucate-Barcares. Future emphasis will be given to medium-priced accommodation rather than expensive second homes. An equally ambitious scheme received government approval in 1971. This involves expanding tourism in Corsica and will complement agricultural revival on the eastern plain. Main objectives involve improving air and sea communications to the mainland, providing better road links between the coast and interior, starting a fifteen-year programme to build accommodation for 300,000 and attempting to conserve the island's fragile ecological environment. The 240 km coastline of Aquitaine is also being planned to receive nine new foci for tourist accommodation, which will be doubled by 1985 to take 500,000 visitors at any one time. Finally, the 'Snow Plan' was announced as part of the Sixth National Plan to provide additional winter-sports accommodation for 150,000 tourists, mainly in new mountain resorts. Management of rural space for alternative or complementary functions to farming will continue to exercise the ingenuity of countryside planners in the remainder of the twentieth century.

Inventory and Prospect

From the preceding discussion it is clear that *l'aménagement du territoire* embraces a complex array of operations, involving not only programmes for the 22 regions but also spatial policies for managing many sectors of French life. It is undeniable that regional planning has made an important impression in modifying the patterns of economic and social activities in contemporary France; nevertheless, fundamental geographical differences remain which can be affected by planning action but cannot be erased. As Parodi (1971) notes, the idea of balanced regional development in France is simply a 'mirage' since resources are distributed unevenly and will continue to be so. In a complementary fashion, recent planning decisions to invest in particular areas rather than others have endowed certain sites with new advantages over their neighbours and thus may have even increased inter-and intra-regional inequalities rather than reducing them. One of the hard lessons that has been learned over the past twenty years is that every region and every settlement in France cannot have everything (Rochefort and coworkers, 1970). Some towns, ports, routeways and regions have undoubtedly been favoured at the expense of others. For example, construction of radial motorways that link Paris with regions generating large volumes of traffic (such as the Rhône valley and the Mediterranean South, the industrial North, and the Lower Seine) has been given priority over other routes that need improvement but carry less traffic. The overwhelming importance of Paris as the hub of national transport networks simply cannot be ignored, no matter how much lip service is paid to developing the provinces at the expense of the capital.

As well as tackling issues of vital concern to domestic interests, *l'aménagement du territoire* has fostered massive projects that reinforce the grandeur of France. By way of illustration, the Gulf of Fos on the Mediterranean and Dunkirk and Le Havre on the Channel coast are being developed as French versions of the Dutch Europoort. Each port installation is being complemented by an ocean-linked heavy industrial focus of truly international scale with good access to densely-developed parts of France and to economic core areas in neighbouring countries. Similarly, the large new cities and suburban nodes on the margins of Paris have important functional roles to play in decongesting if not decentralizing the capital, but they are also prestige developments that will undoubtedly enhance the standing of Paris among world cities and of France among nations.

French experience since the mid-1950s has demonstrated the necessity of modifying ways and means of regional planning as background conditions change. Thus, the objectives of urban management have been extended from very large cities to include regional capitals and medium-sized towns; and in the field of employment, obsession with manufacturing jobs is giving way to concern for tertiary activities. This particular trend poses new spatial problems since Parisian new cities and suburban nodes, *métropoles,* regional capitals the new towns in the provinces and even medium-sized old towns are attempting to capture a share of tertiary growth and decision-making power. The inequality of the competitors is striking and one wonders how towns below *métropole* level and beyond prestige developments in the Paris region will fare in the scramble to become successful office centres. Questions relating to the appropriate size, number and function of French regions have also attracted interest as political conditions have changed and improvements in communications have interacted to 'shrink' geographical space. Much more attention must be paid to these issues in the future. The same is true for the ecological implications of urban and industrial development which at long last have started to arouse concern.

In the past quarter century policies for regional management have dealt with France's transition from an agricultural to an urban-industrial nation. In the next quarter century problems related to urban renewal, industrial renovation and the implications of national transition to post-industrial conditions will come to the forefront. At the European scale France's regional problems are less intense than those of Ireland or the Mezzogiorno. Converting the French from their own relatively successful array of policies for regional development to European Community alternatives from which they stand to gain little will be far from easy. Already there is ample scope for cooperation with Belgian, German, Italian and Swiss planners for managing frontier regions. Whether the fierce independence of France will be subdued to European objectives in this and other realms will remain a matter of speculation in the immediate future at least.

References

Ardagh, J., 1973, *The New France: a society in transition 1945–73,* Penguin, Harmondsworth.

Bastié, J., 1973, La décentralisation industrielle en France de 1954 à 1971, *Bulletin de l'Association de Géographes Français,* **408–9**, 193–209.

Beaujeu-Garnier, J., 1974, Toward a new equilibrium in France, *Annals of the Association of American Geographers,* **64**, 113–125.

Boudeville, J. R., 1966, *Some Problems of Regional Economic Planning,* Edinburgh University Press.

Brayne, M. L., 1972, Rungis: the new Paris market, *Geography,* **57**, 47–50.

Brunet, R., 1973, Structure et dynamisme de l'espace français: schéme d'un système, *L'Espace Géographique,* **2**, 249–254.

Brunet, R. and coworkers, 1974, *La France Maintenant,* Larousse, Paris.

Cazès, G. and A. Reynaud, 1973, *Les Mutations Récentes de l'Economie Française,* Doin, Paris.

Clarke, J. I., 1963, Demographic revival in France, *Geography,* **48**, 309–311.
Clout, H. D., 1970, Industrial relocation in France, *Geography,* **60**, 48–63.
Clout, H. D., 1972, *The Geography of Post-War France: a social and economic approach.* Pergamon, Oxford.
Clout, H. D., 1973, *The Massif Central,* Oxford University Press.
Clout, H. D., 1975, La belle France, *Geographical Magazine,* **47**, 302–309.
Coffey, P., 1973, *The Social Economy of France,* Macmillan, London.
Durand, P., 1972, *Industrie et Régions,* La Documentation Française, Paris.
Elkins, T. H., 1975, Paris transformed, *Geographical Magazine,* **47**, 296–301.
Fielding, A. J., 1966, Internal migration and regional economic growth: a case study from France, *Urban Studies,* **3,** 200–214.
George, P., 1973, *France: a geographical study,* Martin Robertson, London.
Gravier, J. F., 1947, *Paris et le Désert Français,* Flammarion, Paris.
Gravier, J. F., 1964, *L'Aménagement du Territoire et l'Avénir des Régions Françaises,* Flammarion, Paris
Gravier, J. F., 1970, *La Quéstion Régionale,* Flammarion, Paris.
Gravier, J. F., 1972, *Paris et le Désert Français en 1972,* Flammarion, Paris.
Hall, P. G., 1966, *World Cities,* Weidenfeld and Nicholson, London.
Hall, P. G., 1967, Planning for urban growth, *Regional Studies,* **1**, 101–134.
Hansen, N. M., 1968, *French Regional Planning,* Edinburgh University Press.
Lebesque, M., 1970, *Comment peut-on être breton? essai sur la démocratie française,* Editions du Seuil, Paris.
McDonald, J. R., 1965, The repatriation of French Algerians, *International Migration,* **3**, 146–156.
McDonald, J. R., 1969, Labor immigration in France, 1946–65, *Annals of the Association of American Geographers,* **59**, 116–134.
Merlin, P., 1971, *New Towns,* Methuen, London.
Parodi, M., 1971, *L'Economie et la Société Française de 1945 à 1970,* Armand Colin, Paris.
Parry, C., 1963, Une exemple de la décentralisation industrielle: la dispersion des usines de la Radiotechnique à l'Ouest de Paris, *Annales de Géographie,* **72**, 148–161.
Pinchemel, P., 1969, *France: a geographical survey,* Bell, London.
Racine, E. and coworkers, 1975, Planning and housing: France, *The Planner,* **61,** 83–92.
Rochefort, M. and coworkers, 1970, *Aménager le Territoire,* Editions du Seuil, Paris.
Scargill, D. I., 1968, *Economic Geography of France,* Macmillan, London.
Sérant, P., 1965, *La France des Minorités,* Laffont, Paris.
Thompson, I. B., 1970, *Modern France: a social and economic geography,* Butterworth, London.
Thompson, I. B., 1973, *The Paris Basin,* Oxford University Press.
Thompson, I. B., 1975, *The Lower Rhône and Marseille,* Oxford University Press.
Tuppen, J., 1972, Le Vaudreuil: an advance in the French new town concept, *Town and Country Planning,* **40**, 308–311.
Tuppen, J., 1975, Fos: Europort of the South? *Geography,* **60,** 213–217.
Wright, V. and Machin, H., 1975, The French regional reform of July 1972, *Policy and Politics,* **3,** 3–28.
Zetter, R., 1975, Les Halles, *Town Planning Review,* **46,** 267–294.

7

Benelux

François J. Gay

Introduction

The Benelux countries form an excellent laboratory for studying regional problems and schemes for regional development. Admittedly the two principal states display important similarities but the contrasts between them are equally striking. Regional development and planning have a long history in the Netherlands but the inadequacies of *laissez-faire* policies have only recently been recognized in Belgium. Right from the beginning linguistic and political issues have distorted schemes for regional development in that country. One must be careful to avoid falling into the trap of deterministic explanations for differences in emphasis on regional planning in the two main Benelux countries. It would be wrong to use high population density (394/km^2) to explain Dutch concern with regional development, since the Belgian average (320/km^2) is almost as high; and, with the exception of the Ardennes, empty space is just as scarce in Belgium as in the Netherlands. Competition for land is equally strong and environmental problems are equally grave to either side of the frontier. Admittedly the Netherlands took the lead in defending its territory against flooding by seawater and by major rivers (and this may be seen as a starting point for regional management) but Belgium has had to tackle similar problems in maritime Flanders and to the south of the Scheldt estuary.

Even so, there are very clear contrasts between the two countries. A traveller from Lille to Rotterdam, driving along the Flanders motorway, cannot fail to be impressed by the almost continuous sprawl of urban and suburban development from the French frontier northwards to Ghent and even to Antwerp. There are few breaks in the townscape and when they do occur they are rarely more than a few hundred metres in breadth. But once he crosses the Dutch frontier the traveller is surrounded by very different landscapes. Single-family houses predominate in Belgium but give way to multi-family flats and apartment houses in the Netherlands, although single-family houses have become more numerous north of the frontier since 1960. In Belgium the distinction between 'urban' and 'rural' areas is nebulous in the extreme, but in the Netherlands the contrast is more clear-cut and reflects a different background of local politics and planning.

In spite of these differences, common features in spatial organization in the main Benelux countries must not be underestimated. Both nations are densely populated, highly urbanized and criss-crossed by important transport systems. Both have operated cheap workers' tariffs (especially on the railways) and these have encouraged long-distance daily commuting. Suburban sprawl is more noticeable in Belgium but long journeys to work were made possible by railway policies in both countries before the era of widespread car ownership. Finally, both countries face rather similar regional disparities, with 'growth regions' and 'problem regions' being very close to each other in such small states. Hazards of congestion are acute in core areas, while outer regions function as reservoirs of manpower or at best as sites for installing industries that do not require skilled workers. The following paragraphs will examine broad inter-regional differences, will then focus on particular regional problems and finally will outline the distinctive approaches to regional development in the Benelux countries and the varying degrees of success that have been achieved.

Regional Differences in the Netherlands and Belgium

The Netherlands

Dutch regional differences are less important than Belgium's and do not represent a threat to the survival of national unity. Recent policies for regional development have altered details of the traditional east–west contrast but the dichotomy remains, nonetheless. The provinces of Noord-and Zuid-Holland plus Utrecht make up the 'West', which contains 45 per cent of the Dutch population (Table 7.1). The 'East' in

Table 7.1. Total Population and Average Density in the Dutch Provinces, 1972

The West			The East		
	Population	Density (per km²)		Population	Density (per km²)
Utrecht	827,300	623	Groningen[a]	526,600	230
Noord-Holland	2,273,300	856	Friesland[a]	532,500	161
Zuid-Holland	3,013,400	1057	Drenthe[a]	379,600	143
			Overijssel[b]	945,900	249
NETHERLANDS	13,269,600	394	Gelderland[b]	1,558,300	311
			Zeeland[c]	316,100	178
			Noord-Brabant[c]	1,850,500	377
			Limburg[c]	1,022,400	471
			IJsselmeerpolders[c]	20,000	21

[a]Noord region; [b]Oost region; [c]Zuid region.

the widest sense covers the rest of the Netherlands, stretching from Groningen province, with its natural gas fields, to the former coalmining area of Zuid-Limburg. It also includes the southwestern province of Zeeland which is still mainly rural, has low population densities and is less economically developed than the three western provinces.

The contrast between 'East' and 'West' may be shown in many ways, but employment data are particularly useful. Table 7.2 emphasizes the importance of manufacturing industry and especially service activities in the West and of agriculture in the eastern provinces in 1970. The latest estimates show that farming employs less than 4 per cent of the workforce in the West (under half the Dutch average) but still over 15 per cent in the northern provinces and 9–10 per cent in those in the south and east. Even more significant is the growth of service jobs in the West. These account for over 60 per cent of employment in that region, and represent over half (1,300,000) of

Table 7.2. Employment in the Netherlands by the Four Main Regions, 1970, (%)

	Agriculture	Industry	Services
Noord[a]	17·0	38·0	45·0
Oost[a]	11·0	43·4	44·6
Zuid[a]	9·0	47·2	43·7
West[b]	4·6	36·7	58·6
NETHERLANDS	8·0	40·7	51·2

[a]Together comprising 'the East'; [b]'the West'.

all service workers in the Netherlands (2,250,000). Activity rates are higher in the West, which contains 51 per cent of Dutch workers but only 45 per cent of the total population. Income levels in the West are well above the national average and fall below average in the East. Predominance of unskilled or semi-skilled jobs in the latter part of the Netherlands accounts for the difference. Only three provinces have above average income levels and these are in the West. But even so, inter-regional income differences are less pronounced in the Netherlands than in any other member state of the Nine.

The West is a highly urbanized region with four of the most important cities out of the 21 that contain more than 100,000 inhabitants. These four agglomerations housed no fewer than 3,300,000 people in 1972. Cities of more than 200,000 inhabitants are shown in Table 7.3 which confirms the urban dominance of the West. Until recently the most striking index of inter-regional differences was the net flow of migrants to the western provinces. But policies for industrial development in eastern and southern areas have been partially successful and have provided local jobs for large numbers of young people coming on to the labour market and have even attracted some workers from the West. Net outflows of workers from the northern, eastern and southern provinces have slowed down and in the early 1970s Gelderland and Noord-Brabant registered net in-migration. But whether this will continue or not is a matter of debate. In any case, workers from Surinam and the Dutch West Indies have moved to the Randstad to compensate numerically for westerners who moved to new job opportunities in the East.

Table 7.3. Urban Areas with more than 200,000 Inhabitants, 1972

Rotterdam	1,064,000	Eindhoven	345,000	Enschede-Hengelo	237,000
Amsterdam	1,029,000	Arnhem	276,000	Tilburg	208,000
The Hague	702,000	Heerlen-Kerkrade	266,000	Nijmegen	208,000
Utrecht	463,000	Haarlem	239,000	Groningen	206,000

The East–West contrast that one now finds in the Netherlands is the modern version of an important historical distinction. 'Holland' together with Utrecht province remains an important focus of maritime activity and was the heart of an outward-looking, tolerant, trading republic in the past. The rest of the country was ruled by the House of Orange and supported a land-orientated society, tinged with puritan austerity. Money was short in the East, industry was relatively rare, agriculture was less rewarding since soils were poorer than in the West, and the eastern land frontiers were always open to attack. Towns such as Groningen had a glorious past but experienced nothing like the rich and cosmopolitan cultures of the western cities, with their universities, and flourishing intellectual and artistic life. Will it be possible to erase these fundamental contrasts that have been shaped not only by the forces of history and geography but also by external economies and cumulative advantages that have fostered the growth of a post-industrial society in the Randstad?

Belgium

Regional disparities in Belgium do not coincide exactly with the limits of the two main socio-cultural regions. It would be wrong to contrast a uniformly dynamic Flemish North with a completely stagnant French-speaking South (Wallonia). Each of these major regions contains points of strength and points of weakness. The Flemish stronghold of Antwerp has its equal in Liège, the capital of Wallonia; and unemployment affects western parts of Flanders as well as the coal basins of the South. Admittedly there are fundamental differences between the two socio-cultural regions but, with the exception of population, they do not always emerge very clearly from discreetly-presented official statistics. Even so, the relative well-being of the two parts of the country is a matter of great public concern. This contrast between Flemings and Walloons should not detract from what is happening in the Brussels region which exercises a strong grip on the whole country. Only three provinces record above-average income levels, with Brussels being clearly at the head of the league (Table 7.4). (One must remember that regional income differences are much greater in Belgium than in the Netherlands.) These financial disparities are important in themselves but, in addition, they are grafted on to linguistic and social rivalry (if not blatant opposition) and are also reflected by fundamental demographic contrasts.

Table 7.4. *Per Capita* Gross Domestic Product by Belgian Province, 1969, (F.B.)

Oost-Vlaanderen	80,145	Hainaut	77,114
West-Vlaanderen	90,547	Namur	78,709
Antwerpen	106,883	Liège	94,745
Limburg	70,744	Luxembourg	69,155
Flanders	88,827	Wallonia	82,242
Brabant	117,309	BELGIUM	94,099
Brussels Region	135,309		

The French-speaking Walloons are outnumbered by the 5,600,000 Flemings who speak the same language as their Dutch neighbours. The interests and characteristics of Flemings and Dutch people are far from identical, as the rivalry between Antwerp and

Rotterdam shows very well, but the 4,000,000 Walloons are afraid of being submerged by 20,000,000 Dutch speakers in the Benelux union. The Walloons fear, or say they fear, the rebirth of a 'greater Netherlands' which was in existence before 1830—not very long ago. What is more, the elitist groups in Flemish society also speak French and this tends to strengthen their position in the political, administrative and trade-union life of the whole country. Language contrasts are also reflected in political behaviour and social outlook. Roman Catholic interests are stronger in Flanders and the 'Christian Socialist' party is well established there, whilst the socialist party and trade unions are more powerfully developed in Wallonia.

Linguistic contrasts have produced some curious results in Belgian life. One example of this is the creation of a new French-speaking university at Louvain-la-Neuve in addition to the old-established Flemish university of Leuven (Louvain), even at the expense of breaking up an internationally famous intellectual community. After a long history of being the dominated group, the Flemings are now very much aware of their numerical dominance over French speakers. For every ten Belgians, there are now six Flemings, three Walloons, and one resident of Brussels, who in theory is bilingual. In spite of its glorious past, symbolized by Van Eyck and Rubens, Flanders has developed an 'anti-colonialist' reflex action, characterized by '*Wallen Buiten!*' ('Walloons out!'). After many difficulties the language issue has been resolved, in principle at least, by a constitutional compromise which established four linguistic regions whose limits might only be changed by legislation adopted by a 'special majority'. This guarantees the language rights of each community (Table 7.5) and, as we shall see, has given rise to particular problems in the suburban zone around Brussels.

Table 7.5. Distribution of Population in Belgium by Linguistic Region, 1947–70 (%)

	1947	1961	1970
Dutch-speaking region	53·4	55·1	56·1
French-speaking region	34·6	31·1	32·1
German-speaking region	0·6	0·6	0·6
'Brussels-Capital'	11·2	11·1	11·2

The key to the problems of Wallonia is found in the word 'ageing'. The demographic structure and the range of economic activities have become increasingly 'elderly' as they have aged together since their heyday in the nineteenth century. Problems stemming from the decline of the Walloon coal industry, the reorganization of iron and steel production, the contraction of textile manufacturing, and the depopulation in the Ardennes will be examined later. Demographic matters are at the heart of many of the problems of Wallonia. It is true that Flemish rates of natural increase have now slowed down (as they have done in the Netherlands) but the Belgian North benefited from high birth rates until recently. These supplied an abundant labour force that is less militant than its southern counterpart and is eagerly sought by industrialists. By contrast, Wallonia underwent demographic decline similar to that experienced in France until 1945 but since then the Belgian South has not enjoyed the demographic revival that occurred in France up to the late 1960s. Wallonia now contains roughly the same number of inhabitants as it did in 1930, while the population of Flanders has risen by more than 1,000,000. Current trends are simply accentuating the demographic decline of the South, which now contains only 27 per cent of the workforce (compared with 55 per cent in Flanders), offers few attractions for new industrial investment and lacks an adequate range of job opportunities. As early as 1952 the French demographer A. Sauvy had suggested ways of restricting the numerical decline of Wallonia, but to no effect. Belgium now has the second lowest birth rate in the Nine (after Luxembourg) and three-quarters of the total natural increase derives from Flanders. Slight growth of the Walloon population during the 1960s was predominantly due to in-migration (Table 7.6). The national population rose by 5 per cent (+ 460,000) in that decade and it was the Flemish provinces together with Brabant (which contains the national capital) that displayed above-average rates. In fact Brussels received 75 per cent of all inter-provincial migrants during that period.

Table 7.6 Population Change in Belgium, 1961–70, by Province and Linguistic Region (%)

	Natural change	Migratory change	Total change
Limburg	+12·7	+1·3	+14·0
Brabant	+ 1·7	+6·6	+ 8·3
Antwerpen	+ 5·3	+0·9	+ 6·2
West-Vlaanderen	+ 5·9	−0·9	+ 6·2
Oost-Vlaanderen	+ 3·5	−0·5	+ 3·0
Namur	+ 1·9	+1·1	+ 3·0
Liège	+ 0·04	+1·7	+ 1·7
Luxembourg	+ 2·9	−2·7	+ 0·2
Hainaut	+ 0·5	−0·5	0·0
French-speaking	+ 5·7	+1·3	+ 7·0
Flemish-speaking	+ 0·6	+1·2	+ 1·8
'Brussels-Capital'	+ 0·05	+5·05	+ 5·1
BELGIUM	+ 3·3	+1·7	+ 5·0

A more detailed analysis of population change in the 1960s emphasizes the significance of the linguistic frontier (Figure 7.1). All *arrondissements* except one with above-average increases were in the Flemish North. The only French-speaking area to be included was Nivelles, which experienced growth because of the expansion of Brussels. However, rates of increase in Flanders were not uniformly high. The population of western areas is relatively elderly (in response to earlier rural out-migration) and rates of growth are moderate, standing in contrast with very important increases in eastern areas (Turnhout, Hasselt, Maaslik). An 'elderly' population structure is virtually universal in the French-speaking South (Flament, 1974). In short, when natural increase is analysed the contrast between Flanders and Wallonia emerges, with 90 per cent of all natural increase being produced in the Flemish North. But a different picture is produced when migration flows are examined,

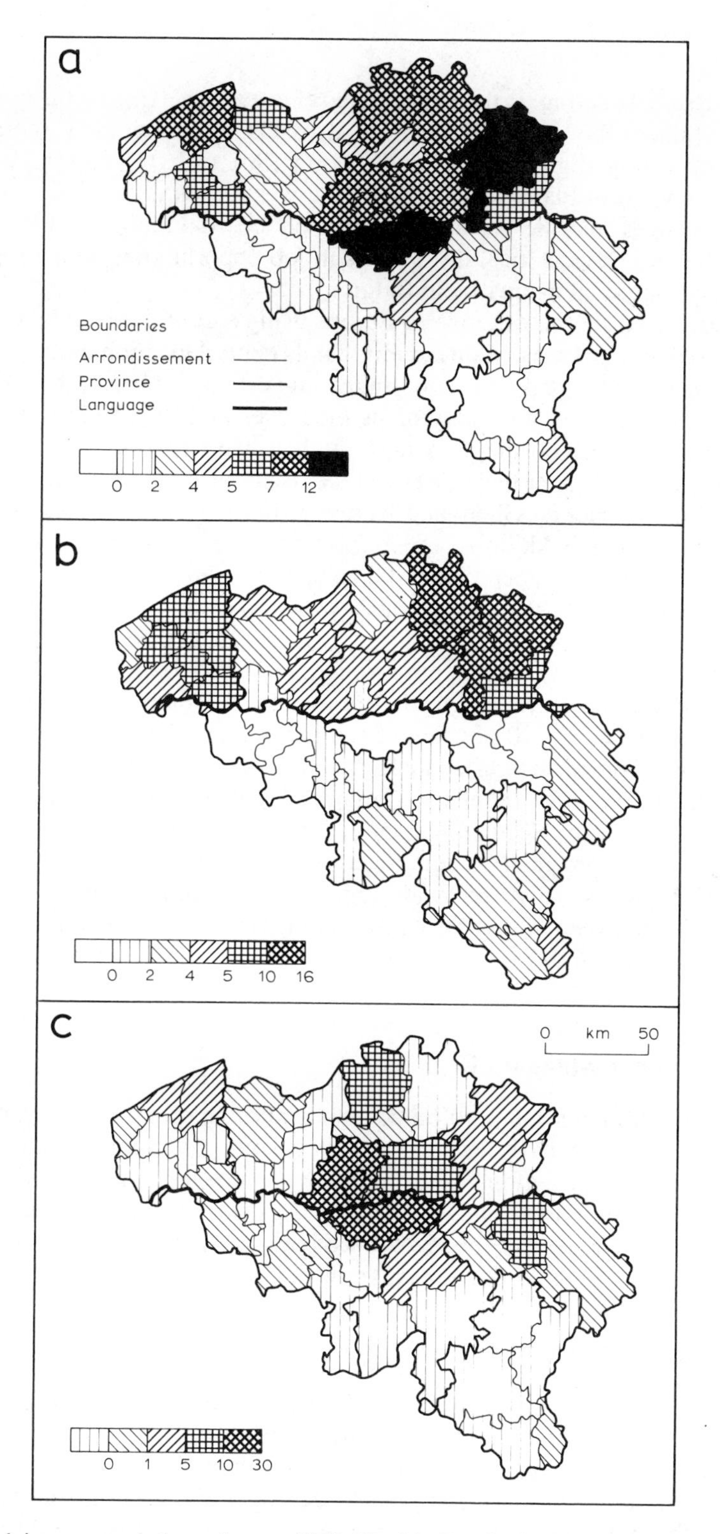

Figure 7.1. Belgium: population change 1960–70. (a) Total change. (b) Natural change. (c) Migratory change. (Reproduced by permission of J. Beaujeu-Garnier and *L'Information Géographique*.)

with greater Brussels acting as the major focus of attraction. This fact is not apparent at first glance, since the officially-defined Brussels agglomeration (comprising 19 *communes*) grew only slightly from 1,023,000 in 1961 to 1,075,000 in 1970. Leading departure areas were in Flanders (especially the western parts) as well as in Wallonia, where the Ardennes and the western sections of the coalmining district were heavy losers. This pattern reflects local differences in job opportunities which form the most serious aspect of regional inequality in Belgium.

Wallonia is the northernmost zone of Europe using a Latin-derived language and has been unable to derive the maximum benefit that its central location in the 'contact zone' midway between France and Germany might suggest. By contrast, the Flemish North has succeeded in modernizing many of its industries and broadening its employment opportunities in a most efficient manner. It has been favoured by two important economic trends. The first of these has been the rise of port-based manufacturing which has had a striking impact on Ghent and Antwerp, in the form of the industrial zones at Zelzate and along the Scheldt downstream from Antwerp to the Dutch border. Space on the right bank is now almost completely taken up and industrial development is pushing ahead on the left bank. Creation of heavy industry in this area has produced many planning problems and the multiplier effect has completely changed the employment prospects not only of the Waasland but throughout Oost-Vlaanderen. The second trend to bring immediate benefit to the North (not to mention future problems) has been the fact that in recent years 75–90 per cent of foreign industrial investment in Belgium has been devoted to Flanders. High-technology industries displaying fast rates of growth have been attracted in large numbers. Certainly Flanders was less industrialized than Wallonia at the end of the Second World War, but she has benefited greatly from State aid for factory development. Flanders received half as much again as Wallonia of the 267 billion F.B. that were voted to aid regional development during the 1960s. The 'golden triangle' between Antwerp, Brussels and Ghent has fared particularly well in the boom years of 1960–74. The fortunes of these and other areas will now be examined.

Growth Regions and Stagnant Regions

The preceding comments relate to broad trends and therefore oversimplify important spatial variations. Now it is necessary to pay some attention to the more or less congested 'metropolitan regions' and the various 'development regions' that include not only backward rural areas but also urban and industrial zones that are stagnating. A range of intermediate areas may be recognized between these extremes but they will not be examined in detail in the following paragraphs.

The Belgo-Dutch Megalopolis

It might appear rather bold to talk of a single megalopolis from Brussels northwards to Zaanstrek beyond Amsterdam. However this zone does display a range of common characteristics even if man-made frontiers and planning regions have worked to subdivide it. The idea of a 'Middengebiet' area of 5,000,000 inhabitants, located to either side of the Belgium/Netherlands frontier, has now become a reality, comprising the two Limburg provinces, plus Antwerpen and Noord-Brabant. In addition, recent industrial growth along the Terneuzen canal has stimulated further urban expansion in this middle area between Brussels and the Randstad. Daily commuting across the border has become important between Noord-Brabant and the Belgian districts of Hasselt and Turnhout. The Delta Plan and development schemes around Bergen-op-Zoom will contribute to an even denser pattern of occupation in this area which even now supports rural population densities of more than 150/km^2. Construction of the Moerdijk canal between Antwerp and Rotterdam will reinforce the complementarity of these two great ports. Finally, the existence of the Benelux union has favoured labour mobility, development of tourism and the recognition of a Dutch-speaking community comprising Flanders as well as the Netherlands. All these trends have contributed toward greater spatial integration in the deltalands of the Maas, Rhine and Scheldt.

The urban region just described stretches in a north–south direction for 200 km, extends for at least 80 km east–west, contains 9,000,000 inhabitants, and includes Amsterdam, Rotterdam, The Hague, Antwerp, Ghent and Brussels. The international trading role of these cities has been strengthened by the world trade centres that have been constructed at Antwerp, Amsterdam and Rotterdam. Belgium and the Netherlands, which occupy vital entry points to North-West Europe, have long had to look beyond their own boundaries for trade. Now they have attracted many European headquarters of international companies and house the 'capital' of the E.E.C. at Brussels. The Benelux countries have strengthened their links with the English-speaking world and were particularly glad to see the U.K. and the Republic of Ireland enter the E.E.C. Massive increases in employment in the tertiary and quaternary sectors, the existence of the Benelux union and the broad use of the Dutch language have all contributed toward further integration between these countries. However the Dutch Randstad and the Belgian 'golden triangle' are quite distinctive parts in the Belgo-Dutch megalopolis and will be examined in turn.

The Randstad

The Randstad, or 'ring city', is not of course a single conurbation, since the towns and cities that compose it have not joined together (Hall, 1966; Lawrence, 1973). However it is difficult to define the Randstad with any precision since its limits are constantly changing. In recent years it has expanded in an easterly direction, with one branch of urban growth extending towards the Gooi area and another towards and beyond Dordrecht. This great horseshoe-shaped region will undoubtedly take on new dimensions in the future, in response to improved forms of communication and new functions that may come to the less-developed parts of the Randstad. The future of the Utrecht–Arnhem–Nijmegen axis and the urban corridor that is developing between Breda and Eindhoven will be particularly crucial. Three other lines of expansion are taking place: northeastwards on to the Zuid-Flevoland polder; southwards into the deltalands; and to the north of Amsterdam across polders drained in the seventeenth century and where the new town of Heerhugowaard is being established (Malézieux, 1973).

The core of this great region remains less developed than the surrounding urban centres and for this reason the Randstad has been called the 'greenheart metropolis'. Urban expansion has been controlled, with varying degrees of success, in the greenheart. Administrative, commercial, educational and other functions are distributed among the towns of the Randstad, even the smaller ones, rather than being concentrated in the largest cities. High-level financial activities are found in

Amsterdam, Rotterdam and Utrecht so no city has overwhelming control of this sector. Transportation links between the Randstad cities are both fast and frequent, with Amsterdam being less than an hour away from Rotterdam by train (Lawrence, 1973). There are few intermediate stopping points on inter-city lines, but local connections are provided by efficient bus and tramway lines. In addition, the location of Schipol international airport 13 km southwest of Amsterdam has further strengthened transport links between that city and other parts of the Randstad. Agriculture on an industrial and highly productive scale also characterizes this region and has developed from ancient Dutch traditions which have been reinforced in recent decades by the need to make farming just as profitable as selling land for urban development. Fruit, vegetables, garden plants, bulbs and flowers are grown in distinctive farming districts, such as the Westland (with its 6000 ha of glasshouses) and the Alsmeer area which produces flowers, bulbs and vegetables.

Planning problems in the Randstad are very much associated with environmental issues. Transport systems need continuous improvement and supplies of drinking water are only just adequate during parts of the summer months in Rotterdam and other major cities. Threats of further atmospheric pollution moved the authorities to respect public opinion and prevent the installation of a steelworks at Europoort (Malézieux, 1971). Increasing separation between place of work and place of residence is a response to strict regulations for zoning land use in and around port-based industrial areas. Transport links beneath major waterways have been installed at great cost, for example, under the North Sea Canal between Amsterdam and the IJmond area (Coentunnel), and under the Nieuwe Maas between western Rotterdam and the Hook of Holland (Benelux tunnel). East–west routes between Rotterdam and its new suburbs and outports have been both costly and difficult to build.

The southern part of the Belgo-Dutch megalopolis is really one great conurbation and therefore has quite a different structure from the greenheart metropolis. Nevertheless, many of the problems experienced in this 'golden triangle' of large cities and ribbon development are very similar to those encountered in the Randstad. Urban development is rather diffuse to the north of Antwerp but becomes more dense as one moves toward Brussels. Both cities have flourished greatly during the Common Market era, with Antwerp acting as a major entry port for iron and steel to North-West Europe and Brussels being the 'capital' of the E.E.C. with a major international airport on its northern margin (Riley and Ashworth, 1975).

The striking growth of Brussels derives not only from its economic and political supremacy but also from its central location at the hub of rail and motorway systems. Urban redevelopment, to accommodate the north–south motorway link, the new underground railway, and extensions to the central business district which houses 400 major European organizations, has brought fundamental changes to the appearance of the city. More than 200,000 people commute to Brussels each day from almost all parts of Belgium, including the Ardennes. Greater Brussels covers a much larger area than the 'Brussels-Capital' *arrondissement* (19 *communes*) that was defined by the so-called 'language laws' and housed 1,075,000 people in 1970. The changing function of innermost Brussels has meant that its residential population has declined but its outer suburbs have expanded rapidly in the Flemish-speaking *arrondissements* of Halle-Vilvoorde and Leuven and the French-speaking *arrondissement* of Nivelles. Greater Brussels now spreads over more than 60 *communes* and contains about 1,500,000 inhabitants (Flament, 1974). Pleasant landscapes and wooded sites in Walloon areas have attracted suburban growth southwards and it is possible that the old industrial areas of Namur and Charleroi may soon be incorporated in the 'golden triangle'. By contrast, the continuing growth of the 'bilingual' (i.e. French-speaking) capital into Flemish territory is viewed with disfavour by the Flemings. Indeed, the creation of a green belt in Halle-Vilvoorde *arrondissement* is in response to the language issues.

Lagging Areas and Peripheral Regions

Areas with relatively low average incomes are found beyond the core areas that have been discussed so far. Such areas may be divided into two groups. The first involves not only mining areas but old industrial regions (such as West-Vlaanderen, Drenthe and Overijssel) which are faced with difficult problems of economic and social revival. The second group includes predominantly agricultural, under-industrialized areas which often suffer from problems of isolation. Groningen, Zeeland, and the Belgian Kempenland fall into this category but they are now experiencing economic modernization in response to improved communications and the distribution of Groningen natural gas.

The Walloon coalfield, running from northern France to West Germany, is a classic example of an industrial 'black country' but contains important internal variations. Early industrial development made use of rich deposits of coal and small quantities of ore for making iron, and also employed local sand for glassmaking. The heyday of industrial development throughout southern Belgium was probably reached by the mid-nineteenth century. The Borinage mining area developed relatively few industries, unlike the Charleroi basin to the south where heavy metallurgy flourished and was eventually joined by manufacture of electrical goods. Urban landscapes are pockmarked with the legacy of the industrial revolution in the form of industrial slums, pit heaps and empty factories. Derelict areas are not rare. By contrast, the Liège–Verviers area had a much broader industrial base (including the manufacture of arms) which was sustained by supplies of high-quality labour and the dynamism of such industrial entrepreneurs as John Cockerill and Solvay. Many important innovations in metallurgy, chemical production and the electrical industries were developed in and around Liège.

The 1930s marked the first phase of the mining crisis in Wallonia, and attempts were made to diversify industrial activities by introducing the manufacture of electrical goods. Poor geological conditions, old equipment and low productivity were points of weakness in the Walloon mining trade. The immediate post-war years brought a new breath of life but many mining companies were scarcely viable and had neglected their installations. Finances were obtained from the E.C.S.C. but these were used to subsidize production rather than to modernize pits and assist mergers. Crisis conditions were experienced with particular severity when cheap American coal started to be imported through Antwerp and oil was piped straight to the heart of Wallonia. More than thirty coal mines were closed between 1966 and 1969, and by 1972 Belgium relied on imports for 88 per cent of her energy supplies, compared with only 33 per cent in 1960. The task of bringing replacement jobs to such areas as Mons and the Borinage, with little or no industrial tradition, has proved particularly difficult. In addition, the steel industry of southern Belgium required modernization following installation of new steelworks with deep-water access at Ghent–Zelzate. This heralded a new episode in the rivalry between Flanders and Wallonia. New developments in southern

Belgium's steel industry, such as the oxygen steelmaking process and the Chertal rolling mills, have not been completely viable. At the same time, the woollen industry of Verviers has encountered serious economic problems.

The coal basins of Belgian Kempenland and Dutch Limburg, brought into production in the early twentieth century, have not been spared similar problems, in spite of much better geological conditions. Production was cut back particularly harshly in Dutch Limburg because of competition from oil supplies and also because the Dutch Coal Board (Staatsmijnen) had interests in natural gas exploitation in Groningen province. Now only 25 per cent of the Netherlands' energy supplies are imported and Dutch policy for employing only 4000 miners in four pits in 1974 and phasing out production by the end of the year contrasted with Belgian support of 20,000 workers in twenty pits. By 1972 coal production had fallen to less than 3,500,000 tons in the Netherlands and under 10,000,000 tons in Belgium. This contraction has been particularly drastic in the two Limburg provinces which until recently had developed few additional activities apart from electricity generation and production of coal-derived goods, such as carbo-chemicals and fertilizers. In addition, the Staatsmijnen and private mining companies in Belgium had created coal-based 'company towns' such as Heerlen-Kerkrade and Geleen-Sittard which had been little more than villages at the beginning of the twentieth century but contained 265,000 and 173,000 inhabitants respectively in 1973 (Riley and Ashworth, 1975).

Rural, Under-industrialized Regions

Dutch Zeeland is perhaps the most characteristic of these relatively isolated, introverted agricultural areas with inadequate systems of transport. This province contains 170,000 ha of islands in the deltaland where protection against the sea and rivers is of prime importance. Historic floods were complemented by grave disaster in 1953 when 1800 people were drowned and over 150,000 ha were inundated. There are many settlements in Zeeland that are of historic interest and are worthy of conservation but this province has suffered serious depopulation and now contains an elderly population living on very small farms. The Delta Plan was introduced primarily to prevent further flooding and avoid landward invasion by saltwater. In addition, new roads were installed along the barrages and sea defences. These roads have improved land communication with nearby urban areas and it is hoped that by opening up the area new industrial activities may be introduced. The Delta Plan has been developed progressively over the past twenty years and is now an integral part of national planning in the Netherlands.

Problems of distance and isolation are particularly great in the northern Dutch provinces of Friesland and Groningen; however, there are two main prospects for economic improvement in these areas. The first is exploitation of large quantities of natural gas, though most of this energy is consumed outside the northern Netherlands. The second is the construction across the IJsselmeerpolders of new road links to the Randstad. But these provinces are still relatively 'blocked' by the German border to the east. For example, the new Dutch port of Delfzijl suffers serious competition from the neighbouring German port of Emden. In spite of the development of natural gas resources, unemployment rates in the northern provinces (5–6 per cent) were almost double the national average (3·5 per cent) at the beginning of 1975. Gas production is highly automated and has created few local jobs apart from in the chemical and fertilizer industries of Groningen and Delfzijl. Particularly poor areas are found in the zone of old peat workings close to the German border which has suffered from decades of out-migration to the U.S.A. and to the cities of the western Netherlands. Housing conditions are poor, settlement is dispersed, income levels are low and service provision is inadequate in this area which has few cities of any size except Leuwarden (88,000) and Groningen (206,000). Now the large student population of the latter city has introduced a new vitality which contrasts with the traditional Calvinist austerity of the hinterland. Between one-half and one-third of the total population of Drenthe, Friesland and Groningen provinces still live in *communes* with fewer than 2000 inhabitants apiece.

The economic marginality of West-Vlaanderen is much more of a paradox, given its glorious past and its proximity to growth centres such as Dunkirk (in France), Antwerp and Ghent. In fact it is the existence of this international border that has prevented the area from deriving maximum benefit from its geographic location (Lentacker, 1973). Industry has not developed along the coast, and while tourism brings seasonal activity it has created few permanent jobs in the Westhoek area which remains strongly agricultural. (Obviously not all coastal areas in the Benelux countries have experienced economic expansion.) The problems of West-Vlaanderen were further exacerbated by contraction in the traditional woollen industry during the 1960s. Until recently birth rates remained high and, in the absence of alternative local employment, considerable numbers of workers commuted daily from West-Vlaanderen (and especially Westhoek) to jobs in France. Up to 80,000 migrants crossed the border each day at the peak of this activity but the volume declined during the 1960s because of improvements in local job opportunities and growth of long-distance commuting in Belgium. Additional incomes are derived by some agricultural workers who take seasonal jobs harvesting beet and hops in Brabant.

Other marginal areas include the Belgian Ardennes and the northern sections of Noord-Holland province. At first glance it would appear that Benelux's regional problems might be solved if only some of the well-being of the Randstad and the Belgian 'golden triangle' could be spread to outer areas. But of course problems are much more complicated than this, being bound up with social and economic inertia and the cumulative processes of external economies. In addition, there is the serious problem of trying to link schemes for industrial decentralization with broader programmes for regional development in the most appropriate ways.

Policies for Regional Development and Planning in Benelux

A distinction must be drawn between policies for expanding employment in particular industrial sectors (which, of course, also have important spatial implications) and schemes for improving use of space and resources in each part of the Benelux countries, and incidentally reconciling actual or potential conflicts between users. The first group of policies has been implemented in a roughly similar way in Belgium and the Netherlands and both countries will be discussed together. By contrast, the second group displays marked differences because of varying forms of legislation and rather different attitudes toward the need for spatial planning. Without decrying what has been achieved at a variety of scales in Belgium, Dutch experience is more interesting in spite of occasional disparities between intentions and results.

Regional Industrialization Policies

The most usual way of attempting to reduce, if not to erase, inter-regional differences is by developing industrial employment in problem areas. Policies are roughly similar in many countries and involve expressions of need for industrial jobs from local organizations (for example, the economic councils for Wallonia and for Flanders), provision of financial aid for creating new and replacement jobs, and measures for equipping industrial estates. Unlike in France or the U.K., relatively little attention has been devoted to decentralizing office jobs in the Benelux countries. Indeed, it is clear that tertiary employment is becoming increasingly concentrated in the Belgo-Dutch megalopolis and in its Luxembourg annexe. However efforts have been made for local decongestion in the western provinces of the Netherlands, for example to benefit such cities as Utrecht. To justify this state of affairs one must look to the international role of the great cities of Benelux and the attitude that they need to be strengthened to withstand competition from other world cities. It is hard to envisage this being reversed and large numbers of tertiary activities being decentralized when one bears in mind the cumulative effects of high salaries, distinctive employment patterns and life styles on the major cities in Benelux. Nevertheless, the Dutch have made a start, with 6500 jobs for civil servants being moved from The Hague to Limburg and the northern provinces before 1977. But it is probable that only low-qualification jobs will be shifted, as was the case with branches of the postal service that were decentralized to Arnhem. That city is in a particularly good position for receiving tertiary activities shifted from the West, possessing good motorway links and a range of research institutions in electronics, atomic energy and land reclamation.

Determining which towns should receive new industries and which zones should be designated 'development areas' always gives rise to serious difficulties. In addition, attempts to spread financial assistance in ways that may be considered too wide, so that inter-regional competition is distorted, arouse opposition from the E.E.C. Commission. (This was certainly the case in Belgium in 1970.) Attitudes also vary regarding the extent to which further growth should be controlled in existing congested areas. In this respect the Dutch have been more environment-conscious than have their Belgian neighbours. Development of the car-assembly industry beyond the Dutch Randstad has been exemplary, with factories opened at Eindhoven and Born (D.A.F.) and at Amersfoort (British Leyland). By contrast, most car-assembly works in Belgium have been located along the Antwerp–Mechelen–Brussels axis, with the exception of Ford at Genk in the Kempenland and British Leyland at Seneffe.

Other difficulties involve the need to take arbitrary decisions to resolve conflicting demands between regions, since there are no universally accepted criteria for defining areas that merit assistance. Thus Wallonia made a case for the need to install new jobs to reduce high levels of unemployment and make use of the existing infrastructure (without accepting the idea of bringing in outside labour to meet the second objective), whilst Flanders pointed to its large supply of young people coming on to the labour market in the Waasland and Westhoek and argued the case for new factories on these grounds. Faced with this dilemma of taking work to the workers or workers to the work, the Belgian government established the principle of balanced development between the two major communities, which could of course only be achieved by a certain sprinkling of aid. Eventually indices of unemployment, inter-regional migration, and dependence on certain industries were agreed on, but not before great political haggling. The map of Belgian regional aid that resulted was much more complicated than the comparable one for the Netherlands (Figures 7.2 and 7.3).

The first Dutch region to receive State aid for industrialization was the peat-digging area in the North-East. Loans and grants were then extended progressively to areas of underemployment within which 40 towns were designated for priority treatment. This legislation was quite effective but became complicated with the passage of time. It was consolidated in 1969 and 'stimulation areas' were distinguished from 'restructuring areas'. The number of development centres was reduced in stimulation areas, which involve northeastern parts of the country, northern Holland, Zeeland and parts of Noord-Brabant. Development aid is more generous in restructuring areas, which include Limburg and Tilburg. Northeastern provinces also benefit from a lower tariff for natural gas consumed locally and this, for example, encouraged the development of an aluminium smelter at Delfzijl.

Systems of aid for regional development evolved more slowly in Belgium and it was not until the coal crisis that development areas were recognized. Volumes of assistance were increased and spread over progressively wider areas in 1959, 1966 and 1970 so that 35 per cent of the country eventually received some kind of aid. Such a state of affairs might be considered excessive (Riley and Ashworth, 1975). In the small state of Luxembourg, with only 340,000 inhabitants in 1975, one can scarcely talk of 'regions' or regional problems. However, the government has provided grants and loans to help diversify manufacturing, reduce excessive concentration on the steel industry and spread more industrial employment to northern parts of the country. American-owned chemical and rubber firms have moved in, being attracted by financial legislation of 1962, and have partially achieved the basic objectives. Canalization of the Moselle and electrification of Luxembourg's railway system has also helped.

For a number of years the Walloon South complained that the greater part of the Belgian motorway network was located in Flanders. But now a motorway has been built linking northern France and West Germany through the industrial corridor of Wallonia. Liège has direct motorway links with Utrecht, Amsterdam and Antwerp. Motorways between Brussels, Namur and Charleroi will link the coalfield area even more strongly to northern Belgium and more particularly to the 'golden triangle'. The Walloons have also campaigned, and with relative success, for their navigable waterways to be improved to European standards so that they might effectively serve south Belgian heavy industry. This has not, of course, prevented development of massive new industrial estates on the coast. Nevertheless, the Sambre system has at last been improved between Monceau and Namur, and the barge-lift at Ronquières has facilitated links with central Belgium and Antwerp.

Another aspect of industrial decentralization is the fact that multinational corporations tend to conform with proposals for regional development. In the Netherlands, the Philips corporation, D.A.F. and the Staatsmijnen have been particularly important. For many years Philips have established factories in rural parts of Noord-Brabant with abundant supplies of labour. Eindhoven, with five large Philips works, employed large numbers of commuters from a hinterland that increased progressively with the passage of time, in spite of the fact that greater Eindhoven was growing fast and was to house 345,000 inhabitants in 1972. Further decentralization was required to reduce long-distance commuting that often exceeded 80 km. Now Philips operates more than 40 factories outside Eindhoven in nearby provinces and in areas as distant as Friesland (Drachten) and Groningen (Stadskanaal). Old textile-

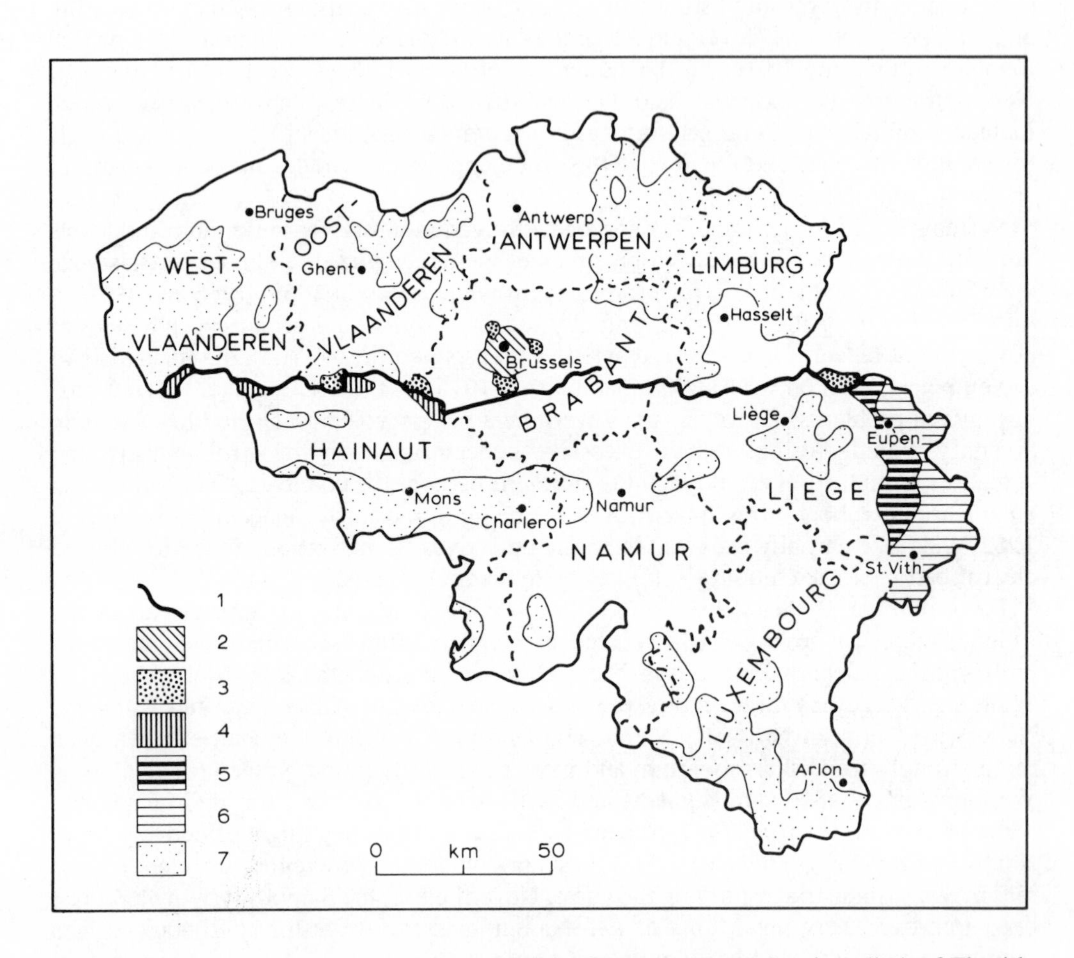

Figure 7.2. Belgium: linguistic communities and development areas. 1 Southern limit of Flemish-speaking region. 2 Bilingual region. 3 Flemish-speaking, with protected French- speaking minority. 4 French-speaking, with protected Flemish-speaking minority. 5 French-speaking, with protected German-speaking minority. 6 German-speaking, with protected French-speaking minority. 7 Development areas (1966 legislation).

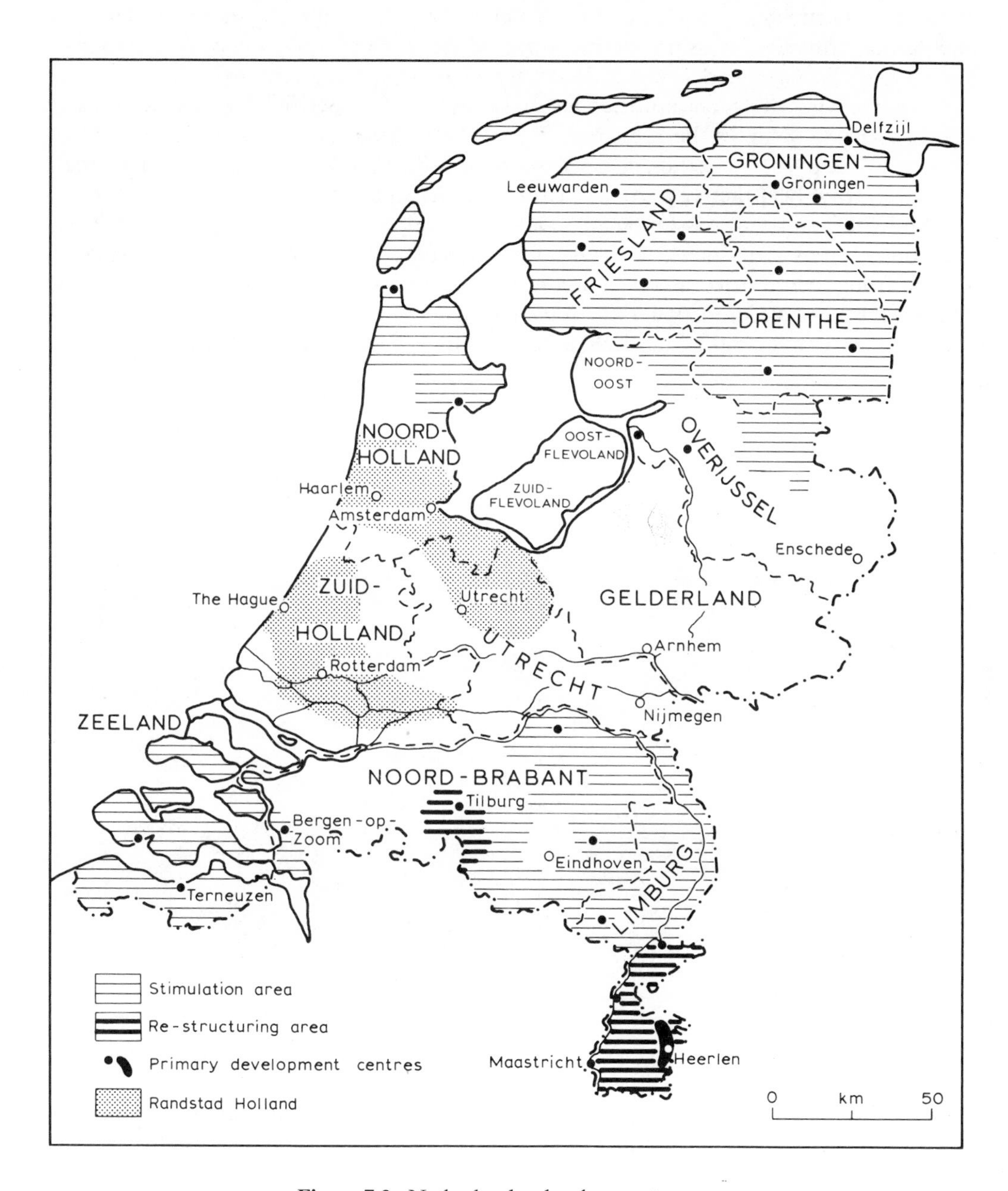

Figure 7.3. Netherlands: development areas.

producing towns in Twente, such as Enschede-Hengelo (238,000) have benefited greatly.

Modern growth industries have often been installed near to the IJssel river, close to pleasant environments such as the Veluwe (which, with its national park, is increasingly functioning as one of the 'lungs' of the Netherlands), and in well-equipped towns like Nijmegen and Arnhem. However, American firms have been less sensitive to the advantages of decentralization and have preferred to sacrifice financial incentives by choosing to settle in the Randstad. By contrast, American firms in Belgium have chosen to adopt the government's schemes for industrial location which, of course, include the highly attractive port of Antwerp with its abundant supplies of local labour.

The final factor that has favoured industrial development in lagging areas has been the policy for creating industrial estates by local authorities using their own finance or capital from central government sources. The Walloons are particularly proud of their industrial estates at Ghlin-Baudour (in the Borinage) and Hauts-Sarts (near Liège), but it is port-based industrial estates in Flanders that have been most successful.

Table 7.7. Increase in Gross Domestic Product by Belgian Province

	1960–65	1960–70	1960–70 (annual average)
Oost-Vlaanderen	49·3	84·3	7·9
West-Vlaanderen	49·4	82·6	7·8
Antwerpen	51·4	94·2	8·6
Limburg	74·2	114·8	10·0
Flanders	52·0	90·3	8·4
Hainaut	41·0	62·9	6·3
Namur	36·0	58·7	6·0
Liège	40·0	55·4	5·6
Luxembourg	40·6	64·2	6·4
Wallonia	40·2	60·3	6·1
Brabant	50·2	82·6	7·8
Brussels region	51·7	83·2	7·9
BELGIUM	48·1	79·0	7·6

It is difficult to draw up an accurate evaluation of regional industrial development in the Benelux countries since suitable statistics are just not available. However, the information in Table 7.7 may be used with caution to show that Wallonia has fared less well than Flanders. Even after the departure of foreign workers during times of economic difficulty, unemployment rates have been consistently above the national average in the southern mining areas during the past 15 years and have reached more than twice the national mean in the Borinage and the Liège area. In 1972–73, 4·5 per cent of the Walloon labour force was out of work, but unemployment was virtually non-existent in Flanders. Southern industries, such as metallurgy and textiles, that have modernized production have also reduced their labour requirements.

Chemicals, electronics, clothing, glassmaking, steelmaking, vehicle manufacture and oil refining are typical activities that have been located on industrial estates in the Borinage, Centre, Charleroi and Liège areas. Admittedly some factories on these estates are completely new operations, sometimes run by foreign firms, but a considerable number are local works that have moved from cramped inner urban areas undergoing renovation (Riley and Ashworth, 1975). Greater Liège, containing over 600,000 inhabitants and having important motorway links, has experienced important urban remodelling and has been particularly successful in attracting new forms of employment. But it must be admitted that diversifying the existing industrial base by new factories has been more successful in Flanders than in the South, with E.C.S.C. finance encouraging electrical and car-assembly plant to settle in the Kempenland.

Dutch data are inadequate to discover the precise impact of regional industrialization but it is known that roughly half of the country's new manufacturing jobs since 1960 have been created in development areas. Results have been most impressive in central-eastern areas, Noord-Brabant and, especially, Zuid-Limburg, where more than 45,000 new jobs have been created since 1961 to compensate for jobs lost in the mines. The Dutch Coal Board, firms like Philips and D.A.F., and the local authorities have played vital roles in reviving the Limburg economy. However, new industries have been attracted to major lines of communication, such as the Juliana Canal and the Maas, rather than to the old mining areas with landscape features inherited from a past era. In fact, the renovation of Zuid-Limburg is a successful part of the total spectrum of Dutch schemes for regional development.

Regional Development Policies

Belgium has been particularly preoccupied with trying to balance out the fortunes of the two main linguistic communities and less attention has consequently been paid to physical planning than has been the case in the Netherlands. Polder reclamation represents an early example of land management, with special drainage associations being in existence since the thirteenth century. From as early as 1901 each Dutch town has been required to prepare its development plan and some cities, such as Amsterdam, have established reserves of land that have been of crucial importance for accommodating recent housing schemes.

Reclamation of the IJsselmeer polders represents an important step towards integration of rural planning as part of the broader operation of regional management (Boyer, 1968b; Clout, 1969). Rural areas in the Netherlands are no longer viewed simply in terms of their potential for agricultural production but rather as ecological environments that must be planned to meet the demands of urban dwellers as well as the needs of the farmer. For example, plot consolidation schemes now incorporate small recreation areas. Interesting practical experiments have been conducted for establishing service centres on each of the recent polders. The earliest was the Wieringermeer polder which was planned to have three or four villages, serving roughly 20,000 ha apiece, plus fifteen hamlets. The idea of building hamlets was abandoned and even four villages proved to be too many. None has managed to function as a successful service centre. The Noord-Oost polder (1948) was planned with a settlement hierarchy ranging from dispersed farms, through ten villages, to the successful market town of Emmeloord. Each village was to serve roughly 48,000 ha. Agricultural workers have generally preferred to live in these nucleations or in Emmeloord rather than in special accommodation built for them on the farms. A new component in the Oost-Flevoland polder is the regional centre of Lelystad that will serve the whole of the IJsselmeerpolders. Target populations for each village centre in Oost-Flevoland have been revised upwards from 2000 to 5000 in response to the declining number of farmworkers required for modern agricultural production and the rising costs of providing schools, churches and other community facilities in triplicate (for each of the main religious groups). Now the number of villages to be built in Oost-Flevoland has been fixed at three, in addition to Lelystad. Rural settlements are only 1·5 to 2·0 km apart in old agrarian areas of the Netherlands but are separated by more than 10 km in Oost-Flevoland, where planners have paid attention to new architectural styles, especially at Dronten. The final settlement structure for the Zuid-Flevoland polder has not been decided but it will probably include two important towns, of which one may receive planned overspill from Amsterdam, in order to reduce pressure on the Gooi area and on old towns such as Naarden. Increased emphasis is being placed on parks, lakes and other environmental facilities, as well as on the role of the polders in improving east–west transport links through the Netherlands. In other words, Zuid-Flevoland is being planned to satisfy various urban needs emanating from Amsterdam. Details of the layout and use of the final, still-undrained polder (Markerwaard) remain undecided.

In the past, Dutch physical planning has operated at the *gemeente* (local) and province levels, but new management problems need to be tackled at different spatial scales. Only recently has it been admitted that fiercely independent municipalities should be linked for purposes of planning; and a number of special sub-regional bodies have been created to manage the southern IJsselmeerpolders (the Rijnmond authority) and the Delfzijl–Eems zone. But no matter how important these special organizations may be they do not supersede the small official 'regions' for which urban-focused structure plans are to be prepared and brought into operation by 1976–77. Conflicts between various levels of authority may be illustrated by the Heerhugowaard polder, where the local administration managed to establish what amounts to a new town even though national and provincial plans had not designated this area for development (Malézieux, 1973). The local authority justified its enthusiasm for urban growth in terms of good communications with Amsterdam, decline of agricultural employment, the need to establish a local service centre, and suitability of this unspoiled area to provide single-family housing for commuters. Now more than 20,000 people live at Heerhugowaard and the success of this venture points to the need to revise certain aspects of planning power.

Coordination of provincial plans, urban plans and projects to develop the eastern provinces has been achieved only recently. Reports were published on problems of the western provinces (1956–58) and on physical planning of the whole country (1960), with a more penetrating analysis of regional problems appearing in a *Second Report on Physical Planning* (1966). Rapid population growth (producing 20,000,000 inhabitants by A.D. 2000) greatly increased car ownership, and a threefold increase in urbanized areas by the end of the century were anticipated (Buchanan, 1969). A policy of 'concentrated dispersion' with improved communication between urban nodes was proposed, so as to avoid the evils of urban sprawl and allow open spaces to be protected. Future prosperity of port-based industrial areas was not to be jeopardized but the report stressed the need to restrict the growth of employment in the West by decentralizing manufacturing jobs to such destinations as Alkmaar, Zuid-Flevoland and Noord-Brabant. The last area already displays important industrial vitality at Eindhoven and in surrounding towns and would be very suitable for receiving 'concentrated dispersion'. The second report was not intended to be a rigid plan but rather a framework within which changing social and economic circumstances might be accommodated and responsibility be coordinated between various authorities.

Detailed research in the late 1960s and early 1970s, plus discussion with branches of administration and the general public, culminated in a third report (1973–74). This recognized the difficulty of implementing plans in a free society and accepted that basic demographic projections had proved to be incorrect. Birth rates had fallen drastically since 1970, with the average rate of 21·1 per thousand (1955–64) contrasting with 17·0 per thousand in 1972. Fertility rates also fell; and a total population of 16,000,000 seemed more realistic for A.D. 2000. Population growth slowed down most strikingly in the western provinces. Short-and medium-term demands on housing space are unlikely to change very much, but a new complexion has certainly been put on long-term needs. Yet in other respects the Dutch have made greater demands on space than was anticipated in the mid-1960s, by acquiring second homes in France, occupying military camps in the Massif Central and, more pertinently, by becoming a suburban society in their own country. In the 1960s and early 1970s many small-and medium-sized settlements reversed their established trend of population loss or stagnation through the arrival of commuters. By contrast, the very largest urban areas were recording population losses.

The average size of newly-built apartments increased from 85 m^2 to 93 m^2 during the 1960s; but the number of occupants per dwelling fell from 4·0 persons in 1960 to 3·45 in 1972. Building densities have also declined to less than 75 dwellings/ha in some new suburbs and even to 25/ha on semi-rural estates in the Gooi area and on the Heerhugowaard polder. The policy of 'concentrated dispersion' seems to be at odds with the growing public preference for single-family dwellings rather than traditional apartment houses. Some inner urban areas (and even parts of new estates such as Bijlmermeer) are becoming 'ghettos' for immigrants from the Dutch West Indies and Surinam. Commuting flows increased beyond all expectation during the last decade, with the number of passenger kilometres travelled in private cars rising fourfold. If all these trends continue, roughly 290,000 ha will be required to accommodate urban growth up to A.D. 2000, by which time the proportion of Dutch territory occupied by urban uses will have doubled from 8 to 16 per cent. Resolution of conflicting demands for land use will become more challenging than ever before.

Finally, the third report noted that development of northern and eastern parts of the Netherlands had not taken place in quite the way that had been expected. Migration had been encouraged from western areas toward the South and East but had been compensated by migration to the Randstad from the West Indies and abroad. With all these trends in mind, the Dutch authorities have placed new emphasis on the need to protect rural areas from urban invasion. The policy of 'concentrated dispersion' continues to operate but with rather greater emphasis on the first word. Greater efforts are being made to protect the greenheart of the Randstad and the green wedges that separate the towns of Noord-Brabant. (If continued attention is placed on developing the Antwerp–Bergen-op-Zoom–Rotterdam axis pressure on Noord-Brabant might well be reduced.)

The key problem is the means of implementing all these plans. Acquisition of land for future development by urban authorities has been fine in principle but has had the effect of pushing the private estates of suburbia even further into the countryside. The Bijlmermeer estate has been built on city-owned land to the southeast of Amsterdam and will house 100,000 people in subsidized apartments by 1980. Its average density is 40 dwellings/ha, which is not much greater than that of many new towns or garden cities. But will massive developments of this kind work hand-in-hand with the policy of 'concentrated dispersion' or will they encourage people to move further afield? Another problem is to convince the various authorities and administrations in each of the Benelux countries to cooperate in terms of planning and regional development. No fewer than 16 directives have been issued to try to plan the 19 *communes* of Brussels-Capital, but whether they will actually be implemented is a matter of speculation. Is it the task of planning authorities to ensure that industrial jobs, small shops and moderately-priced housing are retained in city centres, or to encourage further property development and construction of office blocks which may enhance the city's international status? Planners in Brussels have favoured the latter course whilst their counterparts in Rotterdam have adopted a more balanced approach. This difference once again highlights the conceptual strength and practical importance of physical planning in the Netherlands. Benelux, and especially Dutch, experience shows that urban and regional planning need to be coordinated if they are to be effective; it also demonstrates what may be done in the difficult task of reconciling high levels of economic growth with conservation of the quality of life.

References

Boyer, J. C., 1968a, La notion de région aux Pays-Bas, *Annales de Géographie,* **77**, 323–335.

Boyer, J. C., 1968b, Les centres de services dans les polders du Zuider Zee, *Hommes et Terres du Nord,* 7–16.

Boyer, J. C., 1971, Etudes sur la fonction commerciale des villes néerlandaises, *Hommes et Terres du Nord,* 39–60.

Buchanan, R. H., 1969, Towards Netherlands 2000: the Dutch National Plan, *Economic Geography,* **45**, 258–274.

Burke, G. L., 1966, *Greenheart Metropolis,* Macmillan, London.

Clout, H. D., 1969, Les problèmes de planification rurale aux Pays-Bas, *Information Géographique,* **33**, 114–120.

Clout, H. D., 1974, Economic change in Belgian Limburg, *Geography,* **59**, 145–147.

Clout, H. D., 1975, *The Franco-Belgian Border Region,* Oxford University Press.

Constandse, A. K., 1972, The IJsselmeer polders: an old project with new functions, *Tijdschrift voor Economische en Sociale Geografie,* **63**, 200–210.

Flament, E., 1974, L'évolution récente de la population belge, *Information Géographique,* **38**, 174–184.

Gay, F. and P. Wagret, 1976, *Le Bénélux* (4th ed.), Presses Universitaires de France, Paris.

George, P. and R. Sevrin, 1967, *Belgique, Pays-Bas, Luxembourg,* Presses Universitaires de France, Paris.

Hall, P. G., 1966, *The World Cities,* Weidenfeld and Nicolson, London.

Lambert, A. M., 1961, Farm consolidation and improvement in the Netherlands, *Economic Geography,* **45**, 258–274.

Lambert, A. M.,. 1971, *The Making of the Dutch Landscape,* Seminar Press, London.

Lawrence, G. R. P., 1971, The changing face of south Limburg, *Geography,* **56**, 35–39.

Lawrence, G. R. P., 1973, *Randstad Holland,* Oxford University Press.

Lentacker, F., 1973, La frontière franco-belge, *Information Géographique,* **37**, 46–48.

Lentacker, F. and J. Van Hertog, 1967, Les transformations de la Zélande, *Hommes et Terres du Nord,* 7–20.

Malézieux, J., 1971, Signification géographique d'un projet d'investissement industriel: un centre sidérurgique dans la Maasvlakte de Rotterdam, *Annales de Géographie,* **80**, 428–438.

Malézieux, J., 1973, Une nouvelle ville aux Pays-Bas: Heerhugowaard, *Annales de Géographie,* **82**, 720–731.

Mingret, P., 1962, Quelques problèmes de l'Europe: à travers l'exemple de Liège et de sa région, *Revue de Géographie de Lyon,* **37**, 5–74.

Mingret, P., 1970, Les investissements américains en Belgique, *Revue de Géographie de Lyon,* **45**, 243–278.

Mingret, P., 1972, A factor in the regional evolution of Belgium: the geographical distribution of American industrial investments, *Terra,* **84**, 14–22.

Ministry of Housing and Physical Planning, 1974, *Summary of the Orientation Report on Physical Planning,* The Hague.

Riley, R. C., 1965, Recent developments in the Belgian Borinage, *Geography,* **50**, 261–273.

Riley, R. C., 1967, Changes in the supply of coking coal in Belgium since 1945, *Economic Geography,* **43**, 261–270.

Riley, R. C., and G. J. Ashworth, 1975, *Benelux: an economic geography of Belgium, the Netherlands and Luxembourg,* Chatto and Windus, London.

Snijdelaar, M., 1972, Water management in the Netherlands: the struggle for water, *Tijdschrift voor Economische en Sociale Geografie,* **63**, 211–225.

Sporck, J. A., 1970, La reconversion économique des régions industrielles wallones, *Information Géographique,* **34**, 57–70.

Steigenga, W., 1968, Recent planning problems of the Netherlands, *Regional Studies,* **2**, 105–118.

Steigenga, W., 1972, Randstad Holland: concept in evolution, *Tijdschrift voor Economische en Sociale Geografie,* **63**, 149–161.

Tamsma, R., 1972, The northern Netherlands: large problem area in a small country, small problem area in a large community, *Tijdschrift voor Economische en Sociale Geografie,* **63**, 162–179.

Van Hulten, M. H., 1969, Plan and reality in the IJsselmeer polders, *Tijdschrift voor Economische en Sociale Geografie,* **60**, 67–76.

Wagret, P., 1968, *Polderlands,* Methuen, London.

8

West Germany

Mark Blacksell

Economic Revival

The resurgence of West Germany represents one of the most remarkable aspects of Europe since the Second World War. In 1945 the Third Reich was in ruins, but only thirty years later the western third of the nation has risen to become indisputably the leading industrial country in Western Europe. Indeed, after the United States, West Germany is the second most important trading nation in the world. Its population enjoys one of the highest standards of living and it is likely that the position will continue to improve.

How has such a transformation been possible? Naturally there is no single, simple answer and opinions vary about the relative importance of the economic factors involved. There is one school of thought, headed by the distinguished German economist A. Predöhl, which argues that Germany forms a key part of one of a series of industrial core areas in the world and that, given its location, the post-war successes were inevitable. Grotewald (1973) in a detailed critique of Predöhl's work maintains that this is only a partial explanation. He acknowledges the benefits derived from internal trade within the E.E.C. and the E.C.S.C. but believes that more fluid trade relations between the whole of this industrial core area and its counterparts elsewhere in the world would have benefited West Germany even more. The truth of such speculation, of course, will never be known, but West Germany has certainly profited from European integration and will almost inevitably enhance its relative position, as the effects of present policy-decisions become effective. Clark and his colleagues, in a provocative piece of research published in 1969, showed how the European core area, with industrial West Germany at its centre, had gained in economic strength since the formation of the E.E.C., and then projected into the future to show how its position would be further strengthened if the Community were to be enlarged and if a Channel Tunnel were to be built.

To consider West German progress in such global terms, while in one sense helping to put recent developments into their proper perspective, does less than justice to the resource and ingenuity with which both the society and the economy have been rebuilt. Throughout the country's short history decisions have constantly had to be made about the future and it is these, just as much as any omnipotent global force, that have shaped the face of the nation. This chapter takes a detailed look at West Germany and West Berlin and tries to identify the key events and policies which contributed to the country's present economic preeminence. In particular it examines the internal logic and organization of the one state in Western Europe which can justly claim to be a product of the second half of the twentieth century. For whilst it is true that defeat in two World Wars in the space of a quarter of a century has overshadowed almost every significant political and economic decision made in West Germany, it is equally the case that the country enjoyed a fresh start in 1949 and has been unencumbered by many of the legacies of eighteenth-and nineteenth-century economic imperialism which have dogged some of its European neighbours.

The Nature of the State

West Germany grew out of the three zones of occupation held by the Western Allies (the United Kingdom, France and the United States) after the end of the Second World War. It is difficult to point to a precise date when the state was actually founded, but throughout 1949 a series of events, including free elections on 14th August, gradually transferred political power back to the German people. For the future of domestic policy in general and regional policy in particular, the most important fact about the new state was its federal constitution. The territory was divided into eleven *Länder* (provinces), Schleswig–Holstein, Hamburg, Bremen, Niedersachsen, Nordrhein–Westfalen, Rheinland–Pfalz, Hessen, Bayern, Baden, Württemberg–Baden and Württemberg–Hohenzollern. The latter three were amalgamated in 1952 to form Baden–Württemberg and when the Saarland was returned by France in 1957, it too became a separate *Land*, so that the final tally was ten. The status of West Berlin was and remains anomalous. Stranded in the middle of the Soviet occupation zone, the Western Allies were finally dissuaded from making it an official part of the Federal Republic and a *Land* in its own right. Even today, after the signing of the four power agreement on Berlin in 1971 and the East–West German Treaty in 1972, it is still technically outside the federation. However, the influence it has had upon West Germany and vice versa makes it integral to any analysis.

The *Länder* enjoy a considerable degree of autonomy. While the Federal government is vested with most of the legislative powers, they are responsible for virtually all the administration. They have the power to raise capital through taxation and they are charged with implementing not only *Land* laws, but federal ones as well. It is also the *Länder* which provide most of the services direct to the population. In short the organization of the state is highly decentralized and is a model from which countries with a more centralized organization, such as France and the United Kingdom, could learn much, especially at a time when calls for the regional devolution of power are so strong. Not that the *Länder* themselves conform to any single pattern. They vary widely in both size and population (Figure 8.1) from virtual city states, like Bremen and Hamburg, to huge entities, embracing both urban and rural populations, like Bayern and Nordrhein–Westfalen. However the diversity of form should not be taken as a token of wide regional variations: differences in *per capita* G.N.P. (Figure 8.1) are relatively small, reflecting more or less the relative balance between agriculture and industry. As Johnson (1973) has so clearly pointed out, West Germany is now basically a homogeneous society, its federal political structure notwithstanding.

The initial choice of a federal structure for the new state and its subsequent survival are in many respects surprising, given the high degree of social and economic integration and the relatively small size of the country as a whole. There was, however, a strong tradition of federalism in Germany dating back to the first manoeuvrings for German unity in the nineteenth century. The occupying powers and German people themselves were anxious to foster this tradition in the immediate post-war era as an antidote to the totalitarianism of the Third Reich. The need for a strong central

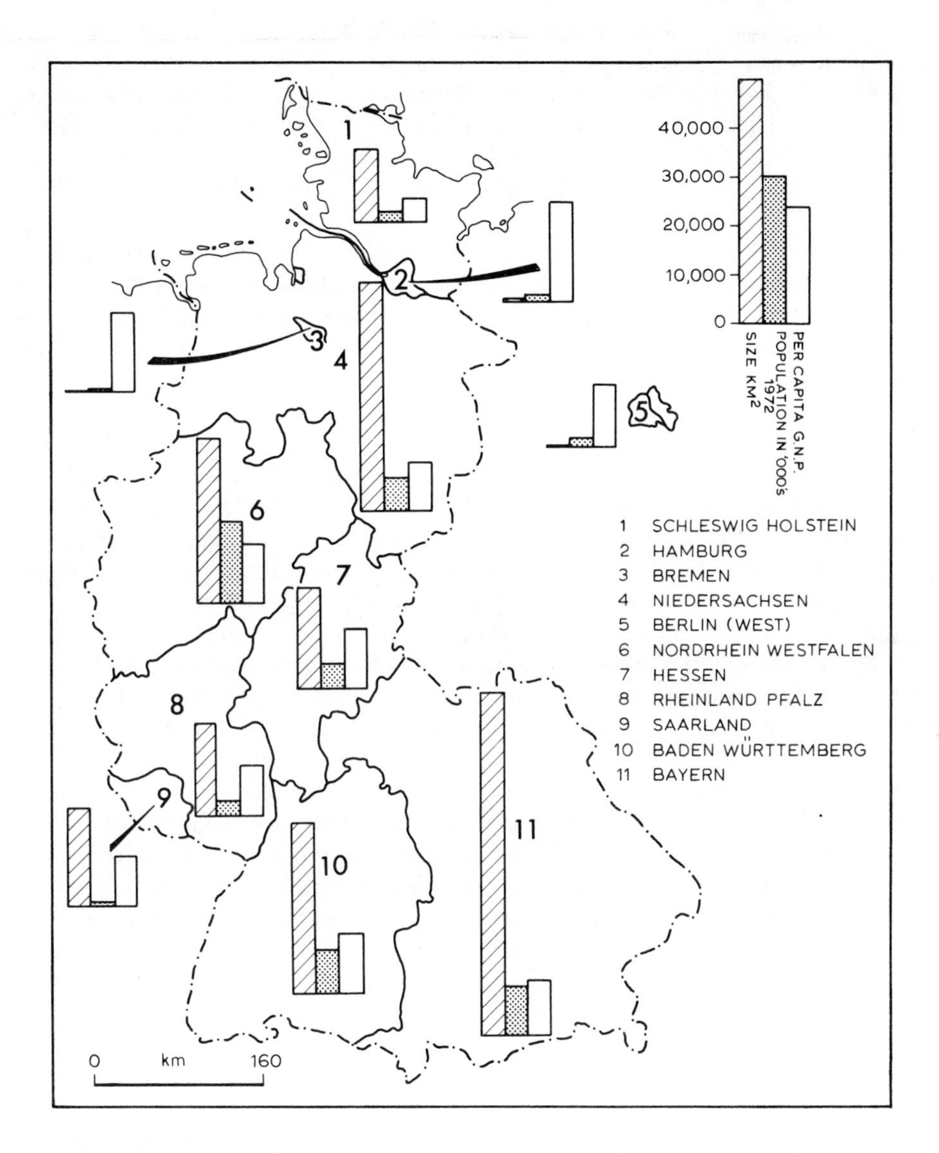

Figure 8.1. West Germany: population and *per capita* G.N.P., 1972 (by *Länder*).

government has also been less than in many other European countries, because of the very limited role that West Germany has until recently played in foreign affairs. It has been concerned mainly with securing the acceptance of the state in the international community and, to this end, its foreign policy has centred upon two essentially domestic issues, the division of Germany and European integration. The E.E.C. was seen as a guarantee of international respectability and West Germany became an enthusiastic member of the E.C.S.C. in 1952 and was a founder member of Euratom in 1956 and of the E.E.C. There has been little interest in influencing world events on a broader scale and hence the need for a strong federal voice was much reduced. The lack of interest in West Germany's world role finds colloquial expression in the way in which the capital, Bonn, is disparagingly referred to as the *Bundesdorf*—the federal village.

The low profile of central government has had its effect on domestic as well as foreign policy. In the 1950s and early 1960s economic thinking was dominated by an almost unshakable belief in the principles of *laisser faire*, interpreted through the now famous *Sozialemarktwirtschaft* (social market economy). The basic philosophy behind this approach was that free competition, rather than state intervention (at an international, national or provincial level) should be the guiding force behind economic development. It was the reverse of the policies adopted in the centrally planned economics of East Germany and the rest of Eastern Europe, and it has been criticized for being too openly anti-communist. As Hallett (1973) has pointed out this interpretation is somewhat unfair and partly due to a misunderstanding in the English-speaking world of the word *Sozialemarktwirtschaft.* Rather than 'social market economy' it is better translated by the phrase 'socially responsible market economy', implying, as is in fact the case, that although state intervention should be kept to a minimum, it should certainly not be eschewed at the expense of public welfare. The key to West German economic policy has been the belief that a steady reliance on the principles of the market economy, linked to improvements in the social services would guarantee a successful reconstruction of the economy as a whole.

The implications of both the decentralized, federal administrative structure and the *Sozialemarktwirtschaft* for regional planning in West Germany have been of fundamental importance. Although in certain limited areas, such as urban renewal and agricultural reform, centralized planning has played a decisive role, these have been exceptions rather than the rule. Regional planning has had but a limited impact on the development of the economy and only since the mid-1960s are there any real indications that attitudes to it are beginning to change. Its organization and achievements are examined in later paragraphs.

Population and Settlement

West Germany has a population of more than 62,000,000 (1973) larger than that of any other European country. Over 60 per cent are fully urbanized, living in towns with more than 20,000 inhabitants and 34 per cent live in cities of over 100,000. Less than one person in twenty is dependent upon agriculture for his livelihood, whereas the comparable figure for both industry and services is one in three. For twenty years the country's economy has been expanding faster than the workforce, something only made possible by a continuous influx of immigrants. In the 1950s and early 1960s they were refugees from East Germany and the former eastern territories of the Third Reich, but over the last fifteen years this source has been replaced by workers from less prosperous countries in southern Europe, drawn by the high wage rates. In general terms it is a pattern familiar to most of the industrialized nations in Western Europe, but there are a number of features, which make conditions in West Germany rather different.

Since the turn of the century all Europe has been in the grip of a 'demographic revolution'. The phrase was coined by the French demographer, A. Landry, to describe the effects of the parallel reduction in both birth and death rates, which seems inevitably to occur as a nation industrializes. West Germany is experiencing just such a transformation, but the impact of two World Wars in the space of a generation has produced important additional imbalances in the structure of population. War deaths have created a very marked preponderance of women among the over-forties. There are nearly 10 per cent more women than men in this section of the population, compared with a slight excess of men (5·16 per cent) among those under forty and, in the 1950s and 1960s, it produced a critical reduction in the active labour force. These were not the only losses: the wars and their aftermath, combined with the impact of economic depression in the 1930s caused a series of short-term falls in the birth rate, thus exacerbating the general downward trend in birth rates, which is a feature of the 'demographic revolution'. In 1953 the birth rate was 15·5 per thousand compared with 18·3 per thousand ten years later. Nevertheless, these figures must be viewed in their proper perspective: the birth rate has declined steadily throughout the 1960s, falling in 1972 to 11.3 per thousand, when for the first time deaths actually exceeded births. The main reasons appear to have been a combination of improved methods of contraception, legalized abortion and general social pressures.

The other side of this rather stagnant picture is the invigorating impact that 9,800,000 refugees have had upon both the economy and the society. Ever since the end of the First World War refugees have comprised an important component of the German population and, since 1945, West Germany has been at the centre of political migrations within Europe. Two-thirds of the 25 per cent increase in the country's population since 1948 is explained by refugees, nearly all of them from East Germany and those parts of western Poland and the Soviet Union which were formerly part of the Third Reich. Immediately after the end of hostilities over 1,000,000 a year were arriving and even as late as 1960 the figure was still 200,000. Most of the new arrivals were young, of working age and gave a vital injection of energy into the labour force at a time when the nation was going through an enforced period of social and economic reconstruction. In the early years the bulk of these people were true political refugees, forcibly driven from their houses. Indeed, one of the outcomes of the Yalta and Potsdam Conferences between the Allied powers in 1945 was an agreement that German minorities should be accommodated within the nation's post-war frontiers. However, as time went on, economic rather than political considerations began to play an increasingly important role. The high wage rates and general affluence of West Germany was a lure that many found hard to resist. It was further encouraged by the shining example of West Berlin. The contrast with the drabness of much of East Germany was unavoidable and the first step in becoming a refugee was a short walk from one side of the city to the other. In many ways it was a fortunate circumstance, for as the Polish-born geographer Kosiński (1970) (who perhaps takes a less rosy view of the movement than many Western commentators) points out, not only were the refugees willing to work hard for relatively low wages, and thus depress wage levels in general in West Germany, they were also an excellent lever for obtaining foreign economic aid, especially from the United States.

In 1961 the East Germans built the infamous Berlin Wall and, overnight, the flow of refugees ceased. In fact for about five years previously it had failed to provide sufficient labour to satisfy the voracious demands of the West German economy and workers from Italy, Greece and Yugoslavia had been attracted to the industrial areas in the Rhine–Ruhr, Frankfurt and Munich regions. The *Gastarbeiter* (guest workers), as the newcomers came to be called, arrived in ever-increasing numbers to a peak of over 500,000 in 1970. By 1973 there were 3,900,000 *Gastarbeiter* in West Germany, three-quarters of them men. The majority of these workers came from five countries, Turkey with over 800,000 was the largest contributor, followed by Yugoslavia, Italy, Greece and Spain. The balance between the various countries of origin has changed somewhat over the years. Initially Italy was by far the largest source but as it too has become more affluent, the countries of southeastern Europe have gradually grown in importance.

It is often argued that a major influence behind this immigration is the freedom of movement guaranteed to workers under the terms of the E.E.C. Rome Treaty, but there is little evidence to substantiate this view (Feldstein, 1968). There was considerable movement of labour in Europe well before the E.E.C. was formed and there is no evidence now to suggest that movement is in any way restricted to member and associate member countries. Economic necessity both in West Germany and the supply countries undoubtedly seems to have been the driving force. In many ways both sides have benefited greatly, but everything has its price. Initially it was never envisaged that these workers would be a permanent section of the labour force, but the realization is slowly growing that most will probably stay permanently. It is a development requiring considerable psychological readjustment on the part of most West Germans. The provision made for the newcomers in terms of housing and social services is inadequate and compares unfavourably with the care and attention lavished on the refugees a decade previously (Burtenshaw, 1974; Hallett, 1973).

It seems unlikely that the upheaval caused by slow, or even negative, population growth and higher standards of living is yet anywhere near its end. A combination of lower birth rates, earlier retirement, the larger numbers in the older age groups, and a longer period spent in full-time education are going to make labour shortages a chronic problem for the foreseeable future. West Germany will have to think beyond its domestic economy for expansion, if the economy as a whole is not to suffer. The losses of capital investment overseas after the two World Wars have made industrialists reluctant to commit resources abroad, but this attitude will have to change if the labour problem is not to become acute.

It is not surprising that a redistribution of population has accompanied such fundamental structural changes. Immediately after the end of the Second World War the population was widely dispersed for what was basically an urban-industrial nation. There were a number of reasons for this, among them the unusually large agricultural labour force (22 per cent in 1950) and the large number of refugees from the East, who initially settled in rural areas such as Schleswig–Holstein, Niedersachsen and Bayern. Most important of all, however, was the wholesale destruction of urban areas, which had resulted from the Allied air offensive. More than 20 per cent of the total housing stock was uninhabitable in 1945 and in general the seriousness of the situation grew with the size of the town (Schöller, 1967). Over half the houses in the large cities were destroyed, a third of those in medium-sized ones and a quarter in small towns. The effect was not only to disperse the population to rural areas, but also towards the smaller towns. Since 1950 the population has been slowly regrouping in the major urban areas, particularly those in the southern half of the country such as Munich, Frankfurt, Stuttgart, Mannheim and Nuremberg. About 50 per cent of West Germans now live in 24 agglomerations, the bulk of them in the Rhine–Ruhr area. As a result the population is now mainly concentrated in a broad arc stretching from Brunswick on the East German border in the north-east, through Hanover, the Rhine–Ruhr conurbation and on to Mainz, Frankfurt, Ludwigshafen–Mannheim and Karlsruhe, before swinging east again to Stuttgart and Munich (Figure 8.2).

The renovation of the major urban areas has been one of the outstanding achievements of West Germany, for not only were most of them severely damaged, many like Hamburg, Erfurt, Kassel and Lübeck were cut off from their traditional hinterlands by the partition of the former Third Reich (Blacksell, 1968; Holzner, 1970). In the early stages of reconstruction, provision of housing took precedence and led to a very rapid extension of the physical size of towns and cities. There was a tendency to choose sites away from the damaged areas, not only because they were cheaper, but also because the state of land holdings in the bombed city centres was so fragmented and confused that it was often impossible to get land released for building. There was also the belief, which had underlain the German approach to town planning as far back as the Weimar Republic, that it was ultimately desirable to separate workplace, residence, tertiary activities and recreation within the urban area (Schöller, 1967). In the 1950s much of the new residential accommodation was of poor quality and design, but subsequently both have rapidly improved and the West Germans are now probably the best housed nation in Europe (Hallett, 1973). This has been achieved by adopting an extremely flexible approach to housing finance, which has prevented the unfortunate polarization that has occurred in the United Kingdom, whereby middle-class housing is financed privately and working-class housing by the state.

After 1955 more attention began to be paid to the fabric of the inner-cities. Slum clearance and general urban renewal, allied to some imaginative city-centre redevelopment schemes, began to transform the face of cities like Hamburg, West Berlin, Cologne, Frankfurt, Stuttgart and Munich. One particular feature was the way many cities adopted trams as the main medium for public transport. The tram was traditionally important and its role has been modernized by underground termini and good links with the other forms of public transport. As will be seen later, the town planning process is more decentralized in West Germany than in almost any other West European country, and the organization and much of the finance for this massive urban renewal programme came from the *Länder* and the cities themselves, but latterly the Federal government has also played an important role (Boudeville, 1974). Urban renewal has been actively encouraged and money diverted to the *Länder* for this specific purpose. As a result, although West Germany has relatively few very spectacular pieces of urban redevelopment, the general standard is extremely high and its urban problems are more tractable than those of any other European country.

Patterns of Movement

For the past twenty-five years the pattern of communications in West Germany has been undergoing an enforced period of radical readjustment. In 1948 the new nation

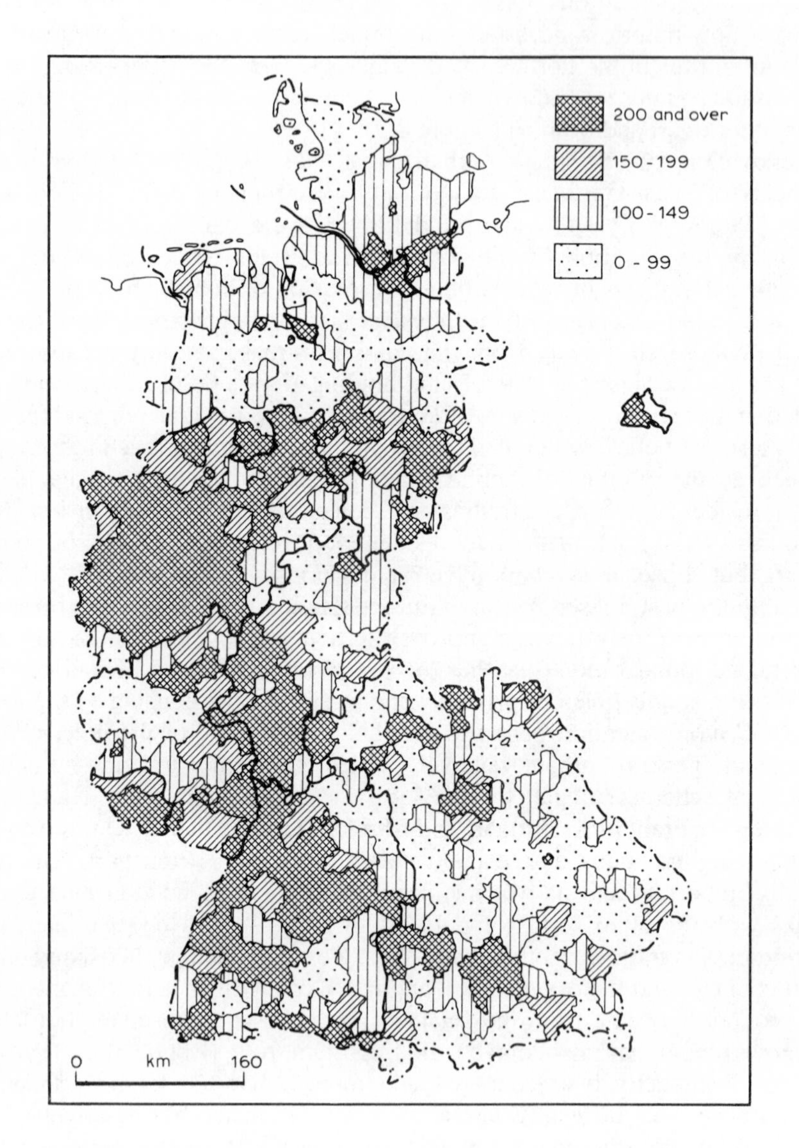

Figure 8.2. West Germany: population density per square kilometre.

inherited a badly damaged network, dominated by the two giants of nineteenth century industrialization, railways and canals, and with the bare rudiments of a modern road system in the shape of the pre-war motorways. The whole system revolved around Berlin, the capital city of three successive German empires and the natural focal point of a nation linking Eastern and Western Europe. Now, of course, the political situation has been transformed: Berlin is a divided city and its western half is a terminus and beleaguered outpost of West Germany, rather than one of the key nodes in the European transportation system. Even those connections which have been maintained are slow and of poor quality in comparison with the internal network within West Germany, for the reorientation has been accompanied by several revolutions in transportation technology. The traditional carriers have been speeded up and had their carrying capacities greatly enhanced and new media, in particular aircraft and pipelines, have assumed a more important role in the movement of both goods and people.

No single centre inside West Germany has assumed the dominant role of pre-war Berlin. The network is now truly polycentric (Figure 8.3) based on several nodes of which the most important are Hamburg, Hanover, Frankfurt-am-Main, Cologne, Munich and Nuremberg. The absence of any preeminent centre is yet another indication of the strength of federation and the importance of the *Länder*; no one city has been allowed to extend its influence to the point where it could completely overshadow the major urban areas in other parts of the country.

The major task in remodelling the transportation network has been to replace the major pre-war axis which ran from west to east, across the North German Plain between the Rhine–Ruhr conurbation and Berlin, by a north to south one based on the Rhine and its tributary valleys. The great need was to draw the southern and southwestern *Länder* of Bayern, Baden–Württemberg and, to a lesser extent, Rheinland–Pfalz into the mainstream of urban-industrial development. Subsequently there has been added the secondary objective of improving transportation links with the E.E.C.

The most flexible and pervasive part of the communications system is the road network. In 1973 there were 166,683 km of classified roads covering every part of the country, and of these more than 10 per cent conform to the rigorous standards laid down by the United Nations Geneva Plan (1949) for a European road network. Pride of place in the network is accorded to the motorways, 5258 km of which are already in service and a further 683 km in the construction or planning stage. It is by far the largest national motorway system in Western Europe, not only in its physical extent, but also in the role which it plays in the general economic development of the country. Motorway construction was the most rapid way of realizing the much-needed north to south axis described above and its success encouraged both Federal and *Land* governments to use the same expedient for opening up other isolated areas, notably the Saarland and Hessen. Recently they have been cast in a rather different role, speeding up intra-urban communications. In the 1930s an urban motorway, called the Ruhrschnellweg, was built through the heart of the Ruhr industrial area and the approach is to be extended, both within the Ruhr itself and in other conurbations. It is a solution to urban congestion, which is hotly disputed elsewhere in Western Europe (cf. the London motorway system); but West Germany appears to have no such qualms about the drastic surgery which urban motorway construction requires.

The other main element in the official drive to restructure communications has been

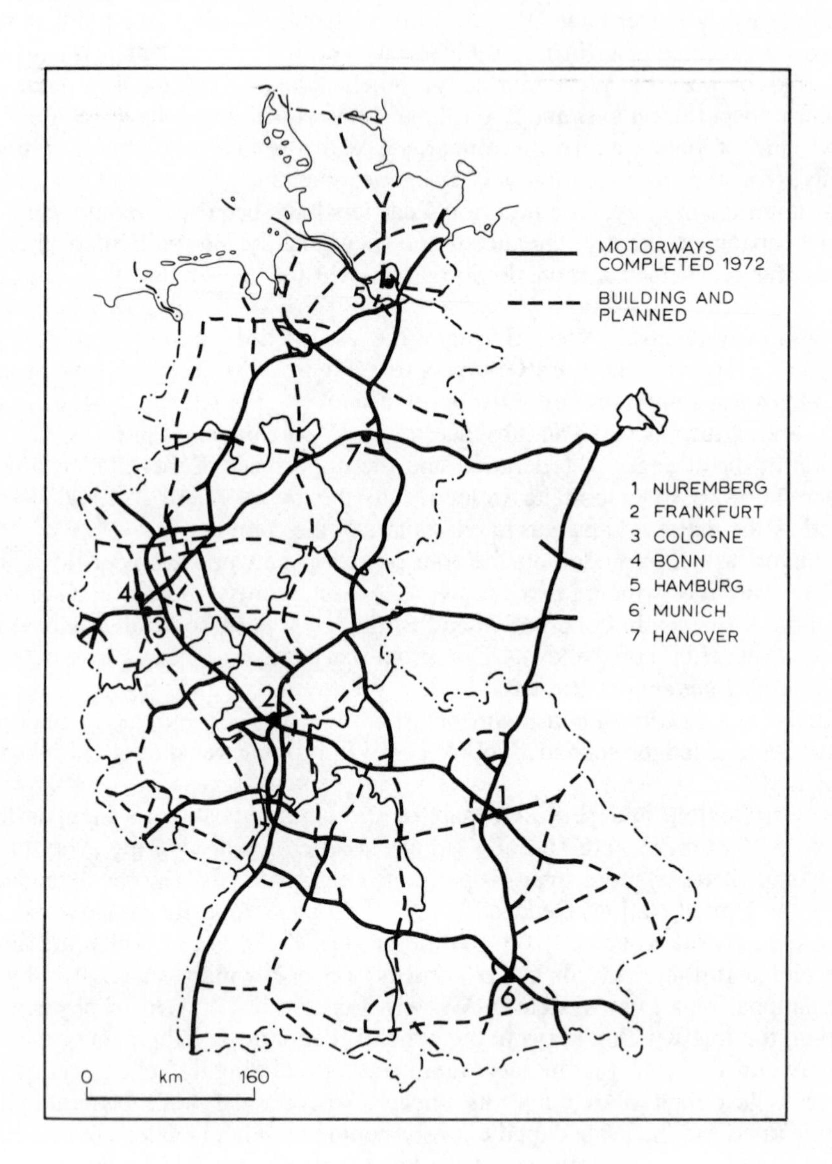

Figure 8.3. West Germany: motorways, 1972.

the renovation and revitalizing of the inland waterways. The Rhine has been a major through route for centuries, but Mannheim was the effective head of navigation and only in this century was it extended to Basel on the Rhine and to Aschaffenburg and Herlbroun on the Rhine's major right bank tributaries the Main and Neckar. Since the Second World War further massive improvements have been made and the Neckar is now navigable for standard 1350 ton barges above Stuttgart. The developments planned for the Main are even more ambitious; it is hoped that a major link will be constructed capable of taking standard barges through to the Danube. The new canal is already open as far as Nuremberg and is due to be completed to the Danube at Regensburg by 1981. Of the left bank tributaries of the Rhine, most of the investment has been ploughed into improving navigation on the Moselle. In 1964, as the result of a scheme sponsored in part by E.C.S.C., standard barges were able to get as far upstream as Thionville in France, thus linking the Lorraine iron ore field with the Ruhr, and subsequently navigation has been extended as far as Frouard north of Nancy. Construction is well advanced on a further section to Toul and is promised to Neuves Maisons and then possibly to the Rhône. Amid all this activity some of the northern waterways have declined somewhat in relative importance. The Mittelland canal now takes little traffic east, beyond Brunswick, and the Rhine–Herne and Dortmund–Ems canals have also attracted little new investment. Nevertheless, all three are still vital links in the Ruhr, which is still West Germany's major industrial region.

The railways are in a similar position. They still carry about 40 per cent of the goods traffic, much more than any other form of transport, and their share shows no real sign of declining. At the same time, they require massive and continuous public subsidy to remain viable and the size of the network, both in terms of track and manpower is constantly being pruned. The pattern of decline is common to all the European industrial nations, but there are signs that, in West Germany at least, the railways are effecting a comeback. New inter-city routes, linking all the country's major cities, have been initiated with great success and more are planned. Expresses already link Hamburg to Munich, Hamburg to Basel, Bremen to Munich, and Hanover to Munich, and others joining Kiel, Regensburg and Passau to the system are planned. Piggy-back wagons which carry long-distance lorries are being used in an effort to increase the railway's share of goods traffic as well.

The most dramatic new development in transportation has, however, been the rapid rise in importance of pipelines. From virtually nothing in 1960, they handled 8·3 per cent of internal traffic and 21·4 per cent imports by weight in 1971. This transformation is of course closely tied to the growth in the use of oil, both as a source of energy and as a raw material, but the effect on transport has been profound. Coal from the Ruhr used to be one of the mainstays of inland waterway traffic, but the increasing importance of oil has reduced the scale of this traffic markedly. Equally pipelines have prevented the traditional goods carriers from picking up the new oil traffic. Geipel (1969) has shown that 79 per cent of oil movements in West Germany are by pipeline and their use is rapidly being extended to other products. In 1968 for example at the Chemischen Werke Hüls A.G., in the new Ruhr town of Marl, 47 per cent of raw materials were received and 10 per cent of products were dispatched by pipeline. Some 21 separate plants were interconnected to the 677 km grid and 21 different materials were moved by it. In addition to being cheap, pipelines have brought a new flexibility to industrial location, the effects of which are already becoming apparent.

Competition between the different forms of transport has always been fierce and the present is no exception, but in West Germany, where economic growth has been generally fast and continuous for thirty years the ill-effects have been minimized. Railways and canals have been able to maintain their volume of traffic, while road transport and pipelines have increased their relative share. By and large, government attempts to manage transport have met with little success, although recent tax changes do seem to have won some traffic back to the roads from the railways. Such action has long been demanded by the railways. They argued that it was impossible for them to compete with road traffic and pipelines because, whereas the latter could concentrate on the most lucrative goods and customers, they had to provide a service and therefore were always at a disadvantage. It is a well known complaint, but in the climate of free competition demanded by the rules of E.E.C. it is difficult for governments to take effective action without infringing the Treaty of Rome.

Patterns of Production

Throughout the first 150 years of industrialization, the Ruhr was the undisputed heart of the economy, but since about 1958 its preeminence has begun to be challenged and there have been some profound changes in the distribution of economic activity. Bartelmus (1970) has divided West Germany into five major regions for the purposes of industrial comparison. Using industrial employment as an indicator, he demonstrates convincingly the relative decline in Nordrhein–Westfalen and southern Niedersachsen, and the relative growth in Hessen, Baden–Württemberg and southwest Bayern. The former, his north-central region, dropped from having 45·9 per cent of the nation's industrial employment in 1954 to 40 per cent in 1966, while the latter, his south-central region, rose from 31·4 per cent to 34·6 per cent over the same period. Subsequently the trend has become even more pronounced, although it must be remembered that the north-central region still remains far and away the most powerful. The other major regions have been static, absorbing their fair share of the general economic expansion, but not altering their relative position. The decisive shift in the balance of industrial power from northern cities such as Essen, Dortmund, Duisburg and Brunswick to Munich, Frankfurt, Karlsruhe, Nuremberg and Stuttgart in the South has been caused by a combination of factors. The new national boundaries and federal political structure; the decline of coal in the face of oil imports; technological advances; the rise in living standards: all have contributed to the new order. The more important of these influences will now be considered in greater detail, so as to demonstrate their impact on different sectors of the economy.

Energy Supply

Industrial vigour in West Germany has traditionally gone hand-in-hand with abundant supplies of cheap, domestically produced energy in the form of coal. In the mid-1970s this is no longer the case. Oil has become the chief source of energy (Table 8.1) and the advantages first of domestic supplies and, more recently, cheapness have been surrendered. The change has had a fundamental impact on the distribution of industry, destroying one of the basic attractions of coalmining areas like the Ruhr and the Saar, and more changes are bound to occur before the stability of the coal-dominated era is achieved once more.

Table 8.1. West Germany: Energy Production from Primary Sources, 1960 and 1972, (%)

	Hard coal	Brown coal	Oil	Natural gas	H.E.P.	Other	Total production[a]
1960	59·7	14·2	22·5	0·5	3·0	0·1	213,219
1972	23·8	8·7	55·3	8·9	3·0	0·3	355,086

[a] In thousand tons coal equivalent.

The vicissitudes of the energy market have been, to say the least, dramatic. In 1945, 35 per cent of German coalmines were either severely damaged or destroyed as a result of the war and, in a world desperately short of all kinds of energy, top priority was given to restoring production to the pre-war level. Money from the European Recovery Programme was poured into re-equipping the pits, miners were allotted extra rations and their houses were the first to be rebuilt. The urgency behind the setting up of the E.C.S.C. lay largely in the fact that it was seen as a means of facilitating the distribution of West German coal within Europe, while at the same time satisfying the strategic restrictions placed on production by the Allies. Despite all the efforts, however, demand still outran supply and West Germany, like all the other members of E.C.S.C., became a net importer of coal, most of it from the U.S.A. The chronic shortage lasted until 1956, when, quite suddenly, it began to turn into a surplus. The cost of shipping a ton of coal from the United States to Bremen fell from $16·31 to $2·94 between 1956 and 1966 (Spelt, 1969) and American imports began to undercut German supplies in both domestic and foreign markets. Huge stockpiles accumulated at most West German pitheads and the High Authority of E.C.S.C., which had put out frequent dire warnings about a continuing energy shortage, began to face up to the implications of a surplus and cut-throat competition for markets.

Cheap coal was only the first symptom of the glut, the underlying cause was oil. The Suez crisis encouraged the major oil companies to bring fields outside the Middle East into production and the price of oil started to fall sharply. The capacity of coastal refineries at Emden, Bremen and Hamburg was increased, and new refineries, fed by pipeline from Rotterdam and Wilhelmshaven, were built on the lower Rhine at Dinslaken and Cologne. By 1960 oil already satisfied more than 20 per cent of the energy market and was competing with coal on its own doorstep in the Ruhr conurbation.

Initially therefore the advent of oil only tended to confirm existing patterns of industrial production, but it was quickly realized both by the major oil companies and the more rural *Länder* that if oil could compete effectively in the Ruhr, its impact elsewhere in the country could be even greater. The first reaction was to extend the North Sea pipelines further south and the construction of a huge refinery near Frankfurt commenced in 1962; but soon the advantages of linking southern German cities to terminals on the Mediterranean coast became apparent. Pipelines connecting Marseilles with Karlsruhe and Mannheim, and Genoa and Trieste with Ingolstadt decisively tilted the balance of industrial power towards the South. Not only was cheap energy now available in Bayern and Baden–Württemberg, it cost the same, or even less than in

Nordrhein–Westfalen. Despite the heavy initial investment required for pipeline construction and the reluctance of oil companies to invest in any but the most profitable sectors, all the *Länder* wanted a share in the bonanza. By 1968 every *Land*, with the exception of West Berlin, had its own refinery and, more important, their associated growth industries (Burtenshaw, 1974).

Any change of such magnitude has its price and the oil boom was no exception. By 1972 oil supplied well over 80 per cent of West Germany's energy requirements and of the 110,000,000 tons consumed, only 5 per cent came from domestic oil fields. It is a situation with many new problems and dangers, neatly summed up by Geipel (1969, p. 123), who said, 'It is clear that the severe devaluation of coal by oil . . . has put the whole West German energy economy at the mercy of world market prices'. In 1974 this is almost a truism, but there were few expressing such fears in 1969. In fact, the worst effects of the rapid rise in oil prices have still to be felt in West Germany, because of its embarrassingly large balance of payments surplus. Indeed the government has been able to use the rapid increase in the cost of oil to resist pressures for revaluation. Nevertheless, it is still the largest consumer of oil in Western Europe and, should there be any weakening of the economy, the effects of increased oil prices will be immediately apparent.

Perhaps the most striking feature of the mass transfer from coal to oil has been the way the rapidly expanding economy took up the slack, caused by unemployment and poor investment in the mining regions. Coal production fell from a peak of 151,300,000 tons in 1956 to 102,500,000 tons in 1972, the labour force has been reduced by more than two-thirds, to less than 170,000, and the number of mines cut by more than half. The worst effects have been felt in the Ruhr, where many of the historic coalmining towns like Bochum no longer support a single productive pit. Yet new jobs and investment have been found and all without the help of a coherent national energy policy, or much effective guidance, certainly in the early years, from the E.C.S.C. The Federal government's belief in the efficacy of the social market economy as the ultimate cure for such ills remained unshaken, and the E.C.S.C. initially had no machinery for dealing with the problems posed by a declining demand for coal. Its main tasks had always been to increase European production and improve distribution in the face of a general shortage.

Nevertheless, the government did make some concessions, bowing to pressure (not least from the ballot box: 30 per cent of the population live in Nordrhein–Westfalen and in 1972 80·4 per cent voted for S.P.D., the highest proportion of any of the *Länder*) and introducing a series of measures to help protect the coal industry. Tax concessions to the industry itself, financial inducements to power stations, which would guarantee to use coal as their main source of energy for a ten-year period, levies on imported coal and restrictions on the level of oil stocks have all been tried, but the decline has continued. The Federal government now hopes that it will be able to hold production at about 100,000,000 tons a year and much of their financial help and the help provided by E.C.S.C. is aimed at easing the transition. Since 1963 there has been a Federal rationalization agency, the purpose of which is to encourage the closure of uneconomic pits, and E.C.S.C. has provided redundancy payments and retraining grants to displaced mineworkers.

Other sources of energy, although of local importance, pale into insignificance by the side of coal and oil (Table 8.1). Lignite is mined extensively at several locations, notably Ville, south of Cologne, but in 1972 production was only equivalent to

28,700,000 tons of coal. Even so, the low costs of strip mining and its usefulness in the chemical industry has meant that production has increased slightly throughout the 1960s. Natural gas production from domestic wells, mainly in the Weser–Ems region of Niedersachsen, has also increased rapidly and it is expected that West Germany will be able to satisfy a much larger proportion of its domestic demand in the foreseeable future. As with oil, the ease with which gas can be transported by pipeline has ensured its wide distribution throughout the nation. Hydro-electric power on the other hand has not enjoyed the growth that might have been expected a generation ago: production has remained more or less constant and its share of the market (7 per cent in 1972) has steadily declined. Atomic energy too has had little impact, owing to the technological problems faced by the industry throughout the world and to a reluctance, both within West Germany and among the nuclear powers, to see heavy investment in nuclear energy in West Germany.

Manufacturing

Manufacturing is West Germany's lifeblood, it employs nearly half the workforce and, in 1972, made up 86 per cent of the country's exports. All the major branches are strongly represented and, almost without exception, have flourished in the past two decades. Mechanical engineering with over 1,000,000 employees and a turnover in excess of 61,000,000 DM leads the field, followed by the electrotechnical, chemical, vehicle and iron and steel industries (Table 8.2). There have however been a number of important changes in both the relative importance of the different branches of manufacturing and their spatial distribution. Broadly speaking, the coal-based industries, such as iron and steel and some parts of the chemical industry, have lagged behind high technology and consumer goods industries, like electronics and vehicles. The contrast has been made all the more stark, since much of the expansion in the latter has occurred at new locations away from the traditional centres of manufacturing.

Table 8.2. West Germany: Manufacturing Industry (Leading Sectors), 1970

	Turnover (million DM)	Employees	Firms
Engineering	71,210	1,176,000	4,611
Food	68,627	530,000	4,243
Chemicals	60,826	642,000	1,727
Electrotechnical	59,538	1,131,000	2,227
Vehicles	48,616	631,000	473
Iron and steel	42,837	546,000	549
Oil refining	31,299	48,000	75
Textiles	26,753	499,000	2,840

The iron and steel and chemical industries clearly illustrate the kinds of pressure that some of the traditional branches of industry have had to face. In spite of structural changes in manufacturing, iron and steel remains one of the cornerstones of the economy. The underlying trend in the production of both pig iron and steel has been upwards throughout the 1960s and, at 32,000,000 tons and 43,000,000 tons respectively in 1972, output is almost as high as it has ever been. Yet the industry has survived not only wholesale destruction through bombing in the war itself, but also a concerted effort by the Allies afterwards to ensure that it would never again dominate the German economy as it did during the Third Reich. They pursued a vigorous programme of decartelization to break up the monolithic organization and created instead a large number of small, independent production units, most of which proved hopelessly uneconomic. With the help of E.C.S.C. and a gradual relaxation of Allied control, the last twenty years have seen economic factors inexorably forcing a slow re-amalgamation on the industry. Vertical integration of production between companies, encouraged by the High Authority of the E.C.S.C., was dictated by the extremely competitive world market for iron and steel and by the huge sums required for investment. New investment was particularly crucial in the late 1950s and 1960s with the arrival of the Linz–Donowitz oxygen steelmaking process, which had the capacity to dramatically improve efficiency, but was extremely expensive to install. Slowly wasteful arrangements, such as existed in Dortmund where the pre-war works was split into three, were eliminated. There is now a single company in Dortmund, the Dortmund Hoerde Hoesch Hütten Union, and recently it has been further enlarged by joining with the Royal Netherlands Blast Furnaces and Steelworks, to form the Royal Hoesch Union. Similar national and European amalgamations have occurred among other companies as well. Since 1967 these moves have been paralleled by the establishment of four marketing consortia covering every plant in the country. The consortia were formed at the instigation of E.C.S.C. to reduce potentially ruinous internal competition, but the marketing brief was found to be too narrow and, in 1971, they were reformed with the wider brief of rationalization.

The Allied programme of decartelization did have some lasting impact, in particular in the stimulus it gave to diversification. Most companies widened the scope of their activities to promote the growth denied them through vertical integration and most West German steel companies are now very broadly based. In some cases, and Krupp in Essen is the prime example, they have abandoned basic iron and steel production almost entirely and become large and varied heavy engineering concerns.

An important aspect of the post-war iron and steel industry has been the relative minor changes in location: Duisburg on the Rhine at the western end of the Ruhr has remained the undisputed heart of the industry and the Saar, Dortmund, Salzgitter, Osnabrück and Siegen, the traditional locations, are still centres of production. It seemed that the pattern might change radically in favour of coastal sites, when overseas ores began to supersede those from Lorraine, but movement would appear to have been something of a short-term aberration. A coastal plant was built at Bremen and opened in 1955, another is being constructed at Hamburg, but a similar project at IJmuiden in the Netherlands has run into difficulties and the company concerned, the Royal Hoesch Union, now plans to concentrate new investment at its main site in Dortmund. As far as West Germany is concerned talk of a migration of iron and steel works to the coast is premature (Fleming, 1967; Warren, 1967).

In the chemical industry decartelization and post-war restrictions on production led to a totally different outcome. At the time of the Third Reich, Germany was the leading producer in the world of chemical products, but after 1945 it was prevented by the Allies from manufacturing most of its basic products, such as synthetic rubber and oil from coal. With almost unbelievable resilience the industry has re-established itself with a completely new range of products mostly derived from imported oil. Ever since oil became abundant after 1958, the industry has been one of the leading sectors of the economy and artificial fibres and textiles have been the key to its success. As with iron and steel the basic location of the industry has not changed significantly: three main areas, the western Ruhr at Marl and Gelsenkirchen, the middle Rhine at Leverkusen and Wesseling, and Hoechst near Frankfurt still dominate, although the last two have grown recently at the expense of the Ruhr. It seems certain however that changes will have to come before very long. The modern chemical plant requires abundant land, a commodity that is both expensive and in short supply near traditional locations in the major urban areas. There is also increasing concern about the dangers of explosion and pollution and these fears have not been helped by the interest the industry is taking in atomic energy as a power source. Since the transport of the main raw material, oil, by pipeline is relatively easy and since economies of scale seem destined to make small plants uneconomic, the future would seem to rest with a small number of large complexes, sited where there is ample land for expansion.

Historical constraints have had a much lesser impact on the recent explosive growth of the electrotechnical and motor industries. Both have been able to turn the greater flexibility in energy supply to their advantage and to develop new locations, which take full advantage of untapped manpower resources and local and federal government grants. In the 1950s the motor industry, now the largest exporter of cars in the world, was concentrated in three zones: Volkswagen at Wolfsburg in the north-east, Opel and Ford in the Rhine–Main area, and a scatter of smaller firms in the south. Since then a disproportionate amount of the growth in the industry has been siphoned off by the southern *Länder*, Bayern and Baden–Würtemberg. BMW at Munich is now the fastest growing, though by no means the largest, of the automobile manufacturers and Volkswagen, Opel and Ford have all set up assembly plants in the south. In the electrotechnical industry the pattern has been similar and in both cases the availability of labour has been one of the main inducements. In fact employees in the southern cities are now as expensive and harder to find than in the Ruhr, but the electrical industry in particular has been very successful at tapping the resources of the rural areas, in Bayern, Baden–Würtemberg and Hessen. Krümme (1970) has shown how the electrical firm of Siemens used Munich as a base to expand into nine other small towns within a radius of 100 km of the city. They also had a complex private transportation system for bringing workers into their factories in Munich itself. These are the developments that have tended to tilt the industrial balance more in favour of the southern *Länder*. The movement is not enormous, but it has produced a more even spread of industrial investment than ever before in Germany, thus counteracting the overwhelming dominance of the Rhine–Ruhr conurbation.

Agriculture

In stark contrast with almost every other branch of industry, the record of West German agriculture has been poor. The organization is antiquated and productivity per employee low. Mayhew (1970, p. 54) summed up the overall situation when he said: 'Agriculture in the Federal Republic of Germany is inefficient in its use of land, capital and labour'. The root cause of this inefficiency is the proliferation of small holdings, rather than any inherent reluctance on the part of the farming community to accept mechanization and other innovations. In 1972 44 per cent of the country's farms had fewer than 5 ha and 83 per cent had less than 20 ha. Nearly half the 12,700,000 ha in cultivation was in holdings of less than 20 ha and only one seventh was in units larger than 50 ha. The problem of small uneconomic farms is compounded by their uneven distribution (Table 8.3): they are particularly heavily concentrated in Rheinland–Pfalz, Baden–Württemburg, western Bayern and Hessen. Such a lopsided regional distribution means that agricultural incomes, which everywhere are low, also vary considerably. Absolute physical size is only one of the structural problems. Many of the holdings are fragmented into widely scattered parcels of land and farm buildings are packed into villages and separated from the farm proper, making the efficient use of labour and capital even more difficult.

Table 8.3. West Germany: Employment in Agriculture, 1970

	Thousands	Percentage of workforce		Thousands	Percentage of workforce
Baden–Württemberg	331·8	7·9	Niedersachsen	327·3	10·9
Bayern	646·9	13·2	Nordrhein–Westfalen	241·5	3·5
Berlin(W)	5·0	0·5	Rheinland–Pfalz	163·7	10·7
Bremen	5·0	3·5	Schleswig–Holstein	97·2	9·4
Hamburg	10·1	1·2	Saarland	9·6	2·4
Hessen	152·4	6·3			
			West Germany	1,990·5	7·5

The poor structure inevitably means that the industry is chronically over-manned. In 1972 7·4 per cent of the workforce was employed in agriculture compared with under 3 per cent in the United Kingdom. Until very recently the proportion was very much higher: in 1950 it stood at 23·2 per cent and even as recently as 1965 was still over 10 per cent. Once again the national average disguises the true picture (Table 8.3), for over 13 per cent of Bayern's labour force is still engaged in agriculture and in both Rheinland–Pfalz and Niedersachsen the figure is over 10 per cent. The farming population is also widely distributed among a large number of small settlements, making cooperation and coordination difficult. Many rural communities are poorly served by public utilities, in particular water supply, making the introduction of modern techniques, such as irrigation, difficult. The wide distribution of the farming population, together with the large variety of different types of production, also compound the problems of concentrating the industry in the more productive areas in southern Bayern, the Rhine valley and the loess *Börde* lands at the southern edge of the North German Plain. Farming rotations and techniques vary enormously, making it virtually impossible to view the industry as an entity that can be planned as a single unit.

Despite these daunting problems, agriculture is very important for the national economy. West Germany still produces nearly 50 per cent by value of its own foodstuffs and the proportion has been rising gradually. However, rapid world inflation is threatening to erode the balance, placing an even higher premium on an efficient and prosperous domestic industry.

Both Federal and *Länder* governments have assumed a responsibility for helping to support agriculture and have recognized the need for continuous financial support. The legal basis for this is the Land Consolidation Act (*Flurbereinigungsgesetz*), passed in 1953, which specifies the roles of both government and the farmer in a national agricultural development plan. In 1954 the first federal aid was forthcoming and since 1956 the details of agricultural assistance have been published annually in the so-called Green Plan, a statutory document, which must be published annually under the terms of the 1956 Federal Agricultural Act.

Federal support alone is currently over 2·5 billion DM annually and takes a number of forms. Nearly half the money is spent on measures designed to immediately raise agricultural incomes, such as subsidized fertilizers, fuel oil and machinery and improved marketing organizations. A further 10 per cent is absorbed by credit schemes, providing cheap loans at fixed interest rates over long periods, and a similar sum is taken by pension schemes, grants for retraining farmers who leave the land and gratuities for early retirement.

In terms of the actual structure of agriculture in West Germany, however, the most important subsidies are those aimed at land reform and the improvement of living and working conditions in the industry. The schemes take a number of forms, but the most important has been land consolidation (*Flurbereinigung*). The purpose of land consolidation has been to merge scattered and fragmented holdings, the result of medieval farming practices and laws of inheritance, into larger viable units. The rate of progress has varied, but has rarely exceeded 300,000 ha annually and recently the pace has begun to slow down (278,000 ha in 1971). Since it is conservatively estimated that there are over 5,500,000 ha which need to be consolidated for the first time and a further 2,500,000 ha where work needs to be repeated in the light of more recent thinking on optimum farm size, it is clear that the problem is beyond the scope of the present measures. It is interesting to note that more than half the land requiring attention is in the three northern *Länder*, Schleswig–Holstein, Niedersachsen and Nordrhein–Westfalen and not in the more agriculturally backward south. The explanation of this apparent discrepancy is that these areas were first in the field with their land consolidation programmes in the 1950s and, in general, the scale of their schemes was too small to be really effective (Mayhew, 1970, 1971).

The two other major elements of structural improvement have been the relocation of farmsteads away from congested settlements (*Aussiedlung*) and the purchase of land for distribution among farmers, so as to increase the absolute size of holdings (*Aufstockung*). Frequently these measures have been introduced in conjunction with a land consolidation programme, but this is by no means always the case. The impact on the landscape of relocated farmsteads in particular has been considerable. In Baden–Württemberg, Bayern, Hessen and the northwestern part of Niedersachsen, isolated farmsteads have mushroomed in areas where nucleated settlements were formerly the rule.

In addition to structural improvements, strenuous efforts have been made, especially by the *Länder*, to encourage farmers to leave the land. Industry has been attracted to the more rural parts of the country by the surplus labour force and, as was pointed out above, has established new factories in smaller urban centres and made elaborate transportation arrangements to carry people from remote villages to work. The new development has had a profound impact not only on the industries, but also on the rural communities. There has been a proliferation of part-time holdings and an increasingly large amount of land is either being left uncultivated, or planted out to fruit trees or timber, uneconomic in themselves but providing a useful income supplement. The Germans have coined the phrase 'social fallow' (*Sozialbrache*) to describe this phenomenon and it is symptomatic of both the decline in the cultivated area and the growing affluence of rural areas.

The most recent threat to the future of agriculture has been C.A.P., the agricultural policy of the E.E.C. Many of the proposals for structural reform in E.E.C.'s Mansholt Plan are very similar to those already in operation in West Germany, yet this is but a small part of the overall policy. The chief goal is a free market in agricultural goods and the antiquated West German industry is at an immediate disadvantage. Initially government subsidies were sufficient to counter the adverse effect of C.A.P., but after revaluation in 1969 import levies on agricultural produce from within E.E.C. were introduced, dealing a severe blow to the free market concept, which West Germany had so fervently promoted at home and abroad in the post-war era. The prospect of a too rapidly declining agricultural industry was unacceptable, both politically and socially.

Regional Policies

In comparison with many European countries, economic policy in West Germany has been non-interventionist and the bulk of public expenditure has been apportioned without specific reference to regional goals. Nevertheless, ever since the late 1940s, there has been a strong sense of collective responsibility, at both federal and local levels, for areas left behind in the general economic expansion and there have often been implicit regional considerations in patterns of public expenditure. In addition, a whole series of measures have been designed specifically to help adjust regional economic discrepancies. Indeed the long-standing commitment to a uniform standard of living in all parts of the country is perhaps the prime illustration of the 'social responsibility' implied in the phrase 'social market economy.'

Achieving this goal has driven successive governments towards more and more comprehensive regional development policies, covering both larger areas and more people. By 1974, well over a third of the population was directly affected by one or other of the programmes, a proportion comparable with the United Kingdom, where management of regional economies has always been a much more explicit objective. Since 1965, under the terms of the Regional Development Act (*Raumordnungsgesetz*), four types of region have been defined, and it is expected that it will be possible to classify every part of West Germany under one or other of these headings. There are 24 urban agglomerations (*Verdichtungsräume*), existing industrial cities, where further growth will be encouraged in the future. There are rural areas, with a balanced and flourishing economic structure based primarily on agriculture, precise definition of which has yet to take place. Then there are two broad categories of problem region: on the one hand, areas which have fallen behind the general level of development or seem in danger of doing so; on the other, the area bordering on East Germany and Czechoslovakia (*Zonenrandgebiet*), together with West Berlin, where the political situation makes balanced growth difficult. The official hope is that remedial measures will raise standards in the two latter types of region to a point where they can compete on an equal footing with the rest of the country.

An important limitation on the scope of regional policy is the strict division of responsibility between the Federal, *Land* and *Gemeinde* (roughly equivalent to a parish) authorities laid down in the constitution (*Grundgesetz*). In certain spheres, such as town planning for instance, the smallest *Gemeinde* enjoys absolute control over the implementation of policy and the upper tiers of government cannot necessarily wield the financial big stick over them with much effect. Taken as a whole, the budget of the *Gemeinden* is equivalent to 65 per cent of the Federal government's, all of it raised directly by them through local taxation. A similar situation exists between the *Länder* and the Federal government, except that with tax receipts totalling 77 per cent of those collected by Bonn, the position of the *Länder* is even more favourable. Differences in tax revenue do of course exist, but they tend to reflect regional variations, rather than political status. There is a fundamental division between urban and rural areas produced by the Industrial Trade Tax (*Gewerbesteuer*). This tax on production makes up 80 per cent of the total revenue of the *Gemeinden* and it exacerbates the differences between the industrial and non-industrial parts of the country. Rural *Gemeinden* tend to become embroiled in a vicious circle, whereby low levels of production reduce their tax base, which limits the extent they can invest in infrastructural improvements, which in turn impedes their chances of attracting new industrial investment. The disadvantages accruing from this antiquated form of taxation are only partially offset by a system of formal income transfers, which requires the richer *Länder* to give a part of their tax revenue to the poorer ones.

The Federal Regional Development Programme dates back to 1951, but until the Regional Development Act (*Raumordnungsgesetz*) in 1965, there was a certain lack of coordination between its various parts. Basically, however, there were, and remain, four distinct types of development programme: the development areas (*Bundesausbaugebiete*), the frontier zone (*Zonenrandgebiet*) and West Berlin, development centres (*Bundesausbauorte*) and integrated regional plans (Figure 8.4).

The development areas have existed in one form or another since 1951. Originally they were called emergency areas (*Notstandsgebiete*) and were designed specifically to cover those rural areas which had taken the brunt of the influx of post-war refugees, mainly the border zone, Schleswig–Holstein and northern Niedersachsen, and those areas in the extreme west which had suffered inordinate destruction during the Allied invasion. In 1953 they were re-named 'renewal areas' (*Sanierungsgebiete*) and, finally, in 1963 their scope was reviewed and they reemerged as development areas. They are defined on the basis of a range of socio-economic criteria, including the level of industrial employment, out-migration, unemployment and *per capita* social product and are now concentrated in five main areas: Schleswig–Holstein, eastern Niedersachsen from the North Sea to the border with East Germany, western Niedersachsen, a strip along the frontier extending from eastern Hessen into northern Bayern, the Bavarian Forest plus the Eifel. All are rural areas, with high levels of employment in agriculture, and the main emphasis has been on attracting industrial investment through direct cash inducements and improvements in infrastructure.

Since 1967 the scope of the development areas has been further extended to include the depressed hard-coal mining areas of the Saar and the so-called Emscher towns in the northeastern Ruhr, Wanne Eickel, Herne and Castrop Rauxel.

The frontier zone is a corridor, about 40 km wide, which runs the whole length of the eastern side of the country. It was established in 1953, to help compensate for the region's isolation from both markets and raw materials. A large part of the frontier zone is rural in any case and so would have qualified for help as part of a development area, but some sections, especially in the north around Lübeck, Salzgitter, Wolfsburg and Brunswick, are highly industrialized and it was thought that special help was required.

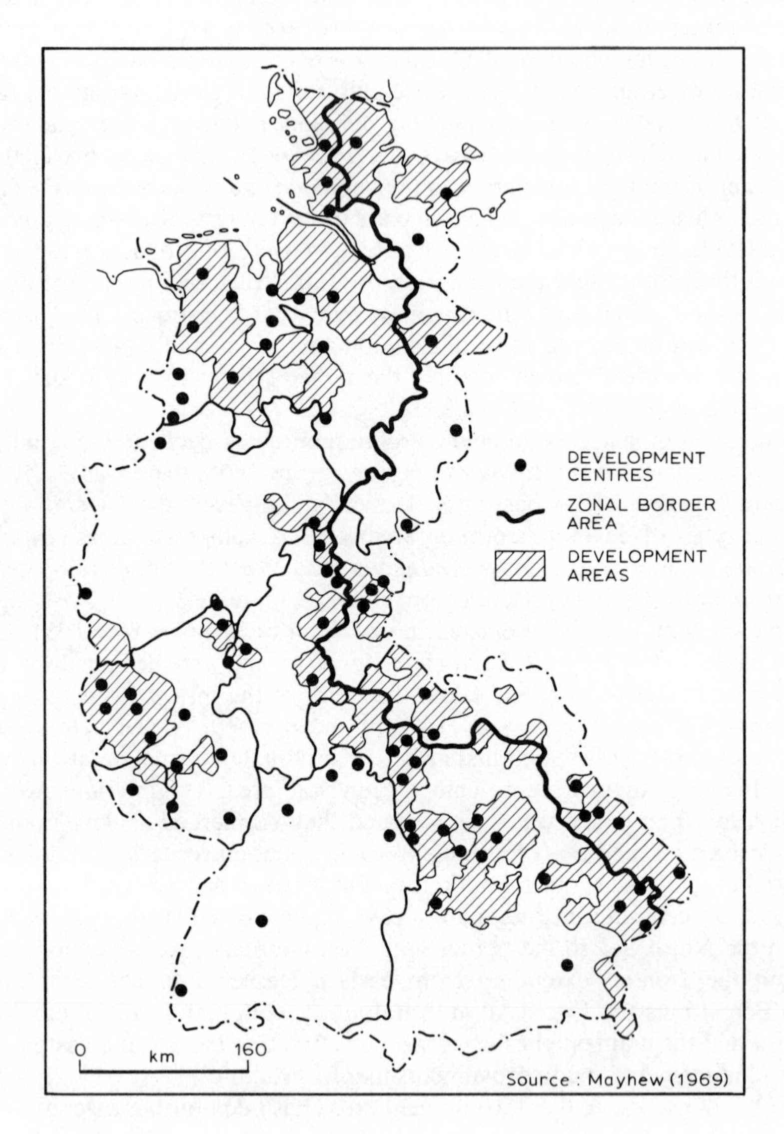

Figure 8.4. West Germany: development centres and areas. (Reproduced by permission of Regional Studies Association.)

At present, with the investment subsidy at 25 per cent, this is the most heavily supported region in West Germany. West Berlin is also included in a similar scheme for the same reasons and both it and the frontier zone enjoy the further advantage of preferential income tax rates.

Development centres grew out of the strong German attachment to applied central place theory in planning. They have very much been a joint project between the *Länder* and the Federal government. The aim is to ensure that the whole country is evenly covered by a fully integrated hierarchy of service centres and the main thrust of the programme has concentrated on small towns in the range 2000 to 35,000. The choice of development centres is based on four main criteria: a high level of unemployment, or underemployment, especially in agriculture; a location giving ready access to the surplus labour force; a minimum of existing social, educational and cultural facilities; and a small amount of established industrial development (Mayhew, 1969). In the first instance, the *Länder* make proposals about which towns should be chosen, and then the list is vetted by the Federal authorities and coordinated with the national list. Nearly 100 development centres have been designated, mostly within, or in close proximity to, development areas; once the structural changes have been achieved, they are declassified and expected to become independent again.

The origin of the integrated regional plans is somewhat different from the other measures, in that they were directly under the control of the Ministry of Agriculture, but in both conception and execution they are clearly a part of the national regional development policy. There are four of these plans, the most important being the Emsland Plan and Programm Nord, started in 1951 and 1953 respectively. The Emsland Plan tried to improve the quality of agriculture in the border region near the mouth of the Ems river. The main stimuli were the unfavourable contrast between the state of farming on the German, as opposed to the Dutch, side of the border, the large amount of land unused because of poor drainage, and the many refugees who settled there after the war. Programm Nord in Schleswig–Holstein was similar, except that refugees did not pose the same problems. In both cases the agricultural objectives were successfully achieved, but, with the Emsland Plan, attracting industry became an important subsidiary aim, and here success was very limited. The region remains predominantly rural and, almost inevitably, lags behind the industrial parts of the nation.

In comparison the Alpen and Küsten plans were much less important. Both were concerned with specific problems: the prevention, in the case of the Alpen plan, of flooding in the Alps and the Alpine foothills and, in the case of the Küsten plan, of marine inundation on the North Sea coast. The objectives were therefore much more circumscribed, but, given the limited brief, they have achieved considerable success.

Since 1969 the whole of regional policy has been undergoing a radical reappraisal in the light of the 1965 Regional Planning Act. In 1969 Regional Action Areas began to be designated under the terms of the 'Joint Task: Improvement of Regional Economic Structure Act', a joint venture between the *Länder* and the Federal Government to further coordinate industrial investment in less developed parts of the country. So far 21 areas (Figure 8.5) have been established and industrial subsidies within them are graded so that development is concentrated into growth poles and the beneficial effects spread to the surrounding area. The level of subsidy varies between 25 per cent and 10 per cent and, since January 1972, it has been virtually impossible to obtain any official assistance for establishing plants outside the official growth areas.

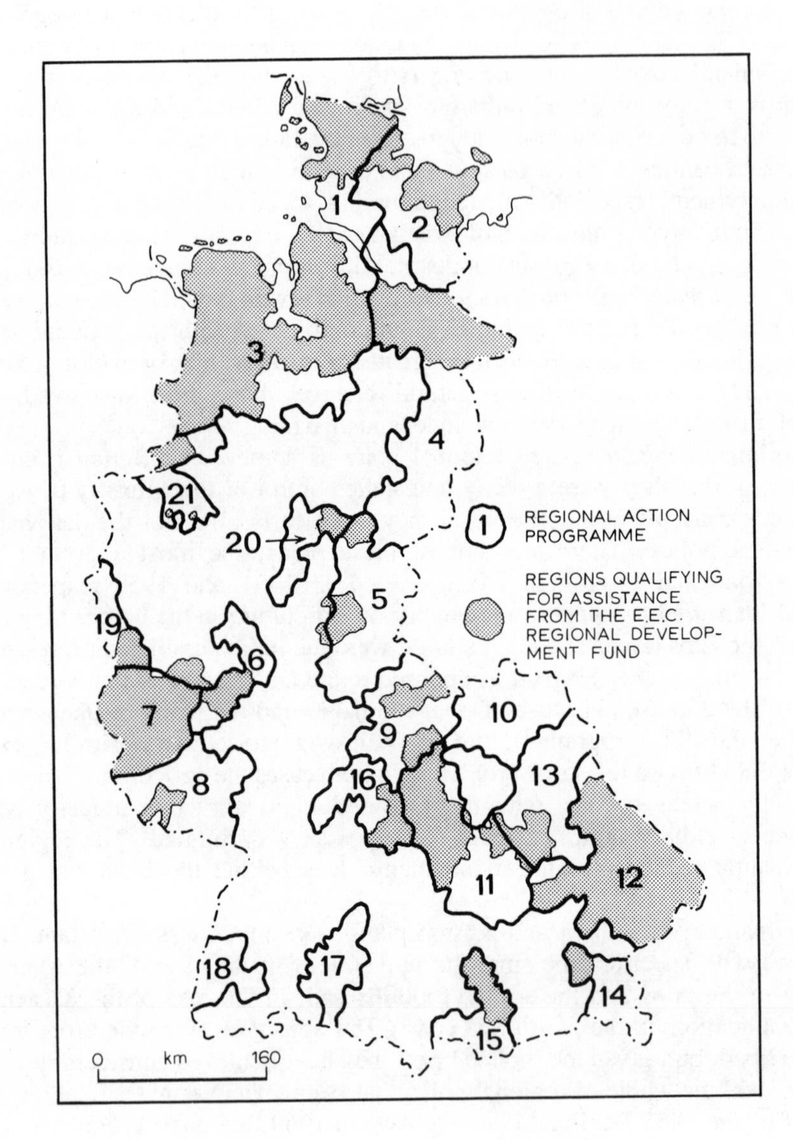

Figure 8.5. West Germany: assisted regions.

Another new element, which seems likely to grow in importance in the next decade, is the influence of the E.E.C. and cooperation with other neighbours on the western frontier. Joint programmes with Switzerland, France, Belgium and the Netherlands were initiated in the 1960s with a view to creating a more coherent economic setting for towns like Freiburg, Saarbrücken and Aachen. In 1973 the E.E.C. published its plans for a Regional Development Fund, in which it defined depressed regions in a Community rather than a national context (see Chapter 1). The findings for West Germany were not dissimilar to the nation's own assessment of her regional problem (Figure 8.5). The only major difference was the more limited extent of the areas which the E.E.C. felt to require assistance, illustrating the small scale of the regional problem in West Germany in comparison with other members of the Community.

To try to judge the success or failure of any regional policy, especially one which has so recently been overhauled, before it has run its full course is a dangerous exercise, but a number of tentative conclusions are emerging. In the south the efforts to encourage industry to expand in rural areas seem to have been a great success from the start, but elsewhere in the country progress was rather slow. However, the new Regional Action Areas have succeeded in spreading the benefits much more widely, so that areas like the Eifel, northern Niedersachsen, Schleswig–Holstein and northern and central Hessen are now beginning to catch up. Nevertheless, the reduced dependence on agriculture is still not necessarily accompanied by real improvements in the structure of farming. The problems of land consolidation seem likely to remain for some time to come, making an efficient, modern industry hard to achieve. Boesler (1974) has also raised two further questions about present regional policy. He is dubious about some of the criteria used to define growth poles and development centres and thinks that some wrong choices may have been made. He also questions the advisability of such an overwhelming concentration on industrial development at a time when service industries are the most rapidly expanding sector of the economy as a whole. Nevertheless, the fact remains that regional disparities are smaller than in any other West European country and it enables West Germany to face such criticism with a certain amount of equanimity.

Conclusion

West Germany has come a long way since 1948. The determination with which the population successfully struggled to reestablish its identity after the war has more recently been directed towards building the kind of society of which it can be proud. Ralf Dahrendorf the distinguished German sociologist and eminent European, in a comprehensive critique of Germany and the Germans in this century has said that 'Germany's curse is not that she did not become a nation, but that she did not become a society' (1968, p. 212). Today she is both: the past quarter of a century has seen the creation of a leading European political force, based on an affluent, egalitarian and, above all, socially responsible society.

References

Bartelmus, P., 1970, Industrielle Strukturwandlungen in Grossräumen der Bundesrepublik Deutschland, *Informationen,* **20**, 443–452.

Beck, R. H., and coworkers, 1970, *The changing structure of Europe,* University of Minnesota Press, Minneapolis.

Blacksell, M., 1968, Recent changes in the morphology of West German townscapes, in R. Beckinsale and J. Houston (Eds.), *Urbanization and its Problems,* Blackwell, Oxford, 199–217.

Boesler, K. A., 1974, Spatially effective government actions and regional development in the Federal Republic of Germany, *Tijdschrift voor Economische en Sociale Geographie,* **65**, 208–227.

Boudeville, J. R., 1974, European integration, urban regions and medium-sized towns, in M.E.C. Sant (Ed.), *Regional Policy and Planning for Europe,* Saxon House, Farnborough, 129–156.

Burtenshaw, D., 1973, Fifty years of the Neckar navigation, *Geography,* **58**, 246–250.

Burtenshaw, D., 1974, *Economic Geography of West Germany,* Macmillan, London.

Clark, C. and coworkers, 1969, Industrial location and economic potential in Western Europe, *Regional Studies,* **3**, 197–212.

Dahrendorf, R., 1968, *Society and Democracy in Germany,* Weidenfeld and Nicolson, London.

Elkins, T. H., 1972, *Germany,* 3rd ed., Chatto and Windus, London.

Elkins, T. H., 1973, *The Urban Explosion,* Macmillan, London.

Feldstein, H., 1968, A study of transaction and political integration: transnational labour flow within E.E.C., *Journal of Common Market Studies,* **6**, 24–55.

Fischer, G. and coworkers, 1971, *Deutschland,* Georg Westermann.

Fleming, D. K. 1967, Coastal steelworks in the Common Market, *Geographical Review,* **43**, 48–72.

Fleming, D. K. and G. Krümme, 1968, The Royal Hoesch Union, *Tijdschrift voor Economische en Sociale Geografie,* **59**, 177–199.

Geipel, R., 1969, *Industriegeographie als Einführung in die Arbeitswelt,* Georg Westermann.

Grotewald, A., 1973, West Germany's Economic Growth, *Annals of the Association of American Geographers,* **63**, 353–365.

Hallett, G., 1973, *The Social Economy of West Germany,* Macmillan, London.

Holzner, L., 1970, The role of history and tradition in the urban geography of West Germany, *Annals of the Association of American Geographers,* **60**, 315–339.

Isting, C., 1970, Pipelines now play an important role in petrochemical transport, *World Petroleum,* **41**, 38–41.

Jensen, W. G., 1967, *Energy in Europe 1945–1980,* Foulis, London.

Johnson, N., 1973, *Government in the Federal Republic of Germany,* Pergamon, Oxford.

Kosinski, L., 1970, *The Population of Europe,* Longmans, London.

Krümme, G., 1970, The inter-regional corporation and the region, *Tijdschrift voor Economische en Sociale Geografie,* **61**, 318–333.

Lindauer, G., 1972, Zum Strukturwandel der ländlichen Gebiete nach dem Zweiten Weltkrieg, *Geographische Rundschau,* **24**, 49–52.

Mayhew, A., 1969, Regional planning and the development areas in West Germany, *Regional Studies,* **3**, 73–79

Mayhew, A., 1970, Structural reform and the future of West German agriculture, *Geographical Review,* **49**, 54–68.

Mayhew, A., 1971, Agrarian reform in West Germany, *Transactions of the Institute of British Geographers,* **52**, 61–76.

O.E.C.D., 1967, *Agricultural policies in 1966,* O.E.C.D., Paris.

Otremba, E., 1970, *Der Agrarwirtschaftsraum der Bundesrepublik Deutschland,* Wiesbaden Steiner.

Ortmann, B., 1971, Delimitation of economic planning regions in West Germany, *Tijdschrift voor Economische en Sociale Geografie,* **62**, 308–317.

Porth, H., 1970, Exploration for natural gas in the Bundesrepublik Deutschland in 1969, *Erdöl und Kohle, Erdgas Petrochemie,* **23**, 546–555.

Perry, N., 1966, The Federal planning framework in West Germany, *Journal of the Town Planning Institute,* **52**, 91–93.

Perry, N., 1967, Recent developments in the West German oil industry, *Geography,* **52**, 408–411.

Reitel, F., 1971, L'aménagement du territoire en république fédérale allemande, *Mosella,* **1** (1), 89–106.

Reitel, F., 1971, La siderurgie de la république fédérale allemande, *Mosella,* **1,** (2), 29–42.

Schmitt, E., 1970, *Deutschland,* Paul List Verlag.

Scholfield, G., 1965, The canalization of the Moselle, *Geography,* **50**, 161–163.

Schöller, P., 1967, *Die deutschen Städte,* Franz Steiner.

Spelt, J., 1969, The Ruhr and its coal industry in the middle 60s, *Canadian Geographer,* **13**, 3–9.

Vinck, F., 1969, Methods of reconversion policy in the framework of the European Coal and Steel Community, *Tijdschrift voor Economische en Sociale Geografie,* **60**, 3–11.
Warren, K., 1967, The changing steel industry of the European Common Market, *Economic Geography,* **43**, 314–332.
Wiel, P., 1963, *Das Ruhrgebiet in Vergangenheit und Gegenwart,* Scharioth'sche Buchhandlung, Essen.
Wiel, P., 1970, *Wirtschaftsgeschichte des Ruhrgebietes,* Essen.
Windhorst, H., 1971, Der Wald de Bundesrepublik Deutschland im Wirtschaftsprozess der Gegenwart, *Geographische Rundschau,* **23**, 432–437.

9

United Kingdom

David Thomas

Historical Perspective

Traditionally, geographers divided Britain into two. To the north and west lay a highland zone, built up of rocks of Palaeozoic age and older. In broad terms the region possessed a certain unity. It was rugged, rainy, had heavy soils, and contained the coalfield–industrial areas. To the south and east lay the lowland zone, composed of rocks of Mesozoic age and younger. In broad terms, this region had relatively unaccentuated relief, light rainfall and freely worked land, it contained a number of major expanding cities on its periphery, and at its pivot was the major metropolitan centre of London. The conventional division between the two zones was accepted as a line which joined the mouth of the River Tees in the North-East and the mouth of the River Exe in the South-West.

But nothing lasts for ever. As time has passed, and as the demographic and economic trends which were the aftermath of the industrial revolution have caused great changes in the distribution of human phenomena throughout the country, a modified conception of the traditional division has emerged—a more modern view based upon the current social and economic geography of the United Kingdom. The North-West is still contrasted with the South-East, but now the emphasis is not principally upon geology, relief, soils, and rainfall, though these factors are at the basis of many of the social and economic differences between the two areas, but upon the levels of prosperity and upon the expectations of advancement in jobs, in housing conditions, and in social infrastructure. The dividing line between the North-West and the South-East has also changed. A more acceptable boundary today would be defined by two lines, one running eastwards and the other southwards from Liverpool to the east and south coasts respectively. To the south and east lie the prosperous parts of the country, to the north and west lie the deprived areas constantly in need of central government assistance. These shifts in the British geographer's perception of his country are, of course, a reflection of powerful forces of change operating within the country over the last 50 or more years. But they are also significant of a growing awareness of differences in regional development which stem not only from the physical fabric of the countryside, but also more and more from enormous divergences in the social and economic momentum of the various regions of the country.

Rapid and differential change causes problems, and problems demand remedies. In the United Kingdom both the identification of the problems and the suggested, and sometimes implemented, solutions have occurred at a number of levels, but two strong themes emerge in any review of the trends through this century. One concerns the problems of land and land use, and the remedies to these problems. They fall broadly within the province of land-use, physical, or town and country planning (the terms are normally regarded as synonymous). The second theme is intimately concerned with economic problems, the solutions to which fall within the scope of economic planning. Sometimes, in the development of regional consciousness and in the growth of regional planning, these themes are quite distinct. At other times and in other places the themes are closely interwoven—a recognition of the fact that physical and economic problems are closely interrelated and, consequently, that physical and economic planning are in many respects indivisible. In the following sections, which trace the growth of regional problems and assess the efficacy of the attempts made to grapple with them, many variations on these major themes emerge.

Early Concern

As the nineteenth century proceeded, the effects of the industrial revolution upon the landscape of the United Kingdom and upon the lives of its people became plain for all to see. Rapid population growth was focused upon the coalfield regions of the north and west, and this converted largely undeveloped regions into swiftly expanding urban–industrial agglomerations. At the same time, existing cities were stimulated by increasing trade and expanding functions, and experienced considerable urban development untempered by any real physical controls. The absence of planning led, particularly in the new industrial settlements, to a seemingly chaotic land-use mix in which dwellings were thoroughly intermingled with the factories and mines, whose workers they housed, the railways, roads and canals which served the urban area, with the shops and other commercial facilities of the towns, with industrial and mining spoil heaps, with churches, schools, and hospitals. Development was unorganized, was quickly to prove inefficient, and was not required to meet, until late in the century, even the most basic of minimum structural standards.

Out of these very unsatisfactory urban forms reemerged the notion of urban planning, the desire to create the perfect city (and sometimes the perfect society also) (Ashworth, 1954). The motivation was a curious combination of the idealistic and the practical. On the idealistic plane there arose a positive desire among a limited number of thinking people—often they were philanthropic industrialists who took an enlightened view of the conditions in which their employees ought to live—to establish urban areas which were both efficient and pleasant places. They were dissatisfied with the standards of housing, health and amenity within existing, rapidly developed urban areas and saw the solution in organized communities where the level of housing design was higher than anything previously thought worthwhile for industrial workers, where public hygiene was consciously accepted as a factor in town development, and where some attempt was made to improve the overall appearance of settlements. For example, Robert Owen, in the early decades of the century proposed, but only partially completed, a model settlement for industrial workers in the Clyde Valley at New Lanark and this was followed by many other model industrial settlements later in the century, such as those at Saltaire in West Yorkshire, at Bournville in Birmingham and at Port Sunlight in Cheshire. These plans culminated in a proposal by Ebenezer Howard, first published in 1898, that new, planned, garden cities should be established to reap the benefits of both town and country life, while avoiding the disadvantages of each. This was to be achieved by creating moderately sized settlements set within an agricultural reservation or green belt. Towns were to be formally planned throughout with an abundant use of recreation land, greensward and trees. Non-conforming land uses were to be strictly segregated, particularly industry from housing. The city was conceived as a complete social unit in which the growth of population and industry was to be as carefully planned as the layout. Once the city had reached its planned size, new growth was to be steered to new garden cities beyond the agricultural zone. In the course of time two such towns were built in Britain, at Letchworth and Welwyn Garden City, countless suburbs imitated the style and, eventually, the new town movement, which grew from the garden city idea, led to the establishment of over thirty new towns in the United Kingdom and to many more elsewhere in the world (Osborn and Whittick, 1963).

Alongside these idealistic notions (a few of which, it is true, actually materialized) there existed in the nineteenth century some very pressing public health problems which had arisen from the rapid and unorganized urban–industrial growth and which called for urgent attention. By the early years of the twentieth century, it came to be recognized tacitly, if not always explicitly by all sectors of society, that to achieve improvements in social, economic and aesthetic conditions it was necessary to interfere with market mechanisms, and also with the property rights of individuals, and to plan the city and its surrounding region. This could only be achieved in practice by a marked extension of statutory controls over land usage, by the introduction, in other words, of statutory physical planning. The protagonists of the ideal or garden cities were certainly important influences in this move towards legal controls. They demanded not only new, model settlements, but also higher standards of amenity in existing towns and cities, and the enlightened development of new suburbs. But it was the immediate need to meet the health and housing difficulties created by the swift urban growth that was uppermost in the minds of most legislators. In this important respect town and country planning in the United Kingdom may be regarded, particularly in its early years, as an extension of public health and housing policy.

Thus it was that the first piece of legislation to include the term *planning* in its title was enacted before the outbreak of the First World War (Cullingworth, 1964). The Housing, Town Planning, etc. Act of 1909 was modest in its aims. It was intended to secure adequate health standards in suburbs about to be built by the introduction of by-laws more rigorous than those allowed under existing late nineteenth-century public health legislation. The local authority could also, if it wished (the legislation was permissive and few local authorities took advantage of its terms), adopt and administer a scheme or plan for the development of an area within and around a town. Provision was made for the payment of compensation to anyone whose property was adversely affected by the scheme, and the Act also allowed the local authority to recover a limited amount of betterment where property increased in value. The Act is important not so much for what it achieved, but because it first introduced the concept of planning legislation and because it laid the foundations upon which so much was to be built in succeeding decades.

The Inter-War Crisis

While the period before the First World War was dominated by a concern for a healthy and well-housed population, the time between the wars became one preoccupied with problems at a scale hitherto not experienced—a new dimension emerged with growing regional imbalances in prosperity and economic potential. Structural deficiencies in the economy appeared which had not previously seemed important and such was their effect that the smaller-scale problems of health and housing were, for a while, pushed into second place. It is important to examine the changes of the inter-war years in some detail.

To an observer at about the date of the First World War, it must have seemed as if the industrial pattern created in the late eighteenth and early nineteenth centuries would continue for ever. London apart, the coalfields dominated the industrial picture, accommodating almost all the heavy industry and often specializing upon particular industries or even branches of industries. The cotton industry was almost exclusively based in Lancashire, the wool industry in the West Riding of Yorkshire, knitwear goods came especially from Nottinghamshire, fine steel and cutlery from Sheffield, the iron and steel industry and the tinplate industry flourished in south Wales, while shipbuilding characterized Tyneside, the pottery industry dominated the small region around Stoke on Trent—and so on. At a very local level, many small parts of coalfields were solely dependent upon the extraction of coal or some other mineral. Though it may not have caused concern when economic conditions were propitious, it is clear that such undiversified regions were always susceptible to shifts in demand as a result of general economic circumstances, to changes in technology or fashion, to cheaper sources of raw materials in new locations, and to changes in power requirements. And the point was proved, rather dramatically, in the late 1920s and 1930s.

The south Wales coalfield, to take one example, was, even up to the late 1930s, totally committed as a region to coalmining and metal manufacture. Little wonder that, in the inter-war period when demand for coal fell sharply and when lower-cost steel production was developed elsewhere, the region became one of the most intractable unemployment blackspots in a period of general industrial contraction. But this one example simply serves to represent general trends within the coalfield–industrial areas. For not only did coalmining decline, but most of those industries associated with the coalfields—producing heavy, capital goods in the main—suffered recession as world economies contracted. At the very same time, the coalfields were losing their locational advantages. New, light, consumer goods industries were developing which widely adopted electricity as their main form of power. They were thus freed from the coalfield regions and could locate wherever advantageous sites offered themselves. Because of the nature of their products, consumer goods industries are characterized by being orientated towards transport routes and towards markets. The Midlands, but especially the South-East of England in and around London, became the favoured locations. In or near London, firms could serve the London market (the largest in the country) and at the same time make use of the road and rail transport nets which focused upon the capital. The Midlands and the South-East therefore acquired most of the growth industries while the coalfield–industrial regions were left with the obsolete plant and worse, with industries having only a small, and sometimes even a negative, growth potential. It is plain why the South and East prospered during the inter-war years at the very same time that the industrial regions of the North and West were suffering stagnation.

Closely associated with these regional differences in economic opportunity were some important new trends in population. Before 1914, London had been the centre of unemployment. Now it was south Wales, the North-East, Scotland and Northern Ireland which became the 'distressed' areas (McCrone, 1969). These changed conditions generated population flows which had the unusual characteristic that they were not mainly from country areas to industrial towns, as they had been in the earlier periods of industrialization, but between existing industrialized urban areas. Thus the 'drift to the South-East' began, though the migration flows were always more complicated than that accepted phrase tends to indicate. Adjustments to population densities within regions were experienced, as well as cross and counter flows between regions. Nonetheless, the predominant movement was towards the South-East, where population density rose at the expense of that in other regions.

Another factor emphasized changing densities. As with rural–urban migration, it was the young adult age group which was mainly involved in the population movements between industrial regions. This had the effect of changing the rates of natural population increase in both exporting and importing regions. Natural increases in population declined in the coalfield regions but increased sharply in the South-East. This made an important contribution to population growth in southeastern England in the inter-war period.

Population pressure in the Midlands and South-East led to the burgeoning of cities; and this was to cause great agonizing among planners in the 1940s and 1950s. City centres were being renewed as their commercial areas expanded. Population was displaced and was accommodated at the city fringe. But also competing for space at the cutting edge of the city were the new factories and also the in-migrants who sought employment in the consumer goods industries. The result was urban sprawl, especially along the main, radial, road arteries of cities such as London.

The tentacles of urban growth along main roads were a result of a number of factors, but among others they were a product of another important phenomenon of the inter-war period—the growth in the number and in the importance of road vehicles. Private cars increased in number from under 200,000 to over 2,000,000 during the period, the total number of vehicles exceeding 3,000,000 by the outbreak of the Second World War. The motor car gave freedom of dwelling location to those who could afford that means of personal transport; at a more limited level the motor bus freed those who could not afford cars from their total reliance upon commuters' railway stations. The motor lorry had equally profound implications for manufacturing industry.

The Planning Response

In the face of these massive problems, mechanisms of planning developed rather swiftly during the inter-war period. At first, the emphasis upon physical planning continued, though the scale of activity expanded to approach a quasi-regional level. Later, as the severity of the problems became clear, a great deal of attention was given to regional economic difficulties. It was a period of great innovation within planning in the United Kingdom and an outline of the more important statutory and other provisions is given below.

It had soon become clear that the Act of 1909 provided an inadequate basis for town and country planning and that a revision of procedure was necessary. The Housing, Town Planning, etc. Act of 1919 was an attempt to meet some of the deficiencies of earlier legislation. It made the preparation of planning schemes compulsory for some of the larger authorities, but more important, it allowed a limited amount of joint action between neighbouring authorities. There were over 1400 local authorities (county boroughs and district councils) in England and Wales alone and they were not large enough for anything but small-scale and short-term planning. The effective planning units could now be enlarged and joint planning studies prepared. The Local Government Act of 1929 allowed the counties to join with the district authorities in joint schemes, while the Town and Country Planning Act of 1932 enabled planning schemes to be applied to all land, irrespective of whether it was likely to be developed or redeveloped.

This legislation cleared the way for a large number of 'regional plans' for counties or parts of counties. They became a feature of the late 1920s and 1930s. They were usually produced by planning consultants and, among other things, frequently recommended routes for new roads, zoned areas for agricultural, industrial and urban development, and designated proposed open spaces and sometimes green belts. But though substantial progress had been made on paper by the date of the outbreak of the Second World War, in practice there had been slower advance in the control of land use and in the solution of the problems posed. It has been estimated that by 1942, while 72 per cent of England and 36 per cent of Wales were subject to interim development control, only about five per cent of England and one per cent of Wales were actually affected by operative schemes. Several major cities of the United Kingdom and many towns were without even an outline plan for present control and future development. The importance of the joint schemes lies not in what they achieved on the ground, but in the experience they provided in drawing up land-use control strategies and in the technical competence they developed for plan formulation. These were to prove vital in the post-war years.

Meanwhile, as the inter-war period progressed, the problems of economic and social imbalance demanded increasing attention. By the early 1930s, unemployment rates had risen sharply, but particularly in the north of England, south Wales, central Scotland, and in Northern Ireland where many individual industrial towns recorded rates in excess of 60 per cent of their workforce. The depression hit the coalmining, shipbuilding, textile and heavy metal industries particularly hard. This caused concern not only because the locations of these industries had a heavy regional component and therefore triggered regional decline and population migration, but also because they were the traditional export industries of the country. Once begun, the economic contraction became cumulative. Depression in the basic industries operated through the multiplier to retard all other sectors of the regional economies where these industries were heavily represented and to create conditions very unsuitable to diversification of the economic structure. In other words, not only were the new consumer goods industries attracted to the South-East, they were positively repelled from those coalfield regions which needed them most to survive the economic crisis. The halting steps towards physical regional planning outlined above were thus matched by an equally hesitant move towards economic regional planning.

In 1934 the Special Areas Act was passed (McCrone, 1969). It designated four Special Areas in south Wales, northeast England, west Cumberland and central Scotland (Figure 9.1(a)), and appointed Commissioners whose task was to promote the rehabilitation of the depressed regions. Powers and financial backing given to the Commissioners at the outset were totally inadequate to meet the enormous problems involved, but between 1934 and 1939, such were the needs of the coalfield areas, that these modest provisions were boosted to provide a battery of physical and fiscal controls the like of which were not to reappear in the United Kingdom until the 1960s. To begin with, the amount of money available to the Commissioners to encourage both small and large firms to move into the Special Areas was greatly increased. Much of

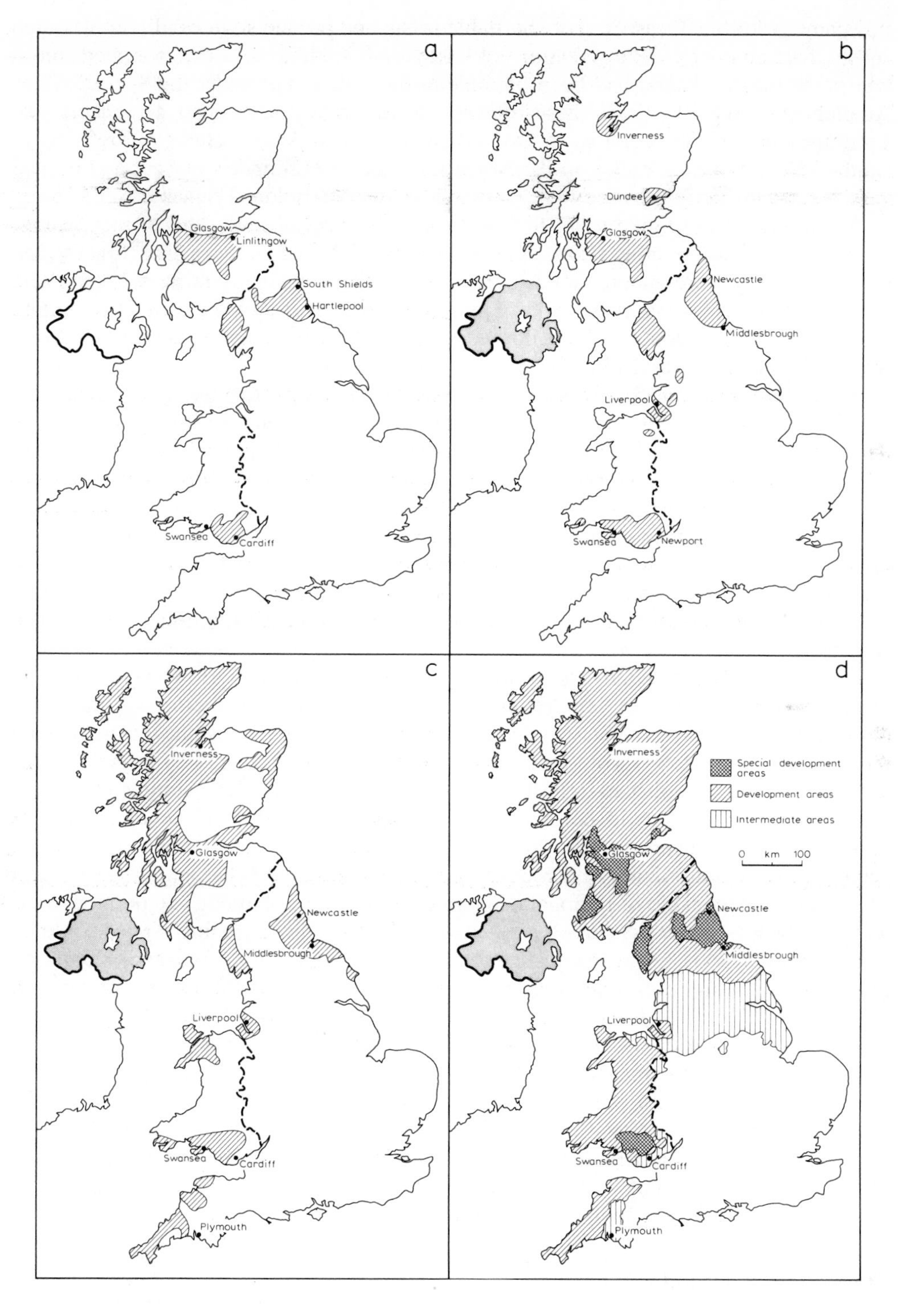

Figure 9.1. United Kingdom: assisted areas.

this came from the Treasury, but charitable trusts and private sources also contributed substantial amounts. The creation of industrial trading estates, operated on a non-profit-making basis, had been possible at the outset, but under the Special Areas Amendment Act of 1937 these powers were widely expanded to enable the Commissioners to let factories in any part of a Special Area. Estates at Treforest in south Wales, at Team Valley in the North-East and at Hillingden in Scotland proved very successful in introducing light industries to the coalfields. The same Act allowed the Commissioners to give relief to new firms in respect of rent, rates or income tax; they could also waive other taxes, such as the new National Defence Tax. Finally, the Commissioners were given facilities for easing the transfer of those workers who moved out of the Special Areas to take jobs elsewhere. This took the form of retraining schemes to convert labour to the skills needed in the growth industries; grants and loans were also provided to assist in migratory moves.

It will be noted that the Special Areas Acts were designated solely to support the depressed regions; they did nothing to control development and ease congestion in the more prosperous areas of the country. That the work of the Commissioners ameliorated conditions in the coalfields there can be little doubt. For example, unemployment rates dropped sharply in the late 1930s. The problem is to decide how much of the improvement was due to remedial action and how much was the result of the upturn in trade which began to revitalize the old basic industries in the few years before the Second World War. Sufficient concern was felt for the future, for the government to set up a Royal Commission to explore the whole problem of the distribution of the national workforce.

The Barlow Commission (its chairman was Sir Montague Barlow) was set up in 1937. It was required to inquire into the causes of the distribution of industrial population, to consider what social, economic or strategic disadvantages arose from industrial concentration and to report what remedies should be applied in the national interest. Its report, published in 1940, was extremely influential. It recommended that a central organization should be created to oversee future industrial development. Its objective was to ensure a regional balance of diversified industry which would be achieved by decentralizing and redeveloping the congested urban areas. The report stressed the relationship of regional economic and physical planning and pointed to the way in which new towns, garden cities and properly planned trading estates could be used to spearhead regional readjustment. The report argued that assistance should be given to local authorities when they tackled regional problems; it also strongly urged the need for research on the location of industry and the use of natural resources in advance of major regional development. The report was not unanimous (there were several dissenting statements), but the reservations were over the detailed measures required to implement the policy rather than over the policy itself.

In many ways the Barlow Report can be regarded as a milestone in the evolution of regional planning. It emerged at a time during the war when attention was focused elsewhere but, as soon as post-war policies began to be formulated, its influence was felt and continues to be felt up to the present day.

The Post-War Period

It is right, at this stage in the discussion, to question the necessity of a lengthy historical introduction to the topic of regional problems and planning in the United Kingdom.

The answer is simple. Without an historical perspective it is not possible to understand fully what happened in the post-war period. The problems that were experienced were essentially a continuation of those which existed in the inter-war years, though the importance attached to those problems altered as the years went by. The solutions advanced to meet the post-war problems developed those of earlier decades. The technical competence to formulate and then to implement plans could only be based on that already available to previous generations of planners. And, more important than any of these, the mistakes of pre-war planners were repeated because, all too often, the wrong exemplars were followed. It would, for example, be attractive to argue that the integrated view of the Barlow Commission persisted and flourished. In fact, the complementary nature of physical and economic planning was never properly recognized in the immediate post-war years and true regional planning did not begin to reappear until the late 1960s. But before turning to planning activities it is necessary to explore post-war social and economic trends in greater depth.

Social and Economic Trends

It had appeared to some observers in the 1930s that once the world trade recession was corrected, the slump in British heavy industry would disappear and the problems of regional imbalance would cause little further concern. It quickly became clear to everyone that this was not to be the case. The economic climate of the inter-war years simply accentuated a long-term trend—widespread readjustment in industrial balance and location which resulted from changing markets and increasing pressure from competitors overseas. The consumer had become king. Hence the dominating position which the Midlands and the South-East had achieved at the expense of the industrial regions elsewhere was perpetuated, and with it the problems which had already emerged so forcibly.

Indeed, as the post-war period progressed, the prospering regions were perceived to have, or to be acquiring, additional advantages. Transport and accessibility, the key to industrial growth, was given added influence by the building of a new system of national motorways. These were first developed where they were most needed—in the Midlands and South-East—but even when the network became widespread, it worked very much to the advantage of the growth areas. Just as the electricity grid had produced not a dispersion of industry but a new concentration upon the major markets (in fact, it released industry from one powerful control—the coalfields—only to release another—the markets) so the new road system gave London and the towns of the Midlands increased accessibility and greatly enhanced their locational attraction.

Another mode of transport, sea transport, gave added impetus to industrial growth in the South-East particularly. Liverpool and the ports of the north faded in importance. They were associated with long-distance trade, often in heavy goods, with Commonwealth and other countries. This trade was in sharp decline, relatively if not always absolutely. The ports of the South-East by contrast—London, Southampton, Dover, Felixstowe, Harwich and others—came increasingly to depend upon trade with the rest of Europe. This was expanding rapidly and became an important factor in industrial site selection. The long prospect of entry and finally the accession of the United Kingdom to the European Economic Community was clearly an important determinant here, but the changing nature of the country's trade should not be discounted in attempting to explain the attractiveness to industrialists of the London region.

The Midlands and South-East also benefited from the nature of their labour resources. Their pools of labour are large and partly because of this, but mainly because of their long industrial histories in craft as well as in diversified factory manufacturing, labour is also extremely versatile. Though labour tends to be expensive, the size and diversity of the resource is such that it more than overcomes these marginally higher costs. The great difference between the inter-and post-war periods is that the one was characterized by a high level of unemployment, while the other has experienced full employment. At particular places and times, and in a number of industries, there have been considerable labour shortages. Thus the importance attached to the character of the labour market in the prosperous regions.

Economies to be gained externally to the firm have increased in recent years and have loomed large in location decision making. All firms regularly purchase goods as well as services from other firms—raw and semi-processed materials, transport, advertising and marketing facilities, legal and financial aid, computer time, and so on. London, in particular, offers such aids to business in great variety, and their proximity not only speeds business but also greatly reduces costs, since specialized components and intermittently-needed special services do not have to be carried by each individual manufacturer. Another aspect of the concentration of services in London is evidenced by the great concentration there of business administration—it was once called the *paper metropolis.* About half the total office space in England and Wales occurs in the South East Economic Planning Region. It is both attracted by, and in turn creates, for other industries, further, external economies.

Of course, there are disadvantages as well as advantages in the location of economic enterprises in the Midlands and South-East. They fall mainly under the heading of *congestion.* This was a major problem identified by the Barlow Commission, but it has been observable since 1945 that the effects of considerable urban-industrial development have been unequal. It is doubtful whether the business units themselves have suffered particularly. If increasing congestion had caused diseconomies then, sooner or later, the benefits of agglomeration would have been overcome and decentralization triggered as part of a natural process of adjustment. In practice, the government and its agencies have experienced great difficulties in bringing about the movement of firms out of the South-East (Luttrell, 1962). It is, of course, the workers in these industries who carry the burden in suffering one or a number of the following: poor housing, longer work-journeys, home and work environments less satisfactory than elsewhere, higher housing costs, greater pollution, social and psychological stress. On the other hand, wages and salaries are higher, educational, cultural and entertainment facilities may be far superior and, given the tendency towards the centralization of administration within London and other towns in the South-East, promotion prospects may be considerably brighter (Coates and Rawstron, 1971).

In the less prosperous industrial regions, opportunities for industrial expansion and in employment appear less attractive. Almost every facet of the social and economic structure which runs in favour of the Midlands and South-East, runs against the North of England, south Wales, Scotland and Northern Ireland. They lack accessibility to the major urban markets for consumer goods, they are badly placed for trade with Europe, the effects of depression have reduced the level of services and, where they existed at all, of component producing firms to a minimum, they are distant from the financial and commodity exchanges, and the level of amenities—in housing, in educational and cultural facilities and in the adverse impressions created by a large amount of industrial dereliction—is perceived to be unsatisfactory. In addition, throughout a period of full employment when labour shortages have often been experienced elsewhere, these regions have maintained small, but significant, unemployment rates and have consistently employed a smaller proportion of the total labour force than the Midlands and South-East.

Given the circumstances, it is not surprising that the population drift to Midland and southeastern England continued strongly in the post-war years. Between 1951 and 1971, the more prosperous regions recorded population increases in the range 12–16 per cent. The less prosperous regions hovered about a population increase of 5 per cent for the same period. The South East Region alone achieved a population increase of 1,900,000 people, well above one-third of the United Kingdom total increase over the 20-year period, and more than 500,000 of these were the result of net migration gains. Population therefore continued to settle at increasingly higher densities in the prosperous regions, causing severe pressure upon land resources and presenting formidable problems to the planners.

As important as the increasing population density was its redistribution within the regions. This caused particular difficulties in the Midlands and in southeastern England where the population pressure was the greatest. As towns and cities grew, their central area functions required more space. Central business districts were further extended and population displaced. Redevelopment of the older, inner, residential suburbs also caused the decanting of population, since the newer building was generally at a lower density. Outer residential suburbs grew rapidly to house this dispersed population and also to accommodate new population (Johnson, 1974). By these mechanisms, the cities and towns of the prosperous regions were 'in flood', causing rivers of bricks and tarmac to inundate the countryside (Self, 1957). The nightmare that London, for example, might soon stretch from Cambridge to the south coast was close to realization, so powerful were the economic and social forces underlying its growth. Overall it was a situation which called for immediate action and this was not long delayed.

Physical Planning

As already indicated, most of the early post-war physical planning had very little connection with the economic planning which was developing alongside it. Despite its great achievements it was the poorer for this. The way had been admirably smoothed, not only by Barlow, but also by two other remarkable state papers, the *Uthwatt Committee Report* into the problems of compensation and betterment and the *Scott Committee Report* on land utilization in rural areas. The wartime preparation was sound and immediately peacetime conditions were reestablished, new legislation transformed physical planning by setting up a comprehensive and obligatory planning code.

The principal means to this end was the Town and Country Planning Act of 1947. Under its provisions planning powers were removed from the districts and placed in the hands of the county and county borough councils. In effect this reduced the number of planning units to 10 per cent of the former number. Each planning authority was required to produce a development plan and to undertake periodic reviews. Control of development—building, engineering, mining, fresh advertising—and control over any material change in the use of land, other than in forestry, agriculture, and in a few other special circumstances, was exercised by the counties, and was consistent with the plan. The Act has since been modified in a number of respects. The sections dealing with compensation payments and betterment charges (the charge levied where the value of land increases as a result of a planning decision) were abandoned in the early 1950s, and other, small additions and alterations have been made, but broadly the Act, as modified in subsequent legislation, still forms the basis of planning in the country. Since 1947, therefore, there has been in existence legislation providing, subject to appeal and ministerial consent, extensive powers for the control of land-use changes which can be, and have been, directed towards the improvement of amenity and to the solution of land-use problems.

An enactment of the same period provided an important legislative control over urban development. The New Towns Act of 1946 (later consolidated in 1965, in which year Northern Ireland first introduced new town legislation) drew its inspiration partly from the garden city movement and partly from the *Barlow Report* and was a further contribution towards balanced and planned land-use development. New towns were an attempt to solve the problems created by overcrowded industrial settlements, by uncontrolled urban sprawl and by the overlong work and pleasure journeys which these urban forms sometimes impose upon their populations. Hence new towns were conceived, ideally, as self-contained communities set in the countryside, where housing, jobs and all necessary facilities were provided. While new towns, and their derivatives—towns expanded under the town development schemes which followed further legislation in the 1950s—may not always have achieved their highest aims, they have certainly made considerable progress in minimizing land-use conflicts and consequently in creating better, happier and more efficient lives for their inhabitants.

There are now more than 30 new towns in the United Kingdom in various stages of development (Figure 9.2). The earliest were designated in the late 1940s to receive overspill from London, Glasgow and other areas, and new towns continue to be founded as the need arises. Their combined population has now reached c. 2,000,000. In addition, about 70 statutory town expansion schemes have been agreed in England and Wales, with another 40 or so in Scotland. A small proportion of these are complete, but the majority are still in progress.

A number of other physical planning measures had implications at a regional scale. One important element in the planning of major cities and their surrounding areas was the introduction of green belts (Figure 9.3). These were zones designated around major urban areas within which it was intended to allow no major urban development (Thomas, 1970). Their purpose was to restrict once and for all the further sprawl of large built-up areas into the countryside—it was plainly undesirable, for instance, to create self-contained new and expanded towns some distance beyond the outer edge of a large city and then allow them to be engulfed in subsequent urban expansion. Green belts were sometimes used to prevent adjacent towns merging and in some cases to preserve the special character of an ancient city. Clearly, green belts modified greatly the mechanisms of the urban fringe. Because they operated through the procedures of statutory development control, they fossilized existing urban uses while preserving the remaining amenities of the open countryside. They tended to push urban pressure further outwards and to create a real urban fringe beyond the fossil fringe. They transferred the problem.

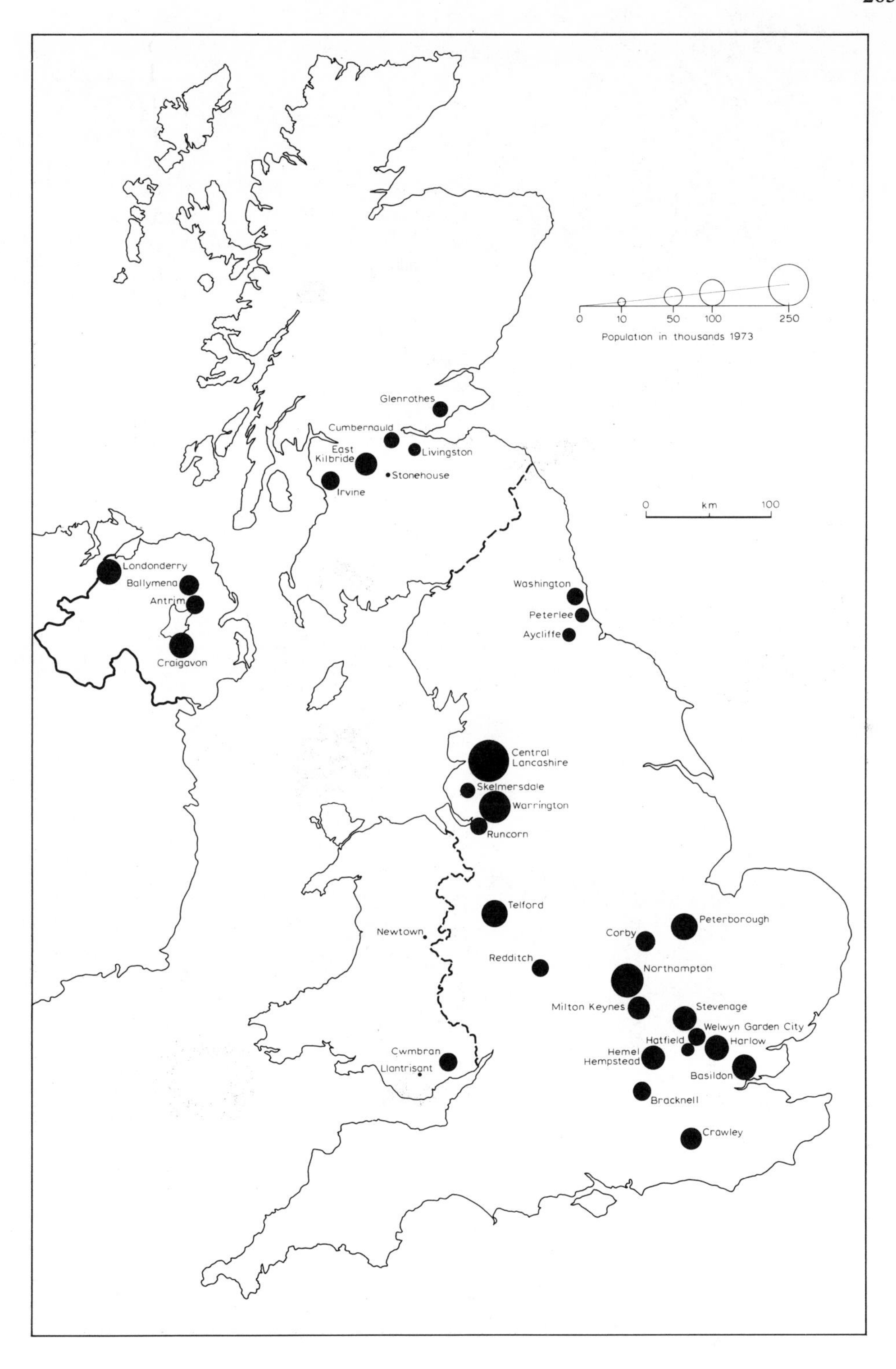

Figure 9.2. United Kingdom: new towns.

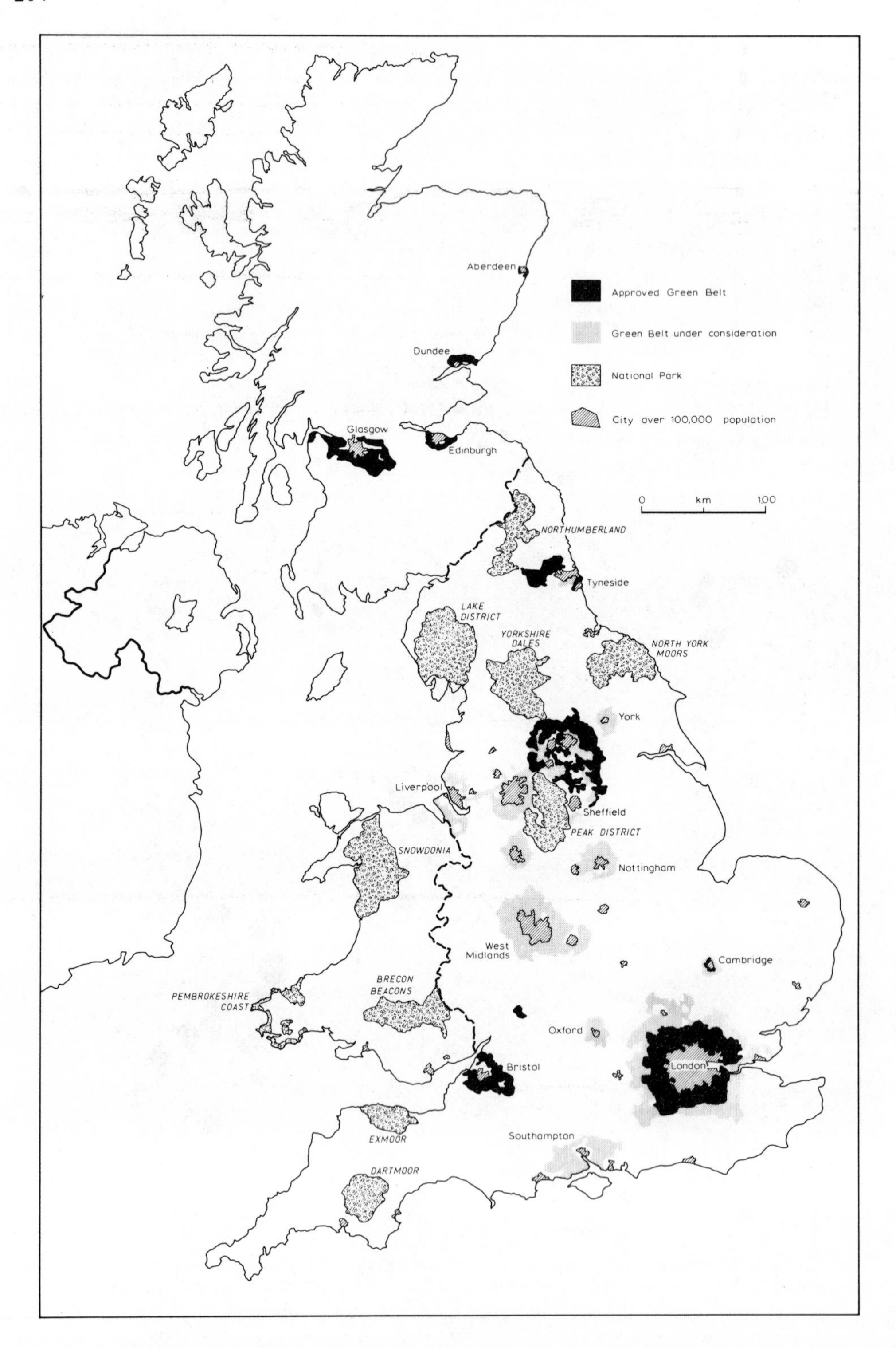

Figure 9.3. United Kingdom: urban areas, green belts and national parks.

Another planning measure to have regional significance was the creation of national parks (Abrahams, 1960). They were designated in England and Wales only, under the National Parks and Access to the Countryside Act of 1949 (Figure 9.3). Their purpose was to preserve the extensive, sparsely peopled, mainly upland areas of the country so as to develop facilities for public open-air enjoyment and to protect beautiful and relatively wild country and its sometimses rare wild life. Meanwhile, established farming, the dominant land use, was to be maintained and supported. The task was not, as can be seen, without internal conflict, but the Commissioners charged with the supervision of the park system have made considerable progress not only in protecting areas of scenic, scientific or historic value, but also in providing regional recreation areas of considerable significance (Burton and Wibberley, 1965). National parks occupy about 9 per cent of England and Wales, an area very similar to that protected by green-belt control.

In addition to the provisions outlined above, a wide range of other planning measures have been adopted to improve the environment and enhance amenity. Urgent attention has been given to the problems of traffic in towns, urban renewal schemes have been strongly supported and in part financed by central government, policies and agencies have been created for the dispersal of office employment from London to other towns and cities, clean air legislation has improved both the appearance and the health of urban areas, historic buildings have been restored and preserved, countryside parks (much smaller than national parks but closer to the centres of population and devoted almost exclusively to recreation) and forest parks have been developed mainly for the use of town and city dwellers—about 90 per cent of the population live in urban areas, half of them in the great conurbations—while nature reserves and sites of special scientific interest have been scheduled to conserve certain types of vegetation, species or landforms of particular value or rarity. By these and other means, towns and cities have been improved internally and their functional, regional linkages with the surrounding countryside have been developed, though regional consolidation has often come as a byproduct of other activities and goals.

Economic Planning

The fear that inter-war unemployment levels would be reestablished gave point and urgency to remedial measures on the economic side also. All political parties were committed to Keynesian economics by the end of the war and there was general agreement upon the aim, full employment, if not always on the means by which it should be achieved. The basis of United Kingdom regional economic policy was the Distribution of Industry Act of 1945. It was later supplemented and modified, but in a broad way its influence remained undiminished through to 1960.

The Act swept away pre-war legislation, redefined new Development Areas in place of the Special Areas (Figure 9.1(b)), adding some entirely new regions in need of special attention, and placed the main responsibility for future policy and for executive action upon the Board of Trade, whose powers were to be similar to those of the Special Area Commissioners at the outbreak of war. The Board of Trade could build factories, make loans (with Treasury approval) both to trading estates and to specific industrial firms, provide basic services and reclaim derelict land. But, in addition, the Board was given another important function. Under the Town and Country Planning Act of 1947 it was provided with the authority to issue industrial development certificates, necessary for all new factory development of any scale or, indeed, for extensions to existing factories. These replaced the building licence system which had operated through and since the war. The Board therefore had the power, by withholding or limiting certificates for the growth regions, to steer industrial expansion into the areas of need.

For a while the new system worked well. The Development Areas were given high priority through a period of general shortage in building materials and in raw and semi-processed goods. New factories were built, the factories which had been located in the less accessible, but labour rich, coalfield regions were converted to commercial use, and new trading estates were established. In some years in the late 1940s, over half the new industrial buildings constructed in the country were located in the Development Areas, though these contained less than one-fifth of the workforce. Unemployment rates, though higher than elsewhere, gave little cause for concern and, partly because of this success, partly because in periods of retrenchment in government expenditure Development Area funds were cut, the policy was greatly relaxed. The pressure from industrialists to locate in the Midlands and South-East was also persistent. With no financial incentives to go to the coalfields, they clearly wished to manufacture where they regarded conditions as optimal. Control through the industrial development certificate mechanism eased considerably. By the mid-1950s regional policy was virtually in abeyance, but after a few years of policy freewheeling, difficulties started to arise once more in the coalfield regions of the North and West.

The post-war period had seen a boom in coalmining and in the traditional heavy industries. Though government action had led to a substantial degree of industrial diversification within the economies of Wales, northern England, Scotland and Northern Ireland, it was not sufficient to have solved the problems of these areas. The boom had simply masked the need for yet further structural changes in the economic framework. By 1958 the boom years were over. The shipbuilding industry had recovered the war-time backlog, and was short of orders, steel output fell, the coal industry began to feel the effects of competition from abroad and from other fuels, the trade cycle moved against the country and deflationary policies were adopted. The problems therefore returned and the Development Areas once more returned to the forefront of attention. In some haste the Distribution of Industry (Industrial Finance) Act was passed in 1958. It kept the existing Development Areas but added to them some additional zones where persistently high unemployment rates had emerged. For the first time some purely rural areas were included, for example, northwest Wales and the Western Isles. The Treasury's power to make loans for industrial development was extended to such places.

The legislation of 1958 was not sufficient to meet the problems which had emerged and it was certainly insufficient to refashion the economies of the less prosperous regions so that in the long term they were capable of healthy and sustained economic growth. Therefore in the 1960s a number of policy developments took place which gave added impetus and greater strength to regional economic readjustment. The first of these advances was enshrined in the Local Employment Act of 1960. The provisions of this Act were greatly stiffened by the Local Employment Act of 1963, and the Industrial Development Act of 1966, and all these measures were bolstered by purposeful government administrative support in the mid-1960s (Manners, 1972).

One of the most profound changes over this period was in the designation of the scheduled areas in need of special support. The old areas were felt to be unselective and the system under which they operated too inflexible. The 1960 Act introduced Development Districts (in place of the old Development Areas which were abolished) based upon local employment exchange areas which could be scheduled (and also revoked) readily by the Board of Trade where high unemployment existed or was likely. In practice, 4·5 per cent of the insured population was regarded as a critical unemployment rate, though other factors were also taken into account. Almost all the old problem areas re-emerged as in need of attention, but so also did some major regions not hitherto given special treatment, such as the South-West of England and the Highlands of Scotland (Figure 9.1(c)). The difficulty, which was soon perceived, was that unemployment rates did not necessarily indicate the economic potential of an area, though they were an adequate guide to social deprivation. This was offset to some degree by defining growth points in central Scotland and a growth zone in northeastern England, an important conceptual advance in economic planning. Another difficulty was that marginal districts had an unsettling tendency to enter and later to drop out of the scheduled list, making long-term planning very difficult for firms located within them.

The 1966 Act put an end to these administrative inefficiencies by abolishing the Development Districts and reinstituting Development Areas, but drawn far more liberally than those used previously. Almost all of Wales, Scotland, and the North of England were scheduled and also a sizeable section of the South-West of England. Together they covered over 40 per cent of Great Britain and contained about 20 per cent of the population. They could be amended relatively easily and were not based upon a strict application of unemployment data. A year later the government identified some smaller areas (they were the cores of the pre-war Special Areas) which required special treatment—the Special Development Areas. Here, particularly favourable levels of assistance were available, though the growth pole philosophy had certainly received a setback. It was further eroded in 1969 when the government implemented some of the proposals of the Hunt Committee (Chairman: Sir Joseph Hunt) which had investigated the 'grey' or intermediate areas, the less prosperous regions which were not sufficiently deprived to enter the Development Area category. Subsidies and other types of government assistance were offered in these areas too, though not at the level attainable in the Development Areas. In effect, the government was refusing to endorse the growth zone idea, in which funds would be channelled towards areas of opportunity and potential, in favour of allocating its assistance more widely on the basis of need. In the course of time, both the Special Development Areas and also the Intermediate Areas were expanded substantially (Figure 9.1).

Integrated Regional Planning

From the foregoing it is plain that the Barlow Commission's prescription for successful planning, in which policies for regional economic balance were to be fully integrated with physical planning measures, had not been followed. Regional organization and administration had never attracted much public attention and, in government, physical and economic planning had been strictly segregated, despite the lip-service that was conventionally paid to their closely complementary natures. In practice, physical planning had always been lodged under the planning ministry (it had several different names in the post-war period) while economic planning had been administered by the

Board of Trade. It was not until the mid-1960s that real moves were made to set this matter right.

The adoption of a regional organization did not emerge in a vacuum. Inter-war physical planning legislation had already been moving in this direction and it had taken place in the context of a very healthy debate. While planners had, on the whole, approached regionalism pragmatically, geographers and other scholars had found regional analysis a convenient descriptive and analytical device and a few had applied these notions to planning and administration in the United Kingdom. Though much of this work dates from the early inter-war period, it needed the outbreak of war in 1939 to bring about a formal regional arrangement. Civil Defence regions were hastily devised. Wales, Scotland and Northern Ireland were established as individual regions. England was subdivided into ten regions, each with its own capital. Most government departments employed these regions when dispersing their activities from London and so, though the regions themselves were far from ideal geographical units, they grew to have some administrative importance. With some small post-war modifications they became the Standard Regions for statistical and other purposes and persisted until 1965.

At that time it had become essential to plan on a regional basis for the use and development of both physical and economic resources. Since the existing regions were unsuitable for the purpose, new economic planning regions were devised, each with its own economic planning council and board. The councils are composed of individuals selected by the government for their range of knowledge and experience of the region (it is not impossible that in the near future the councils may be composed of elected members). The boards are composed of senior civil servants, specialists in economic and social matters, who supply the technical expertise. Together the councils and boards assess the economic potential of each region, draw up regional plans (at the moment these are of an advisory nature since the regions have no executive authority) and make recommendations for the implementation of the plans in such a way that they contribute to national objectives. Wales, Scotland and Northern Ireland have a council and board apiece while England is divided into eight regions (Figure 9.4). Though the majority report of the Redcliffe–Maud Commission on local government reorganization in England, which reported in 1969, suggested new Provinces with augmented regional powers and fresh boundaries, no major changes occurred. The coordination of the work of the regions was placed in the hands of a newly created Department of Economic Affairs. It has since ceased to exist and now physical planning matters and the work of the economic planning regions are dealt with by the new super-ministry, the Department of the Environment, while industrial policy is handled elsewhere, within the Department of Industry. Imminent legislative and administrative devolution to Wales and Scotland will enable a considerable amount of coordination in physical and economic planning in the future in those two countries.

Many regional and sub-regional plans and studies have now been produced. These have guided local authorities in their detailed day-to-day planning as well as in the preparation of their structure plans. They have also provided basic data for central government about the regions' physical, social and economic resources, and hence have guided policy-making on the balance to be achieved in development between the regions. At last it can be claimed that a proper start has been made at integrated regional planning in the United Kingdom. It has not yet proceeded far and doubtless the system is capable of considerable improvement and further integration in the next

Figure 9.4. United Kingdom: economic planning regions.

decade. Its success cannot now be judged, since the regional problems of the country are immense and can only be solved over a period of many years. But a structure exists, the techniques of analysis are available and a growing number of professional planners and other specialists are aware of the needs and also of the means by which those needs can be satisfied. The real hope is that within the United Kingdom a planning response has evolved which is as powerful and as focused as the problems it seeks to cure. But only time will tell.

References

Abrahams, H., 1960, *Britain's National Parks,* Country Life, London.

Ashworth, W., 1954, *The Genesis of Modern British Town Planning,* Routledge and Kegan Paul, London.

Burns, W., 1963, *New Towns for Old,* Leonard Hill, London.

Burton, T.L. and G. P. Wibberley, 1965, *Outdoor Recreation in the British Countryside,* Wye College.

Cameron, G. and D. B. Clark, 1966, *Industrial Movement and the Regional Problem,* Oliver and Boyd, Edinburgh.

Cherry, G. E., 1974, *The Evolution of British Town Planning,* Leonard Hill, London.

Chisholm, M., 1962, *Location of Industry,* Political and Economic Planning, London.

Chisholm, M. and G. Manners (Eds.), 1971, *Spatial Policy Problems of the British Economy,* Cambridge University Press.

Chisholm, M. and H. B. Rodgers, 1972, *Studies in Human Geography,* Heinemann, London.

Chisholm, M. and J. Oeppen, 1973, *The Changing Pattern of Employment,* Croom Helm.

Coates, B. E. and E. W. Rawstron, 1971, *Regional Variations in Britain,* Batsford, London.

Cullingworth, J. B., 1964, *Town and Country Planning in England and Wales,* Allen and Unwin, London.

Daniels, P. W., 1969, Office decentralisation from London—policy and practice, *Regional Studies,* **3,** 171–178.

House, J. W. (Ed.), 1973, *The U.K. Space,* Weidenfeld and Nicolson, London.

Howard, R. S., 1968, *The Movement of Manufacturing Industry in the United Kingdom 1945–65,* H.M.S.O.

Johnson, J. H. (Ed.), 1974, *Suburban Growth,* Wiley.

Luttrell, W. F., 1962, *Factory Location and Industrial Movement,* N.I.E.S.R.

Manners, G., (Ed.), 1972, *Regional Development in Britain,* Wiley.

McCrone, G., 1969, *Regional Policy in Britain,* Allen and Unwin, London.

Needleman, L. and B. Scott, 1964, Regional problems and location of industry policy, *Urban Studies,* **1,** 153–173.

Osborn, F. and A. Whittick, 1963, *The New Towns—the answer to megalopolis,* Leonard Hill, London.

Rhodes, J. and A. Kan, 1971, *Office Dispersal and Regional Policy,* Cambridge University Press.

Self, P., 1957, *Cities in Flood,* Faber and Faber, London.

Senior, D. (Ed.), 1966, *The Regional City,* Longmans, London.

Thomas, D., 1970, *London's Green Belt,* Faber and Faber, London.

10

Republic of Ireland

James H. Johnson

Introduction

The Republic of Ireland, comprising 26 out of the 32 counties that make up the whole island of Ireland, is a politically independent state which has had power to shape its own regional development since 1922. Political independence, however, should not be allowed to mask the very strong economic, social and demographic connections with the United Kingdom. Certainly it can be said that the single most important factor influencing many aspects of the human geography of Ireland is the presence nearby of the larger island of Great Britain, while the existence of the province of Northern Ireland as a political unit independent from the rest of Ireland also complicates the operation of various aspects of policy-making in the Republic of Ireland, particularly in the North-West, where the border weaves a tortuous path through areas with similar demographic and agricultural problems and almost isolates Donegal from the rest of the State.

The key to many of the State's social and economic attributes lies in the size, distribution and history of its population. At the last census, in 1971, the total population of the Republic was just over 2,900,000, making it one of the smallest countries in Western Europe in terms of its total population. The resultant limited home market makes many of the economic problems of the Republic difficult to solve within the confines of the national economy, since the small domestic demand restricts the kinds of manufacturing industries that can flourish without recourse to the greater uncertainties of an export market. The small population also affects the scale at which regional problems are perceived and solutions attempted. Judged by the planning regions commonly found elsewhere, all of Ireland might well seem an appropriate 'regional' area for planning but, leaving on one side the political separation of Northern Ireland, there is sufficient awareness of local problems among the citizens of the Republic to have demanded the recent development of a system of allocating resources to different parts of the country on the basis of regional plans, with the result that some of these regions have very restricted populations. For example, the North-West Region was the smallest with only 78,549 people in 1971, comparable with a small British city, and it might be argued that in dealing with such restricted numbers it is difficult to bring precision to the planning of employment, since at this scale the quirks of fortune affecting individual firms can assume unpredictable importance.

The small total population may be one reason why Irish policy-makers often appear to have been particularly obsessed with national population numbers. This concern has been reinforced by the history of Irish population during the nineteenth and twentieth centuries, since, following the Great Famine of 1845–47, the population of Ireland fell dramatically and this decline continued until the outbreak of the First World War

(Commission on Emigration, 1955). The most important cause of this decline was emigration: between 1853 and 1900 over 3,300,000 people of Irish stock emigrated from the British Isles to places other than Europe, 84 per cent going to the U.S.A. To this must be added the flow of permanent emigrants to Britain, which is more difficult to quantify since there were many journeys made in both directions across the Irish Sea and there are no records of gross or net movements. Between 1841 and 1901, however, the total number of permanent emigrants to Great Britain cannot have been less than 700,000 and is likely to have been more.

In spite of great outward movement the total population might not have been reduced if it had continued to reproduce itself at the rate found before the Famine. In the second half of the nineteenth century, however, Ireland was notable for a declining marriage rate and a consequent reduction in the birth rate. The falling marriage rate was caused by an increase in permanent celibacy and also by a later average age of marriage. This curious phenomenon was directly and indirectly associated with emigration. Directly the process of emigration involved young unmarried adults, and thus reduced the marriageable population of the country. Indirectly its results were equally important. Because it was easier for a single person to emigrate and because emigration became the socially accepted means of economic advancement, many people who eventually remained in Ireland delayed marriage until all hope of emigration had faded. The eldest sons of farmers who stayed at home also delayed marriage until they could inherit an undivided family holding—a system made possible by the emigration of most of their brothers and sisters.

As a result of the combined effects of reduced fertility and emigration, the population of what is now the Republic of Ireland fell from 6,500,000 in 1841 to about 3,200,000 in 1911. Urban growth during this period of population decline was extremely modest, rural depopulation devitalized life in the countryside and, during a period of rapidly developing nationalism, emigration was seen as a process by which the nation's life blood was being drained away, although the economic advantages that migration brought to the individuals involved were too often overlooked.

Economic Policies after Independence

These many years of intense population decline left an indelible mark on Irish political consciousness and, in particular, influenced the economic policies that were adopted in the early 1920s (Meenan, 1970). These new policies were designed to promote economic independence from Britain to parallel the political independence that had already been achieved. As a result a tariff wall was erected, designed to provide an economic environment where new manufacturing industries could take root and serve a predominantly home market. A protected home market also appeared to be a mechanism for encouraging the growth of employment in smaller towns. National self-sufficiency was also developed in other fields. The State intervened in the establishment and running of organizations like the Electricity Supply Board and Bord na Mona (the Peat Board) with the aim, among others, of exploiting indigenous sources of energy, like peat fuel and hydro-electric power. The government-owned Irish Sugar Company aided import substitution by encouraging farmers to grow sugar beet. State forestry, begun by the British Government, was developed further to maintain population on the land as well as eventually to reduce imports of timber.

Institution of national independence and the establishment of this inward-looking economic policy coincided with a new period in the changing population of Ireland, which extended from the first census of the Republic of Ireland in 1926 until 1951. During this period the rate of population decline slowed down greatly. Between 1926 and 1951 the total population of the State fell by only just over 11,000 people (compared with over 167,000 in the immediately preceding intercensal period from 1911 to 1926). At first sight this improvement would suggest that the new economic policy had been dramatically successful. Certainly some new industries which had previously been unable to survive in the face of British competition became established. Footware, cutlery, crockery, textiles and cement are only a selection from a longer list of such products. The electricity industry eventually achieved a situation in which it produced over 80 per cent of its total output from indigenous sources of peat and water-power. Employment in manufacturing rose by over 63 per cent from 164,000 in the 1920s to 268,000 in the early 1950s. The tertiary sector of the economy also expanded, if somewhat less rapidly.

Table 10.1. Republic of Ireland: Population Change, Natural Increase and Net Emigration, 1871–1971

Period	Population change	Natural increase	Net emigration
1871–81	−183,167	327,681	510,848
1881–91	−401,326	195,999	597,325
1891–1901	−246,871	149,543	396,414
1901–11	− 82,135	179,404	261,539
1911–26[a]	−167,696	237,333	405,029
1926–36	− 3,572	163,179	166,751
1936–46	− 13,313	173,798	187,111
1946–51[b]	+ 5,486	125,054	119,568
1951–56[b]	− 62,328	134,434	196,762
1956–61[b]	− 83,561	132,080	215,641
1961–66[b]	+ 65,661	146,266	80,605
1966–71[b]	+ 87,228	148,148	60,920

[a] 15-year period; [b] 5-year period

The success of the policy, however, is much less impressive if it is evaluated in the context of its original goals and of the general demographic situation in Ireland immediately before and afterwards (Walsh, 1968). Even before Irish independence the rate at which population was declining was being reduced (Table 10.1). That slower rate of decline can be detected between 1901 and 1911, and although there was an increased fall in population between then and the next census in 1926, it should be recalled that the intercensal period was 5 years longer in this particular case and that also the upheaval of Irish independence during this period encouraged the withdrawal of British citizens to Great Britain or Northern Ireland and represented an exceptional event in the process of population change. Hence the period of underlying demographic stability can be pushed back earlier than the institution of new policies for industrial expansion. At the other end of the period similar difficulties arise. Without any change in economic policy, population decline in the Republic became apparent again after the census of 1951 and during the decade to 1961 reached nineteenth-century levels of decrease.

Even with the period of demographic stability from 1926 to 1951 there is an element of illusion in the statistics. In spite of changes during the nineteenth century the rate of natural increase was high compared with the levels now becoming established elsewhere in Western Europe. Although the percentage of married women of fertile age remained relatively low, their fertility was high and showed remarkably little fall during this period. Any drop in fertility was much more than offset by a sharp fall in infant mortality, which dropped from over 105 per 1000 live births in 1901 to 45 per 1000 in 1950. As a result, unlike other countries, the Republic showed no reduction in natural increase during this period and the stable total population merely indicates that emigration was continuing at a high level. During this period, in fact, it appears that the level of natural increase was an important factor influencing the volume of out-migration. What had been created was a relatively static economic situation in which, although new jobs were created in manufacturing, there were in fact no additional jobs, because of the continued decline in the agricultural labour force.

The policy was also a failure if internal contrasts within the Republic of Ireland are considered, since hidden behind the virtually stable total population were important variations in different parts of the country. The population of rural areas was decreasing almost everywhere. The decrease was relatively lower in the South-East: in Leinster rural population declined by 13 per cent between 1911 and 1951 compared with 22 per cent in the whole country. This area was agriculturally more prosperous than many other parts of the Republic and accessibility to Dublin must have played some part in arresting population decline here.

The forces encouraging rural depopulation were also resisted somewhat more firmly in the more prosperous parts of the province of Munster, but in other areas the decline was grave, particularly in the North-West where many areas suffered a higher than average reduction in population during the first half of the twentieth century, sometimes reaching more than 10 per cent per decade. This region has a long tradition of emigration and is roughly coincident with the area occupied by the smallest farms. Population decrease was most severe where these small farms were being operated under difficult agricultural conditions, as, for example, in most of Leitrim and parts of Donegal. Here, agriculture was often less intensive than in other parts of Ireland, as a result of an increasing emphasis on the production of store cattle. Farming no longer provided an acceptable standard of living for the smallest farmers and their families, but there was no alternative employment. Decades of constant emigration also meant that most rural craftsmen and agricultural labourers had been removed from the population.

In those areas where depopulation was most severe, the reduction in numbers between every census was relatively constant. Where rural population totals were less severely reduced, the rate of change was more likely to vary (Duncan, 1942–4). What seems to have been happening was that those areas with a smaller reduction in total population were more sensitive to the changing economic climate from period to period. In areas of considerable depopulation, on the other hand, the relative state of the Irish and British economies was of little immediate importance, since between every census local opportunities compared unfavourably with those in Britain or, for that matter, in Dublin. In these areas too, the migrants included small farmers as well as agricultural labourers and rural craftsmen, whereas elsewhere the number of actual landholders often remained relatively stable in the twentieth century.

Urban population exhibited similar internal contrasts within the Republic. Certainly the proportion of the total population that lived in towns increased. In 1901, for example, 28 per cent of the total population of what is now the Irish Republic lived in towns and 72 per cent in the country. By 1951 41 per cent of the population was found in urban areas and 59 per cent in rural. But during this period the rural population of the Republic had fallen by over 602,000 so this changing proportion was as much the result of a decrease in rural population as of urban growth.

In fact, the growth of urban population was almost all concentrated in Dublin, and the smaller and more remote towns more often than not had a declining population. Not surprisingly, the city of Dublin and its immediate evirons dominated long-distance movements of population within the Republic. Although Dublin enjoyed a stately expansion during the nineteenth century, the restoration of its function as a capital city in 1922 stimulated its economic growth both in manufacturing and in tertiary industries. Dublin also provided the most attractive location for new employment in both manufacturing and services because it represented a large and increasing proportion of the national workforce and home market. In addition, the city lay at the point of maximum access to the rest of the country and it was by far the largest port providing direct access to Britain—still the dominant source of imports and the destination of most agricultural and industrial exports.

It is clear that the policy of economic protection failed to revitalize the small towns, particularly those in the West, since even those industries producing for the local Irish market were often unable to resist the pull of Dublin and of the larger towns like Cork and Limerick. The policy of industrial protection did nothing to counteract rural depopulation by providing alternative urban employment in the West and North-West of the Republic, and it did little to steer economic expansion away from Dublin and its immediate zone of influence, which grew rapidly at the expense of the rest of the country.

The need for a revision of economic policy became increasingly clear in the 1950s. After the recovery of the British economy from the upheaval of the Second World War, the Republic began to experience emigration at the highest levels that had been recorded in the twentieth century. As a result, the total population of the State also began an accelerated decline and those areas where population had been decreasing in the 1920s and 1930s experienced a remarkable reduction in population, particularly in the light of the fact that rural depopulation had been under way in these districts for a century (Johnson, 1963). Between 1951 and 1961, for example, nearly every rural district in Eire north of a line between Galway Bay and Dundalk Bay suffered a decline in population of more than 10 per cent. In parts of Leitrim, Sligo and Donegal this decline was over 15 per cent. Even in the South-East, where agricultural incomes had been increasing most, the population was sometimes reduced by more than 10 per cent and often by more than 5 per cent. Only three rural districts showed an increase of more than 5 per cent and their location adjacent to Dublin and Cork suggests that the increase represented the spread of suburban population beyond the legal limits of these two cities, rather than any genuine increase in rural population. Towns of less than 5000 people commonly showed a reduction in population during this period except where they were within the ambit of Dublin. In other words, the policy of economic protection had visibly collapsed in the changed situation after the Second World War, with the growing attraction for Irish workers of the British labour market, to which they had free access. Hence, the dominating factor behind this severe population decrease was the contrast between the relatively dynamic economy of Great Britain during this period and the more or less stagnant economy of the Republic.

This changed demographic situation revealed starkly the limitations of the policy of economic protection, since a total population of less than 3,000,000 could only provide a restricted base for employment growth to meet the demands of the home market. Similarly it was more obvious that the absence of any explicit regional component in the policy of protection was politically, socially and economically unacceptable. The contrast in the economic and social environment of a great city like Dublin with the rest of the country was so great that a very uneven result from a nationally orientated policy was inevitable in any situation, but this contrast emerged much more sharply during this period of alarming national population decrease.

A concern with these internal contrasts within Ireland can perhaps be traced back to the work of the Congested Districts Board, which was established by the British Government in 1891 and gave special assistance to those areas where the rural population was particularly numerous in relation to the resources available to support it. One of the Board's most important activities was the restructuring of agriculture, although it could be reasonably argued that in recasting the rural landscape the Board tended to allocate holdings that were too small for commercial survival. The work of the Congested Districts Board was taken over after independence by the Irish Land Commission, which concentrated on improving the viability of holdings in much of the West; but again the sizes of the compact farms which it created tended to lag behind the commercial realities of farming, not surprisingly since one of the implicit aims of the Irish Land Commission was to maintain as many people as possible on the land.

The Congested Districts Board had also encouraged rural craft industries, but a more recent and more thoroughgoing policy of increasing manufacturing in the western counties was introduced by the Underdeveloped Areas Act of 1952, which attempted to steer more employment to the West and North-West of the Republic by offering grants for buildings and equipment for small-scale rural industries (O'Neill, 1971). This first step had little effect in reversing depopulation, but it was expanded by later measures to develop the tourist industry and to attract new manufacturing by the provision of grants and tax concessions. For example, the Industrial Grants Act of 1956 gave higher grants to firms locating in the Underdeveloped Areas. It should be noted, however, that this piece of legislation applied to the whole country, so that it also initiated a policy of encouraging national development generally by means of increased industrial grants. The policy of encouraging national economic growth emerged as a dominant one in the late 1950s as the grim population statistics of the decade began to penetrate the consciousness of administrators.

At this time, too, an interesting, if minor, step in the direction of regional development was taken with the establishment in 1957 of Gaeltarra Eireann—a state-sponsored board under the direction of the Department of the Gaeltacht, which was founded to encourage the extension of Irish-speaking in those areas where native Irish speakers still formed a substantial proportion of the population (Figure 10.1). It was the intention of Gaeltarra Eireann to increase employment in these Irish-speaking rural areas, thus restraining out-migration from the nuclei where Irish was still a functioning language; but as there were only about 80,000 people in the Gaeltacht, these developments affected a very small proportion of the total population.

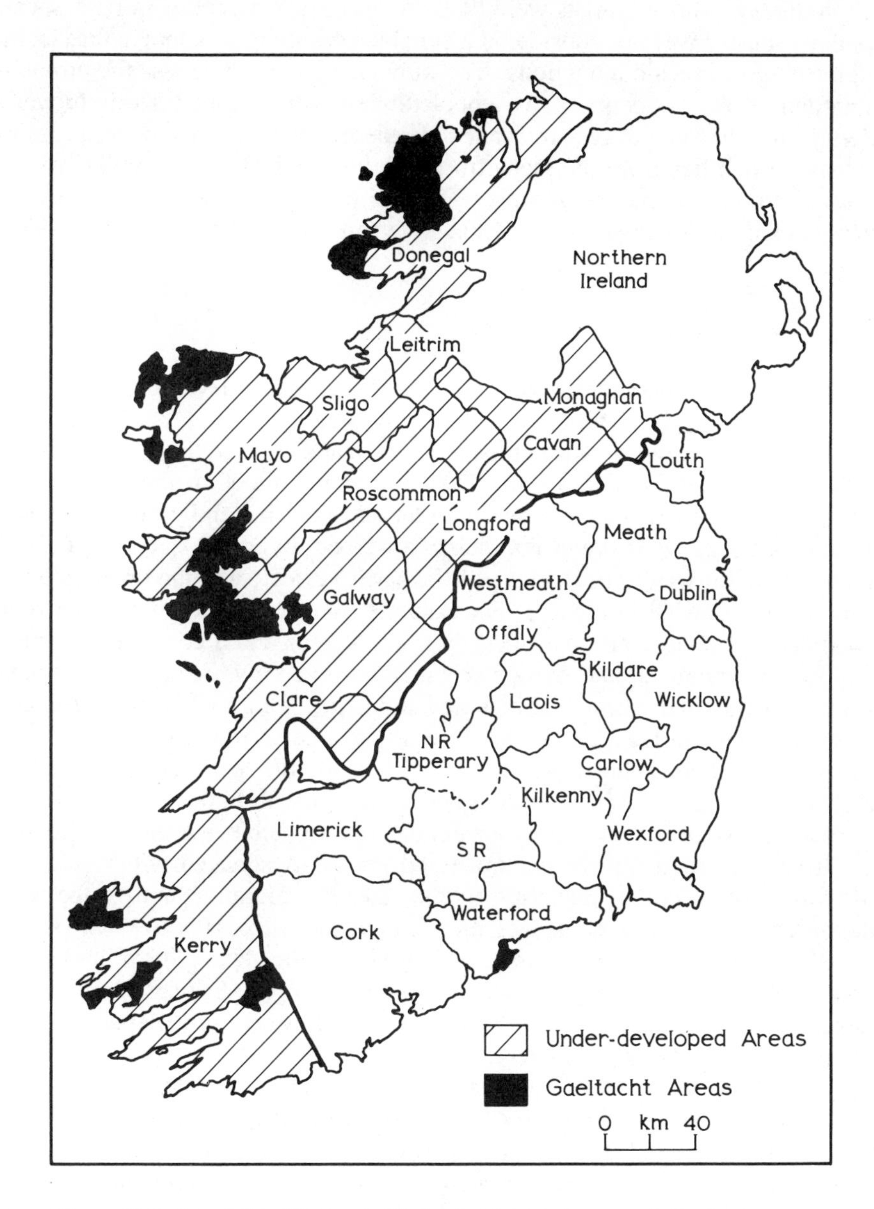

Figure 10.1 Republic of Ireland: Underdeveloped Areas and Gaeltacht areas.

Gaeltarra Eireann was empowered to give financial assistance to new industries and also established its own industrial activities, with the result that a number of small factories, producing such items as tweed, knitwear, toys and plastic goods, were sited in the Gaeltacht. In fact, these factories had only a limited success, and some have ceased production. Since tourists were likely to encourage a decline in Irish-speaking, the holiday industry was not considered a suitable activity to promote in the Gaeltacht, in spite of its great scenic attractions. The only exception to this was the provision of accommodation for students and schoolchildren who wished to improve their knowledge of the language. In spite of these efforts, and certain grants to Irish-speaking families from the Department of the Gaeltacht, the population of these areas has continued to decline sharply. The one exception is a small area of county Waterford, of doubtful linguistic usefulness, where the total population in 1971 was only 866.

Expansion in the 1960s

The new policy of encouraging general national economic development, hinted at in the Industrial Grants Act, was more explicitly stated in the First Programme for Economic Expansion, which was published in 1958 and covered the years from 1958 to 1963 (Government of Ireland, 1958). The Programme aimed at the revitalizing and transforming of the Irish economy in order to halt the emigration of displaced agricultural workers who were not being absorbed by other sectors of the Irish economy. This aim was to be achieved by what, in effect, was the opening up of the Irish economy to outside influences. For example, the Government no longer required that all industries should be controlled by Irish citizens and a concerted attempt was made to attract foreign capital, particularly if it was to be invested in export-orientated industries. To this end credit facilities and industrial grants were improved and tax reliefs given on export earnings. As a result, a greater proportion of government investment was channelled towards directly profitable activities, particularly in manufacturing. A target of a 2 per cent per annum growth in the G.N.P. was set—in the event a modest aim, since more than double this was achieved, most of the increase coming from the production of manufactured goods. Associated with this economic growth was a considerable reduction in the rate of emigration and in the resultant population decline. In its early stages the demographic success of the new policy was reinforced by the sluggish British economy of the late 1950s and by new Commonwealth sources of migrant workers attracted by the British labour market. In the longer run, however, the new outward-looking policy certainly played an important part in reversing population decline.

The First Programme was followed by a Second Programme for Economic Expansion, which operated from 1964 to 1969 (Government of Ireland, 1964). Entry to the E.E.C. was expected during the operation of the Second Programme, with the resulting economic changes that this would bring. As a first preparatory step into a much more competitive environment, an Anglo-Irish Free Trade Treaty was signed in 1965 (and an application for membership of the E.E.C. was submitted in this same year). The Anglo-Irish trade agreement abolished duty on Irish exports into the United Kingdom and, as a result, foreign firms were given an additional stimulus to establish plants in the Republic of Ireland because of the entry which such a location gave to the British market. In turn, and equally important, the agreement involved the progressive removal over a ten-year period of the protective duty on Irish imports from Britain. Thus, in effect, the treaty indicated a willingness to dismantle the protected home market and confirmed the changed orientation of Irish economic policy.

The failure of Britain to achieve entry to the E.E.C. during the 1960s meant that inevitably the Republic also could not proceed alone, because of the importance of the British market for its exports, particularly agricultural products. As a result it was necessary to introduce a Third Programme in 1969 to cover the years to 1972 (Government of Ireland, 1969). But although the Second Programme failed to reach its detailed targets, the 1960s as a whole represented a period of remarkable economic success for the Irish economy as a whole.

Table 10.2. Republic of Ireland: Growth of G.N.P., 1961–71

Year	G.N.P. at 1968 prices (£m)	Percentage change over previous year
1961	1002·4	
1962	1035·2	+3·3
1963	1077·8	+4·1
1964	1117·8	+3·7
1965	1148·8	+2·8
1966	1164·2	+1·3
1967	1227·1	+5·4
1968	1330·4	+8·4
1969	1384·4	+4·1
1970	1422·0	+2·7
1971	1464·0	+3·0

The economic growth during this period is summarized by the expansion of the G.N.P. The period of expansion which had been introduced in the later 1950s by the First Programme was continued in the 1960s. Between 1961 and 1971 the G.N.P. increased by 46 per cent; and although there had been a setback in 1965–66 when a growth of only 1·3 per cent was attained, the achievement throughout the decade was remarkably consistent (Table 10.2). This success was reflected in levels of emigration. During the 1950s emigration had been running at over 40,000 per annum, but between 1961 and 1966 annual emigration was reduced to just over 16,000. Between 1966 and 1971 emigration was further reduced to 12,000 per annum. Natural increase during this same decade was slightly higher than in the 1950s, with the result that the total population increased by over 65,000 (2·3 per cent) in the period from 1961 to 1966 and by over 87,000 (3 per cent) from 1966 to 1971, thus reversing an almost unbroken downward trend of over a century. (The only intercensal period between 1851 and 1961 at which there was an increase was 1946–51, when population increased by only 5486: 0·2 per cent.)

The number of people employed within the Republic did not show the same dramatic increase as national productivity. In fact between 1961 and 1971 the total labour-force at work increased by only 1·8 per cent (just over 18,000 people). If allowance is made for unemployment, the total labour force increased by 2·8 per cent (31,000). More

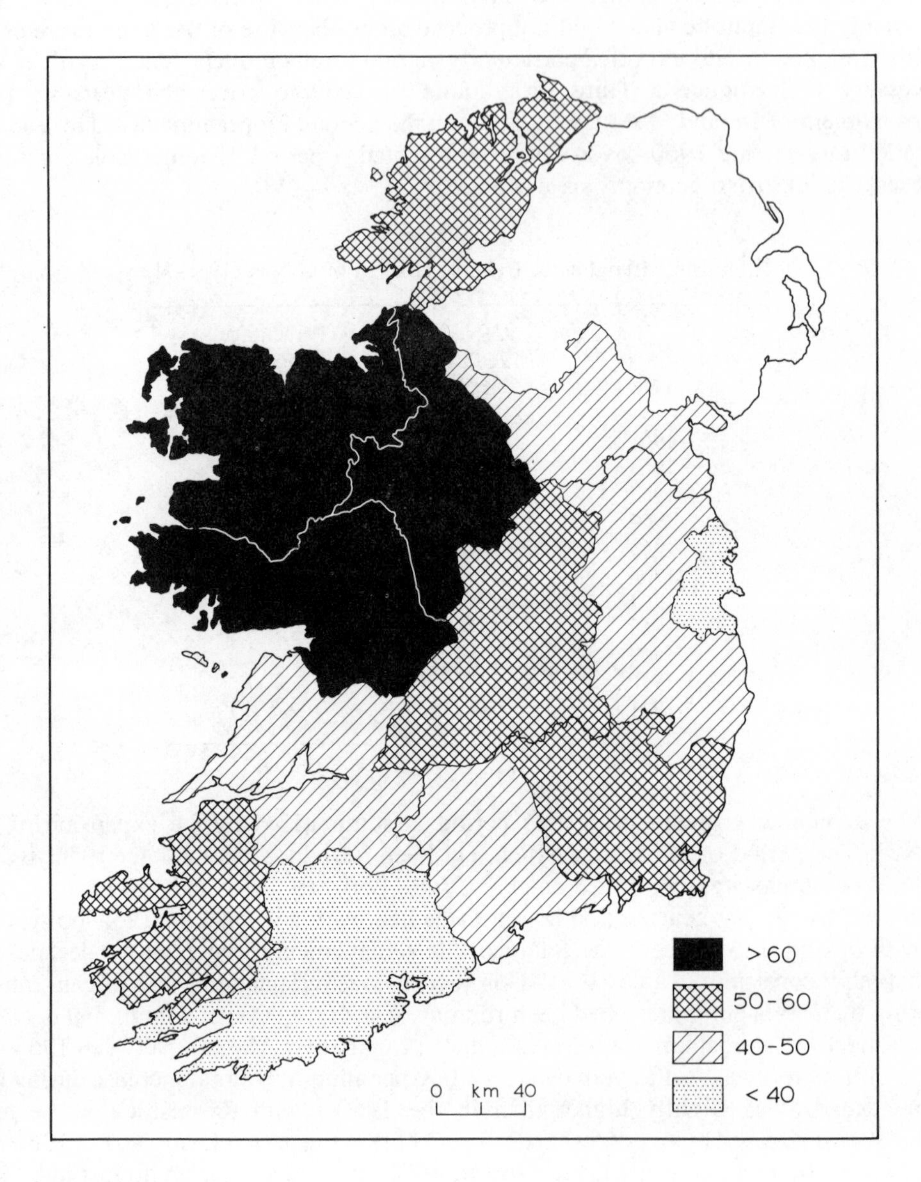

Figure 10.2. Republic of Ireland: percentage of total employment in slow growth sectors, 1966.

important than any increase in the total number of jobs was a reorientation of the relative importance of the various sectors of the economy. Employment in agriculture, forestry and fishing decreased by over 97,000 (25·7 per cent). During the same period employment in services (including tourism) increased by over 45,000 (nearly 11 per cent). Finally the industrial sector (which includes mining, quarrying, turf production, building and construction, and gas, water and electricity supply as well as manufacturing) showed an increase of nearly 71,000 (27·5 per cent). 44,200 of these additional jobs were in manufacturing industry. This change in emphasis was also accompanied by a remarkable growth in exports, which increased by 198 per cent (at current prices). This overall figure hides a 593 per cent increase in the export of manufactured goods, which by 1971 had become the driving force and principal component of Irish exports. The key to this rise lay in 550 new industrial enterprises established during the period. Some 68 per cent of these were the result of foreign participation, largely by firms from Britain (41 per cent), the U.S.A. (23 per cent) and West Germany (20 per cent) (I.D.A., 1972, part I).

The improvement in the general economic well-being as a result of these more expansionist policies is clear; but in spite of this success the detailed regional pattern of population change and employment is less impressive, with the slow growth sectors of the economy having a marked regional distribution (Figure 10.2). Donegal, the Midlands, the West and the North-West Regions all showed a continuing decline in population and high levels of out-migration. Although somewhat different maps are produced when individual criteria are examined in detail, such indices as population change, employment rate, level of personal income (Figure 10.3), population employed in industry and of total government expenditure for industrial development produce basically the same picture, with these less-favoured regions suffering consistently unfortunate conditions (I.D.A., 1972). This situation was not necessarily the result of a lack of good intentions by central government agencies: for example, in terms of spending on industrial development per head of the population, the North-West Region was ranked second in the country. This region, however, has the gravest problem. In agriculture it is burdened by small farms, which have survived in spite of the long history of emigration, by less profitable agricultural activities and by particular difficulties in the reorganization of holdings because of the retention of landownership by emigrants. The soils and climate of the North-West also impose inflexibility on the viable types of agricultural output that can be produced: the degree to which agricultural officers can promote more intensive farming is limited and hence a further decline in the agricultural labour force is inevitable. As in other rural areas of the Republic of Ireland the solution for this area must lie in the stimulation of manufacturing industry—although the degree to which this is necessary is perhaps greatest here. So too are the difficulties, since in this region (and also in Donegal) the smallness of the base for manufacturing industries is a severe limitation, impeding any substantial development of industrial employment, except in small independent units. In some favoured localities tourism is providing a source of income which could be further developed, but the very short season makes large investment in conventional tourist facilities a doubtful economic proposition and the very temporary increase in employment during the summer can do little to restrain permanent emigration. It is not surprising therefore that population decline continued here relatively unabated between 1961 and 1971, in spite of population increase elsewhere.

By way of contrast the situation in the East Region (dominated by Dublin and its

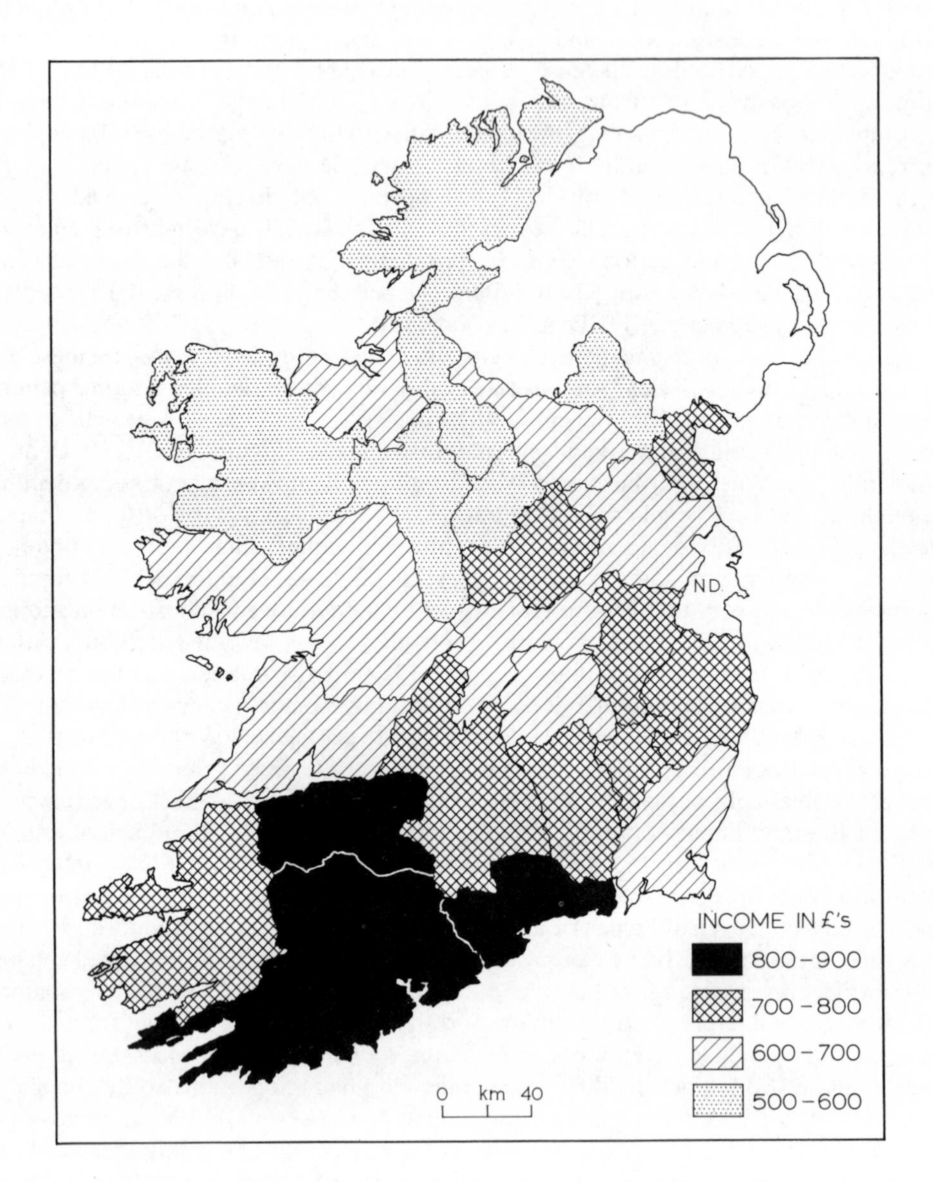

Figure 10.3. Republic of Ireland: personal income per head of the workforce, 1965–66.

environs) exhibits a different range of problems. This area emerged, demographically and economically, as the most favourable area of the Republic, with County Dublin experiencing nearly 29 per cent of the total population growth in the Republic between 1961 and 1971. During the 1960s the proportion of manufacturing jobs located in and around Dublin declined, but this masks an absolute increase in the amount of manufacturing. At the same time there was a continued expansion of service employment. Under Irish conditions, where Cork is Dublin's nearest rival with a population of about 125,000 and Limerick (the third urban settlement) has only 58,000 people, employment in the tertiary sector (largely in offices and shops) would be difficult to move away from the metropolitan centre. Manufacturing might be more susceptible to relocation, although the non-economic attractions of Dublin for entrepreneurs and managers should not be underestimated.

In this context, too, it is important to notice that, unlike the situation in earlier decades, much of the population growth of Dublin was the result of natural increase. Although the population expanded by over 129,000 between 1961 and 1971, net immigration was less than 13,000. It follows that the solution of the problems of the less-favoured regions does not lie in steering too much economic growth away from the East Region, since the 670,000 inhabitants of greater Dublin have as much right to expect the provision of employment close to home as the citizens of any other region of the country.

The Growth of Regional Policies

Running parallel with the survival of regional problems during this period of economic expansion was a continuing and strongly growing effort by the government to resolve more localized problems in the Irish economy. These attempts suffered from the lack of any coherent overall strategy, but nevertheless they represented an important series of steps in the development of current policies and in the evolution of a more all-embracing approach to the regional planning problems of the Republic.

One important early step associated with the drive for more rapid economic growth was the establishment of the Shannon Free Airport Industrial Estate in 1959. The importance of this development was that it illustrated the feasibility of creating employment for 5000 people (more than half of them men) in a location adjoining the undeveloped areas of the West. There is no doubt that the success of the Shannon Estate influenced later policies, but it is important to note the special circumstances that existed there.

This industrial estate was partly designed to boost the declining traffic of Shannon Airport, which was being increasingly bypassed by longer-range aircraft (Soulsby, 1965). As a result special efforts were made to exploit the facilities of the international airport for the handling of air-freight and for the easy transit of visiting managers, giving the estate advantages that were unique in Ireland. In addition, the estate was designated a free trade area to reduce the complications of import and export—again a unique feature, difficult to repeat in many other locations. Finally, it could be argued that a disproportionate amount of effort was directed into making this estate a showpiece for the new expansionist industrial policies.

These advantages attracted British, Japanese, South African, Dutch, American and German firms, producing goods as diverse as pianos and transistor radios; but it remains a matter of considerable doubt whether such a success could be repeated elsewhere, particularly in the West. The industries on the estate in fact established few linkages with each other or with other firms within Ireland, although the concentration of activity certainly improved the educational, shopping and training facilities available in the nearby town of Limerick and probably made this area more attractive to further industrial growth. Yet the estate remains an enclave, which assists the Republic's balance of trade, but hardly induces self-sustaining growth in the national economy or extends an influence beyond the local labour market area centred on Limerick.

In spite of these reservations the success of the Shannon estate at a time when the idea of growth poles was beginning to make an impact on regional planning theory was an important factor in the evolution of a new policy of concentrating aid in a limited number of centres. The early rumblings of this new policy can be detected in the findings of the Committee on Industrial Organization (C.I.O.), appointed by the government in 1961 to steer industry into a more competitive world. In 1962 the C.I.O. suggested the abandonment of special aid to the underdeveloped areas and the concentration of funds in a small number of growth centres. The suggestion that policy should be reorientated in this way was confirmed by later reports from the same committee. The Committee on Development Centres and Industrial Estates established by the Minister for Industry and Commerce in 1963 to investigate the problem further, concluded that development centres would provide an effective strategy for the attraction of new industry, and this view was also supported by the National Economic Council in 1965 (O'Neill, 1971). As a result a government policy statement was issued in 1965 accepting the concept of development centres as a means of promoting further expansion of economic activity, but doing nothing further at this stage.

In the meantime the Second Programme for Economic Expansion had been launched and had included a proposal that a national physical planning programme should be prepared to guide the social and economic changes associated with economic growth. To implement this proposal the State was divided in 1963 into nine planning regions, formed by grouping together administrative counties in a rough and ready way (Figure 10.4). In the next year this growth of interest in regional administration was confirmed by the establishment by Bord Failte Eireann (Irish Tourist Board) of another set of similar, but not identical, areas associated with eight regional tourist organizations, which were to take responsibility for developing and promoting regional tourist facilities. (In passing, it is alarming to note that yet another regional division was established in 1970 for the operations of Regional Health Boards (Figure 10.4).) In 1964 also, reflecting this more practical concern with regional development and physical planning, An Foras Forbartha (The National Institute for Physical Planning and Construction Research) was established, with assistance from the United Nations. In the next year the government set up a regional development committee, representative of all relevant government departments, to advise on regional policy; and consultants were appointed for the preparation of three regional studies based on the new planning regions, some of the necessary staff being provided by An Foras Forbartha. In effect, the three reports which emerged translated the growth centre idea into more tangible proposals.

The first, a study of the Limerick region by Nathaniel Lichfield and Associates was published in 1966 (Lichfield, 1966). This study attempted to marry economic and physical planning for a region which included the rapidly growing Shannon Estate. Within the region Lichfield wrestled with the strategy problem which also applied at a broader scale for the country as a whole: whether to concentrate all the investment eggs

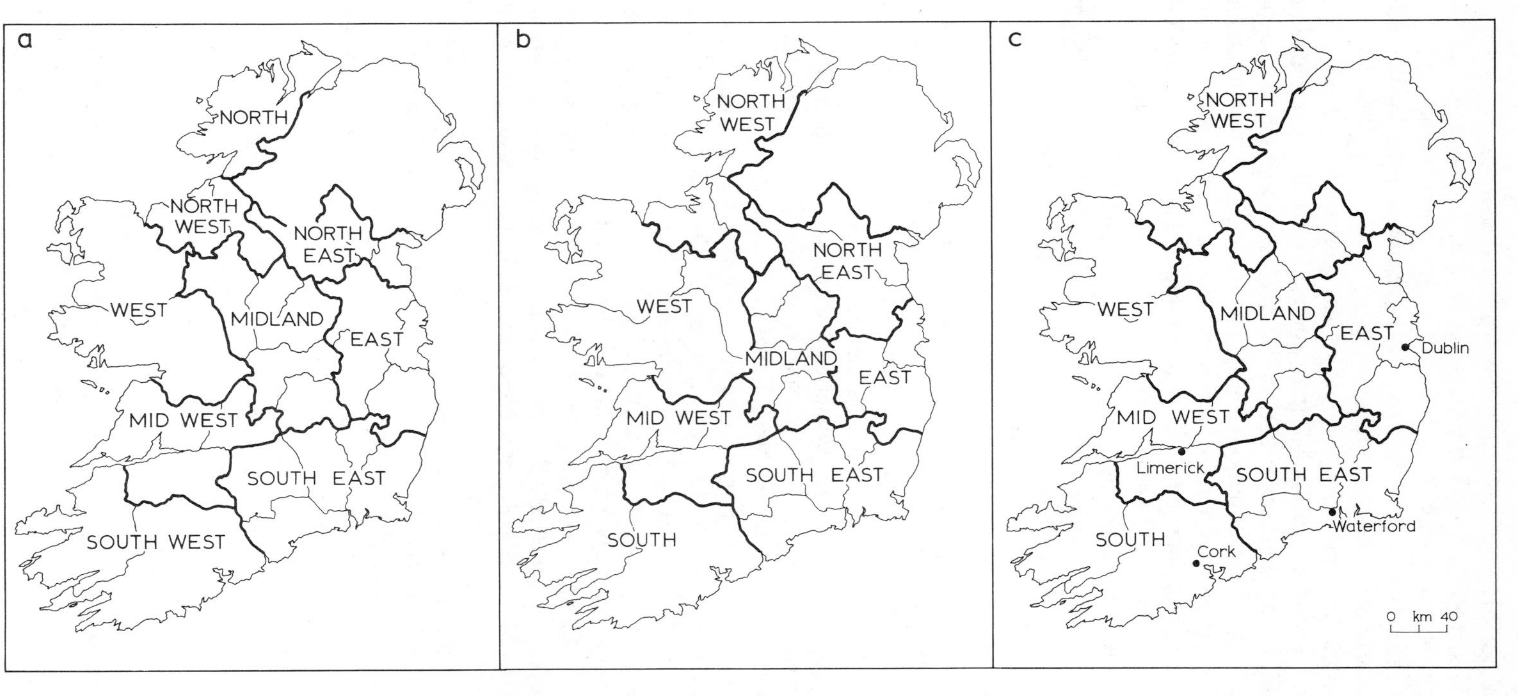

Figure 10.4. Republic of Ireland: regions defined by three authorities. (a) Industrial Development Authority Regions (these are currently regional planning regions, based on regions established in 1963); (b) Tourist Board Regions (established 1964); (c) Health Board Regions (established 1970).

in one basket or not? His sensible solution was to designate the nearby towns of Limerick and Ennis, together with the Shannon Estate, as forming a central growth area, with Nenagh, Roscrea and Thurles as subsidiary centres. The study as a whole, however, suffered from the lack of any overall government plan for the allocation of investment to the various regions of Ireland.

The second plan, for the Dublin Region, was prepared by Myles Wright and published in 1967 (Wright, 1967). Wright assumed that the continued growth of metropolitan Dublin was inevitable; and in his plan the necessary expansion of the metropolitan area was to be accommodated by a westward growth of the built-up area, fingering out along the road pattern. There were few substantial towns nearby which could form substantial counter-attractions to Dublin, but it was proposed that the populations of Drogheda, Naas/Newbridge, Arklow and Navan should be approximately doubled to produce outlying subsidiary growth centres.

The third and most influential document was prepared by Colin Buchanan and Partners in association with Economic Consultants, and published in 1968 (Buchanan, 1968). This project was originally directed at the remaining seven regions of the Republic, but in fact the plan proposed a strategy for the whole State and regional details were not worked out with any real precision. The major influence of the report was to articulate the argument for growth centres in a persuasive manner and to propose specific locations for their development. Economic growth was to be concentrated because it could be backed by good training facilities, effective communications and a reasonably large labour market. The greater ease of movement which the expansion of private transport was bringing to the Irish countryside would allow these growth centres to develop more extensive journey-to-work hinterlands around them.

In all, nine possible centres were designated (Figure 10.5). Three of these centres already possessed virile economic growth: Dublin, Cork and Limerick (including Shannon and Ennis). Dublin had such dominance in the urban hierarchy that its role as the most important administrative, commercial and industrial centre in the Republic could not be interfered with: the report indicated that no attempt should be made to stop Dublin's growth (although industrial expansion should not be given any special inducements). Cork and Limerick–Shannon were also designated as national growth centres, to be allocated a major share of the country's industrial growth, with a special effort being made to ensure their success.

According to the Buchanan strategy there would be advantages in 'a certain amount' of industrial development taking place in other centres as well, although from this choice of words it appears that these towns were only to have a minor role in national economic growth. These other towns, which were designated 'regional growth centres', were Sligo, Galway, Drogheda, Dundalk, Athlone and Waterford. There were still gaps in the regional coverage and hence it was suggested that Letterkenny might be chosen as a 'local growth centre' serving Donegal and that Tralee would fill a similar gap in the South-West. Castlebar was more tentatively recommended as a local centre for Mayo, and Cavan (really because there was nowhere else) was suggested for the difficult small-farm area embracing Counties Longford, Cavan and Monaghan.

The Buchanan strategy suffered from two basic defects. One was to do with the concept itself. The national growth centres might well be viable, but all the other centres were very small by West European standards. It is very unlikely that these smaller centres could become genuine centres of self-sustained growth. Their local labour markets would always be limited, there could only be a restricted number of

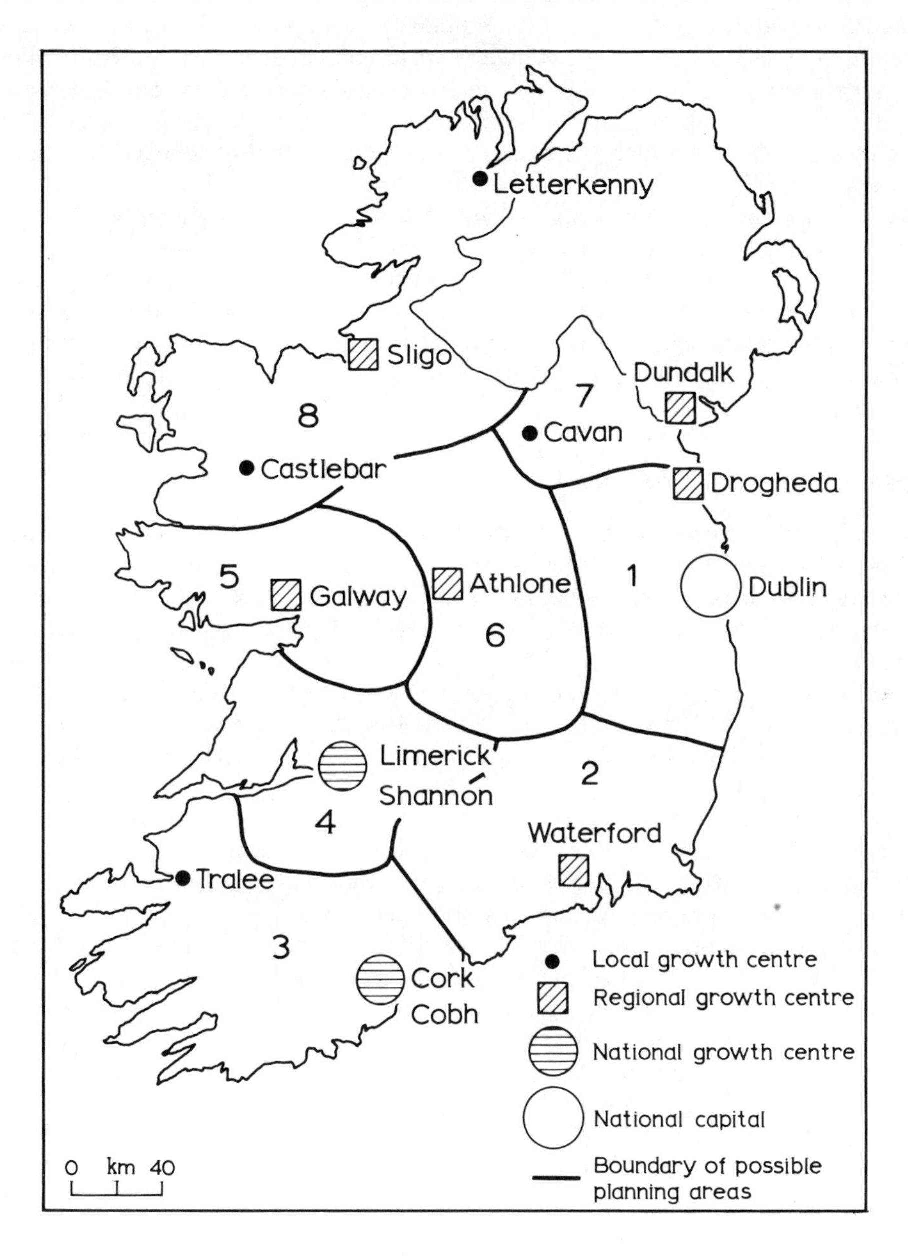

'igure 10.5. Republic of Ireland: growth centres and planning regions recommended by Buchanan.

industrial units in each and, even if it were possible to create external economies by concentrating industrial growth, this effect would always be of small importance. On cool reflection it also appeared doubtful if economic benefits would trickle down to the small towns and the rural population, particularly in the West. The industries that were likely to flourish in these centres would be small independent units, probably aimed at the export market. If so, then critics of the proposals wondered whether there were any real advantages in discouraging the establishment of new industries in other centres, which would otherwise miss the benefits of increased employment (O'Farrell, 1974; Newman, 1967).

The second difficulty was simply the problem of what is politically possible in a small nation, where the combined resistance of those towns not chosen for development could well be sufficient to overthrow the whole policy (e.g. Chadwick and coworkers (1972)). As a result the government, although it appeared to welcome the report, found difficulty in facing up to its more unpleasant implications and in due course a somewhat more equivocal position has become established in governmental attitudes.

Modifications to the Buchanan Proposals

The decision made immediately after the publication of the *Buchanan Report* tended to accept its findings implicitly: for example, industrial estates were sanctioned for many of the regional centres and industrial expansion was seen as the principal means of reshaping the economic map. As the main method of implementing the Buchanan recommendations the Industrial Development Authority (I.D.A.) was established in 1969. This authority is a state-sponsored organization which has been given responsibility for industrial development in the Republic. Its functions are to administer grants and incentives for industry, to develop industrial estates, to provide housing where necessary to attract employees, and to foster the national objective of regional development. Apparently in order to take account of the existence of a vested interest—albeit a successful one—the I.D.A. was made responsible for only eight of the planning regions, with the ninth region (the Mid-West) being administered for the I.D.A. by the Shannon Free Airport Development Company.

Buchanan had suggested a more logical recasting of the planning regions of the Republic, but in fact the I.D.A. planning regions were very similar to the 1963 regions, presumably to avoid upsetting county authorities. With the passage of time, other recommendations were also watered down. This became manifest with the publication in 1972 of regional industrial plans for the period from 1973 to 1977 (I.D.A., 1972). These plans are concerned with industrial employment only, but this limitation is not so grave as might be thought, since the plans also examine the contribution to regional development of agriculture and services, although proposals for developments in these fields lay outside the competence of the Authority.

The planning objectives and strategies suggested by the I.D.A. combine a mixture of the sensible and the fanciful. The realistic demographic objectives the Authority has taken upon itself include the increase of population in all regions in the longer term (by 1986), the reduction of emigration rates in all regions, and the restriction of Dublin's growth to the equivalent of its natural increase. In addition the Authority has also encumbered itself with more difficult social goals. Not only is the existing pattern of communities to be maintained and improved, but at the same time the clustering of geographically and functionally related towns is to be encouraged to form a stronger base for industrial and social development, and in each region a number of centres providing a wide range of higher order services is to be developed. The development of smaller centres is to be spurred by the location there of smaller industrial units, while additional employment is also to be provided in the Gaeltacht. Selected towns are to be made growth centres, but employment opportunities are to be provided within a travel-to-work time of not more than 30 minutes. These aims may well be admirable; but listed together they introduce a note of unreality, since not all of them fit together coherently. The amount of new employment available is not unlimited; and it is difficult to see how growth centres, short journeys to work, and the creation of manufacturing employment in rural areas and smaller towns can all be successfully achieved. Similarly, where and how are higher-order service centres to be conjured up in at least some of the Planning Regions? The clustering of related towns may just be possible in Limerick–Shannon–Ennis—but where else in the real geographical world of the Republic of Ireland?

The Authority's specific industrial objectives reveal similar unrealities. Suitable industries, according to the I.D.A., would have low capital intensity, use Irish materials and services, employ men, and link with other Irish industries. They would also be highly profitable, be likely to grow, demand highly-skilled labour, and be involved in international trade. Besides these excellent features the companies to be encouraged would be Irish-owned or have substantial participation by Irish principals, or they would be international companies which allow their Irish subsidiaries a high degree of independence. They would originate outside the E.E.C., but would wish to trade with the enlarged Community. They would be of high commercial stability and strength. It is not clear how many of these attributes a suitable firm would possess, but the prescription seems a counsel of perfection, especially when the effort involved in attracting any employment to the more remote parts of Ireland is recalled.

The I.D.A. regional plans propose that 38,000 extra jobs should be created in manufacturing between 1973 and 1977. This represents almost as many jobs as were created in the prosperous 1960s (42,000 between 1961 and 1971), and these new jobs are planned to be more widely distributed throughout the Republic. In fact, the current state of the world economy already suggests that these totals will be particularly difficult to achieve: it is somewhat ominous that the level of unemployment in the early 1970s has been running about 1·5 per cent higher than in the mid-1960s, in spite of the real achievements in creating new jobs. The test of the plans, however, will not simply be the totals that are attained, but the extent to which these jobs are distributed in accordance with a specified regional pattern. In this important sense the work of the I.D.A. represents a significant new departure for regional planning in the Republic in that there is an explicit quantified target for all the regions of the State (Table 10.3).

While the I.D.A. plans were being elaborated and published, a government interdepartmental committee completed an examination of regional policy. This led to an official government announcement in May 1972 of all overall regional strategy to cover the next 20 years, including a set of population targets for the main urban centres (Department of Labour, 1972). Not surprisingly, this pronouncement coincided with the most important and realistic details of the I.D.A. plans. In fact, now that the dust of controversy is clearing, the current official strategy emerges as not very different from the original Buchanan proposals. Four main strands of policy can be distinguished. First, the future development of Dublin is to accommodate the natural increase of its

Table 10.3. Industrial Development Authority: Employment Growth Targets in Manufacturing Industry, 1973–77

Region	Gross job creation target	Possible redundancies	Net extra jobs	Percentage increase in total manufacturing employment
Donegal	2,800	800	2,000	31·3
North-West	1,700	400	1,300	41·9
West	5,300	1,100	4,200	48·8
Midlands	3,400	600	2,800	30·8
South-West	9,400	2,400	7,000	19·8
South-East	4,700	1,500	3,200	15·9
North-East	5,300	1,900	3,400	20·2
East	17,000	6,700	10,300	9·4
Mid-West	5,400	1,600	3,800	21·2
National Total	55,000	17,000	38,000	16·7

existing population. Second, there is to be expansion centred on eight other urban areas. Third, there is to be the development of county or other large towns in each region. Finally the special measures for the development of the Irish-speaking Gaeltacht areas are to be continued. In short, political constraints and economic realities have been blended to form what is probably a workable, if not an ideal, compromise—a situation not unknown in other countries where a regional dimension has been given to economic planning.

References

Baker, T. J. and M. Ross, 1970, The changing regional pattern in Ireland, *Economic and Social Review*, **1**, 155–166.

Black, R. D. C. and J. V. Simpson, 1968, Growth centres in Ireland, *Irish Banking Review*, **Sept. 1968**, 19–29.

Buchanan, C. and Partners in association with Economic Consultants Ltd., 1968, *Regional Studies in Ireland*, Stationery Office, Dublin.

Chadwick, J. W., J. B. Houston and J. R. W. Mason, 1972, *Ballina: a local study in regional economic development*, Institute of Public Administration, Dublin.

Commission on Emigration and other Population Problems, 1955, *Reports*, Stationery Office, Dublin.

Department of Labour, 1972, *Manpower Policy in Ireland*, Stationery Office, Dublin.

Donaldson, L., 1967, *Development Planning in Ireland*, Praeger, New York.

Duncan, G. A., 1942–4, Movements of the rural population in southern Ireland, 1911–36, *Proc. Royal Irish Academy*, **48B**, 1–14.

Forbes, J., 1970, Towns and planning and Planning in Ireland, in N. Stephens and R. E. Glasscock (Eds.), *Irish Geographical Studies*, Queen's University, Belfast, pp.291–311.

Government of Ireland, 1958, *First Programme for Economic Expansion*, Stationery Office, Dublin.

Government of Ireland, 1964, *Second Programme for Economic Expansion*, Stationery Office, Dublin.

Government of Ireland, 1969, *Third Programme for Economic and Social Development, 1969–72*, Stationery Office, Dublin.

Industrial Development Authority, 1972, *Regional Industrial Plans, 1973–77*, I.D.A., Dublin.

Johnson, J. H., 1963, Population changes in Ireland, 1951–61, *Geographical Journal,* **129**, 167–174.

Lichfield, N., 1966, *Report and Advisory Outline Plan for the Limerick Region,* Stationery Office, Dublin.

Meenan, J. F., 1970, *The Irish Economy since 1922,* Liverpool University Press.

Newman, J., 1967, *New Dimensions in Regional Planning,* Stationery Office, Dublin.

O'Farrell, P. N., 1970, Regional development in Ireland: problems of goal formation and objective specification, *Economic and Social Review,* **2**, 71–92.

O'Farrell, P. N., 1971, Regional development in Ireland: the economic case for a regional policy, *Administration,* **18**, 342–362.

O'Farrell, P. N., 1971, The regional problem in Ireland: some reflections upon development strategy, *Economic and Social Review,* **2**, 453–479.

O'Farrell, P. N., 1972, A shift and share analysis of regional employment change in Ireland 1957–66, *Economic and Social Review,* **4**, 59–86.

O'Farrell, P. N., 1974, Regional planning in Ireland: the case for concentration: a Reappraisal, *Economic and Social Review,* **5**, 499–514.

O'Neill, H. B., 1971, *Spatial Planning in the Small Economy: a Case Study of Ireland,* Praeger, New York.

O'Neill, H. B., 1973, Regional Planning in Ireland—the case for concentration, *Irish Banking Review,* **Sept. 1973,** 9–20.

Scully, J. J., 1971, *Agriculture in the West of Ireland: A Study of the Low Farm Income Problem,* Stationery Office, Dublin.

Soulsby, J. A., 1965, The Shannon Free Airport Scheme: a new approach to industrial development, *Scottish Geographical Magazine,* **81**, 104–114.

Walsh, B. M., 1968, *Some Irish Population Problems Reconsidered,* Economic and Social Research Institute, Dublin, Paper no.48.

Wright, M., 1967, *The Dublin Region: Advisory Plan and Final Report,* Stationery Office, Dublin.

11

Denmark

Christopher Elbo

Denmark's Post-War Industrial Revolution

Denmark is the smallest of the Scandinavian countries, having a population of 4,950,600 in 1971. The land area of metropolitan Denmark (43,069 km^2) is a little less than one-fifth of that of the United Kingdom and consists of the peninsula of Jutland plus 406 islands, 97 of which are inhabited. Two other territories are attached to the Danish Crown. The Faroes (38,680 inhabitants) have formed a self-governing region of the Kingdom since 1948, and Greenland (46,300 inhabitants) has been upgraded from its colonial status and was first represented in the Danish Parliament in 1953. The present chapter will concentrate on the regional management problems of metropolitan Denmark.

The traditional image of the Danish economy as being almost wholly reliant on a sophisticated highly efficient agricultural sector has recently undergone a drastic transformation such that could hardly have been predicted in 1945. Being almost devoid of indigenous fuel and mineral resources, Denmark's earlier economic development had been closely linked to the land. The eastern half of Jutland and the archipelago are endowed with fertile glacial till deposits and supported a network of coastal central places and market towns which served agricultural communities inland. Copenhagen reigned supreme and unchallenged as the legislative, administrative, manufacturing and transportation centre of the nation. Dislodging some of the well-being of Copenhagen to the benefit of the country has proved the most demanding issue in regional management. Denmark's population had grown at a fairly steady rate of 0·7 to 0·9 per cent per annum in the inter-war years. During the Occupation and the immediate post-war period, that rate exceeded 1·0 per cent but fell back to 0·6–0·8 per cent after 1950. At present the crude death rate is 9·0 per thousand and the birth rate 15·0 per thousand. The latter figure is low, resulting from increased use of the contraceptive pill and the 'deferment' of having children among many members of the 18–25 age group. Between 1920 and 1960 the volume of immigration roughly balanced emigration; but since 1960 many foreign workers have entered Denmark mainly to occupy industrial jobs. A total of 54,000 was recorded in the country in 1972.

Much more important are the changes that have occurred in the internal distribution of Denmark's population. Table 11.1 shows that the general pattern was constant between 1921 and 1940, but that in more recent years the significance of Copenhagen has declined in relative terms (if not in absolute numbers). Provincial towns have experienced important growth in the past 30 years and the extension of commuting areas into the peri-urban countryside raised the proportion of Danes living in administratively 'rural' *kommunes* between 1950 and 1970. Rates of mobility are high, with approximately 10 per cent of the total population changing its place of residence each year since 1950. Regional trends of population change are shown in Figure 11.1.

Table 11.1. Population of Denmark, 1921–70

	Copenhagen (%)	Provincial towns (%)	Rural *kommunes* (%)	Total (millions)	Density (per km^2)
1921	22·6	22·9	59·8	3·10	76
1930	21·7	22·2	56·1	3·55	83
1940	23·1	24·3	52·6	3·84	89
1950	22·7	26·4	50·8	4·28	99
1960	20·2	27·1	52·7	4·59	106
1970	16·7	28·3	55·0	4·91	114

This emphasizes the dominance of Copenhagen but also shows that since 1960 a new growth area around Aarhus on the eastern coast of Jutland has begun to emerge. In 1964 and 1965 a temporary reversal of trends occurred and Copenhagen lost population. The reasons for this sudden reversal remain unknown but have made Danish planners very cautious about attempting to forecast future migration flows.

Changing patterns in population and economic activity are linked to Denmark's post-war 'industrial revolution' and rapid increase in tertiary employment. The nation's economy was quick to recover after the Second World War and when the period of worldwide economic growth appeared in the late 1950s Denmark was ready to claim its share. Between 10,000 and 15,000 full-time farmworkers left the land each year during the 1950s (Bunting, 1958, 1968). Welfare and living standards rose sharply and an increasing demand for labour in urban industries accelerated the exodus from farming (Table 11.2). The volume of decline in the agricultural labour force was reduced in the 1960s, with 8–10,000 leaving the land annually. At the same time the importance of part-time farming increased so that 13 per cent of Danish farmers held off-farm jobs in 1968. In addition, many workers moved straight from the farms to tertiary jobs in the late 1960s and 1970s. Agricultural employment will undoubtedly experience a further decline in the immediate future since more than half of Danish farmers were over 55 years of age in 1968. However, increased capital intensiveness has allowed Danish agriculture to raise its volume of output in spite of a rapidly contracting labour force. Between 1960 and 1970 the annual volume of investment in manufacturing industry more than doubled with an increasing share being devoted to machinery. As a consequence the industrial sector was able to increase production more than two and a half times and to increase exports more than three times during the decade.

Evolution of the Danish Planning Machine

Town Planning

The economic progress that Denmark experienced in the late 1950s and 1960s was associated with enormous increases in the quantity and quality of her urban fabric. Urban land uses spread over roughly one additional square kilometre each week. The total volume of housing in Western Europe rose by roughly one-third between 1938 and 1970, but an increase of 75 per cent occurred in Denmark over the same period. Under

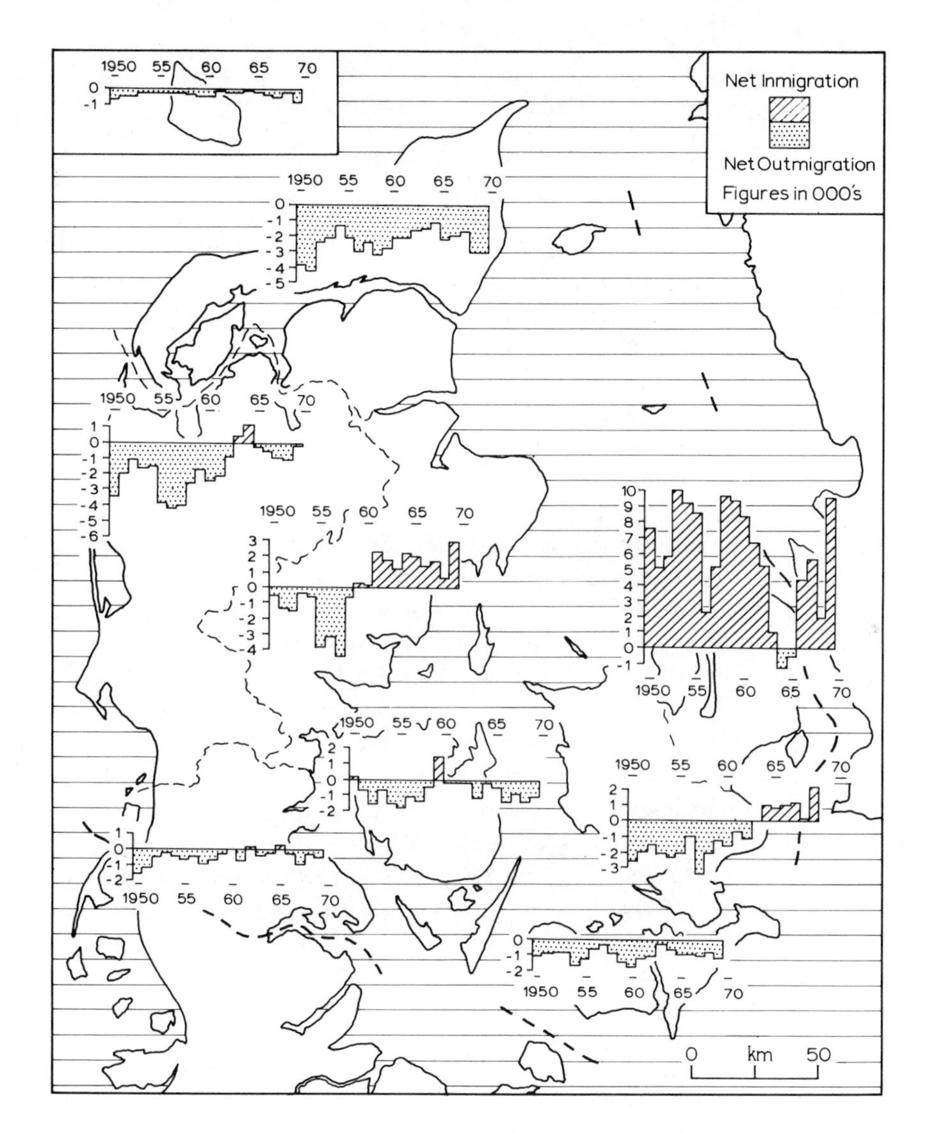

Figure 11.1. Denmark: annual migration, 1950–70.

Table 11.2. Full-time Employment in Denmark (%)

	Agriculture	Manufacturing	Tertiary activities
1950	25·6	33·8	40·6
1960	17·8	37·0	45·2
1965	14·6	38·0	47·4
1970	11·2	37·2	51·5

such circumstances of growth urban planning was absolutely essential. Denmark's first real town planning legislation was introduced in 1938, improving and extending a law of 1926. Now virtually all municipalities with any form of urban settlement have prepared town plans. In 1949 a town development act was introduced which required Copenhagen and the three largest provincial towns to prepare urban development plans. In addition, 38 other town development areas were defined which together covered all urban centres of any significance in Denmark. In 1970 the legislation was revised and required that planning should operate in all urban settlements with more than 1000 inhabitants. Simultaneously, the *kommunes* (local authorities) were able to adjust their plans to align with the local government reorganizations that were completed in 1970.

Copenhagen is of particular interest in the realm of town planning, since it is almost large enough to be recognized as a region in its own right. Planning of the capital began in 1937 when a Green Plan defined a continuous system of open spaces. Twenty *kommunes* had to prepare their own development programmes but there was no coordination between them. In 1948 Copenhagen's 'Finger Plan' proposed a hand-shaped pattern of future development spreading away from its coastal city core. As a result of the 1947 town development act, a plan was approved in 1951 which divided the capital into inner, middle and outer zones. This continued to operate until 1966 but endowed local authorities with only 'negative' powers for preventing undesirable building schemes. They were not able to propose positive alternatives for development. In 1960 the 'Draft Outline for a Regional Plan for Urban Development until 1980' proposed that Copenhagen should have more than one major central area. In the same year a new law required coordinated planning along the southern tentacle of the earlier Finger Plan and represented a modest attempt at preparing a kind of new town act (Diem, 1973; Hall, 1967). Since then Copenhagen has developed increasingly as a multi-centre city, with three large central areas and thirty smaller ones. The three large centres (Rødovre Centrum, Lyngby Center and Høje Taastrup Storcenter) are designed to focus future development and to avoid the spread of straggling dormitory suburbs. At the same time the supremacy of inner-city areas is being protected by, for instance, rejecting proposals by foreign retail chains for opening out-of-town shopping centres on the margins of Copenhagen to cater for a motorized clientele. In 1967 a regional planning council for the capital was formed and outline regional plans were issued in 1971 and 1973. The authorities received enhanced powers under the National and Regional Planning Act (1974) and consequently the future management of Copenhagen would seem to be under firm control.

Regional Planning

Non-mandatory plans were produced for several parts of Denmark after the Second World War (for example, the Copenhagen 'Finger Plan' of 1947, the Aarhus Plan of 1954, and the Outline Plan for Copenhagen of 1961), but Danish regional planning really came into being with the first regional development act of 1958. This was designed to stimulate new employment opportunities in backward rural areas and in other zones that suffered special economic difficulties. A regional development council was set up under the Ministry of Commerce, with ten members representing ministries, trade and industrial organizations and local government bodies. The Council's main objectives are to act in an advisory capacity and provide financial aid for firms in problem areas. Its effectiveness has subsequently been increased and the areas qualifying for assistance have been rationalized and reduced in number. For example, in 1967 one third of Denmark, containing a quarter of its population, had been designated for help (northern, western and southern Jutland, Lolland, Falster, Bornholm, Langeland, AErø, Samsø, the Faroes, and a few other areas). But in 1972 management schemes were amended and five major regional development boards were set up.

Two main forms of financial aid are given, in the form of grants and government guarantees for loans. These help meet initial costs of setting up a new factory in a development area and other expenses associated with removal, extension, rationalization and redeployment. Government aid is extended to local authorities that develop industrial premises for sale or for lease. Grants are also available for up to three years for new businesses which relocate in development areas but experience a temporary fall in efficiency or profitability because of their new location. Finally, a special law of 1969 intensified aid for creating new jobs in the provinces. Following entry into the E.E.C. further changes in Danish regional development policy are to be expected, but up till December 1974 programmes for regional development might be applied without restriction to Bornholm, AErø, Samsø, Langeland and the special development area of northern Jutland. Since 1 January 1973 aid to other areas has been restricted to less than 20 per cent of costs.

Conservation

Running parallel with its schemes for regional development, the Danish government has enacted a conservation policy on a regional basis. A nationwide conservation programme was first suggested in 1937 but it was not until 1959 that fourteen regional conservation committees were established to propose areas for conservation and others suitable for second home development. A Nature Conservancy Commission was instituted in 1961 and issued a report six years later that provided the foundation for the Town and Land Zoning Law (1969). Using the evidence of a 'Green Zone Plan' Denmark was divided into urban zones, areas suitable for summer cottage development, and more purely rural territory (Holmes, 1973). Radical changes, such as a powerful policy of nature conservation, were not proposed but the need for creating regional parks was emphasized. The Zoning Law (1969) also introduced a range of building restrictions, to prevent new construction less than 100 m from the coast, 300 m from woodlands and 150 m from major roads, large lakes and watercourses. These restrictions may be lifted only by the appropriate regional conservation board. In practice, nature conservation and land-use zoning were combined under the law of 1969. A number of pieces of legislation were amended in association with E.E.C. entry because they either conflicted with Common Market rules of free competition or because they had proved to be ineffective. For example, in 1972 commercial letting of country cottages and camping grounds required obtaining special licences from the Ministry of Housing. In the same year another law authorized the public acquisition of land for recreational use. Young farmers were given added protection at law from competition from townspeople in the use of land. Now, any purchaser of a farm of more than 5 ha is required to live on the holding and to exercise agriculture as his principal occupation. Finally, new legislation in 1972 required the permission of the Minister of Public Works before further exploitation of stone, gravel and other deposits might take place.

Danish Problem Areas

Northern Jutland and the small islands form Denmark's major problem areas. A special regional plan (1966) for northern Jutland emphasized the lower quality of life in the area and called for special planning measures and provision of aid to improve infrastructures. In the past ten years the new fishing port of Hanstholm has been built, a new university opened at Aalborg (August 1974), and the national organization that handles trade between metropolitan Denmark and Greenland decentralized from the capital to Aalborg. In addition, northern Jutland receives aid by virtue of its 'special development area' status. The problems of 65 of Denmark's small islands were studied by the Ministry of Cultural Affairs. In 1970 it reported that development of important manufacturing or summer cottage activities was not feasible because of the small size of many of the islands. But subsidies for improving ferry services were recommended as was special government aid for developing hotels and facilities for boating and camping.

National Planning

Although various forms of physical planning in Denmark had been initiated as early as 1926, it was not until the period of rapid economic and urban growth in the late 1950s that a general awareness arose of the need for sophisticated planning. The dramatic reshaping of Danish society opened the floodgates to a series of problems that the Danes had not faced before; towns were spreading rapidly, new industry required more land (often coming into conflict with established interests), the summer house 'explosion' severely threatened the hitherto unspoiled Danish rural landscape, and the more backward parts of the country wanted a share of new industrial growth. In brief, conflict situations arose between a series of groups in competition for scarce land resources. Statutory measures were essential for land to be allocated in ways that were most likely to meet national objectives. Geographers and planners produced schemes on a national scale. Johannes Humlum proposed a north–south motorway through the centre of Jutland; Erik Kaufmann Rasmussen suggested in his 'Stjernebyplan' that a number of provincial towns should be upgraded as regional centres; and the Regional Planning Secretariat put forward the idea of an 'H-Plan', in which two north–south lines of development should run through eastern Jutland and eastern Zealand with the crossbar being formed by an axis from Esbjerg to Copenhagen. In 1960 the political climate was right, but the protagonists had little experience of national planning and in any case the necessary machinery did not exist. Consequently Prime Minister Viggo Kampmann announced in 1961 that an interministerial committee (the Landsplanudvalg, or L.P.U.) should be established on the Dutch model to propose guidelines for national planning. The L.P.U. consisted of civil servants from thirteen ministries and had a secretariat of planners. Its first major work was the 'Zone Plan 1962' which outlined areas suitable for future urbanization, industrial development, conservation, agricultural production and summer cottage development (Figure 11.2). The government, press and public gave the report a warm reception, since the call for a 'plan' had been answered. In the following year the government attempted to introduce a package deal of land laws in order to contain urban sprawl. This was rejected in a referendum and had the unfortunate side effect of damping political interest in planning activities in the following years. By 1965 the L.P.U. had faded out of existence, probably as a result of the split loyalty that its members had for their respective ministries and for the Committee. However, the secretariat of the L.P.U. has survived and has had considerable influence in recent years on developing a working machine for national planning.

Regional and National Planning in Concert

Following the quiet death of the L.P.U., its secretariat issued a report on 'Regional Planning and Regional Divisions' (1966) which became part of the preparatory evidence for local government reform. In 1958 there had been 1400 *kommunes* but by 1966 the total had declined to 1000 as a result of voluntary mergers. A Kommune Reform Commission was instituted in 1966 to organize a reform leading to legislation in 1970. The subsequent local government structure comprised 277 new *kommunes* which made up 13 new counties. By establishing this structure a major obstacle to combining national and regional planning objectives has been overcome. In addition, data collection and analysis by the secretariat of the old L.P.U. had made good progress and in 1966 the government passed a law giving the National Statistical Bureau greater powers for collecting essential data required for planning purposes. New programmes will lay down guidelines for siting future urban growth: delimiting urban zones, identifying major traffic zones, siting key public institutions, polluting factories and mineral resources, and steering the siting of country cottages and other recreation facilities. Eventually, in April 1974, a law on national and regional planning and on managing Greater Copenhagen provided the first statutory authority for planning Denmark on a truly nationwide scale. Copenhagen and the twelve other counties (in association with their respective *kommunes*) now have to submit regional plans to the government. If they receive approval their implementation is mandatory. Copenhagen has been regarded as something of a special case and its plans deal not only with land-use issues but also with economic growth. Capital-city planners have greater legal powers than their provincial counterparts. Schemes for Copenhagen have to be approved by the Metropolitan Council as opposed to the central authorities. Thus Denmark has finally achieved the goal proposed in the Zone Plan of 1962. This plan undoubtedly came too early, but provided Denmark with a useful educational exercise without which the present type of coordinated planning machine probably would not have been created. The implementation of a national planning policy has been

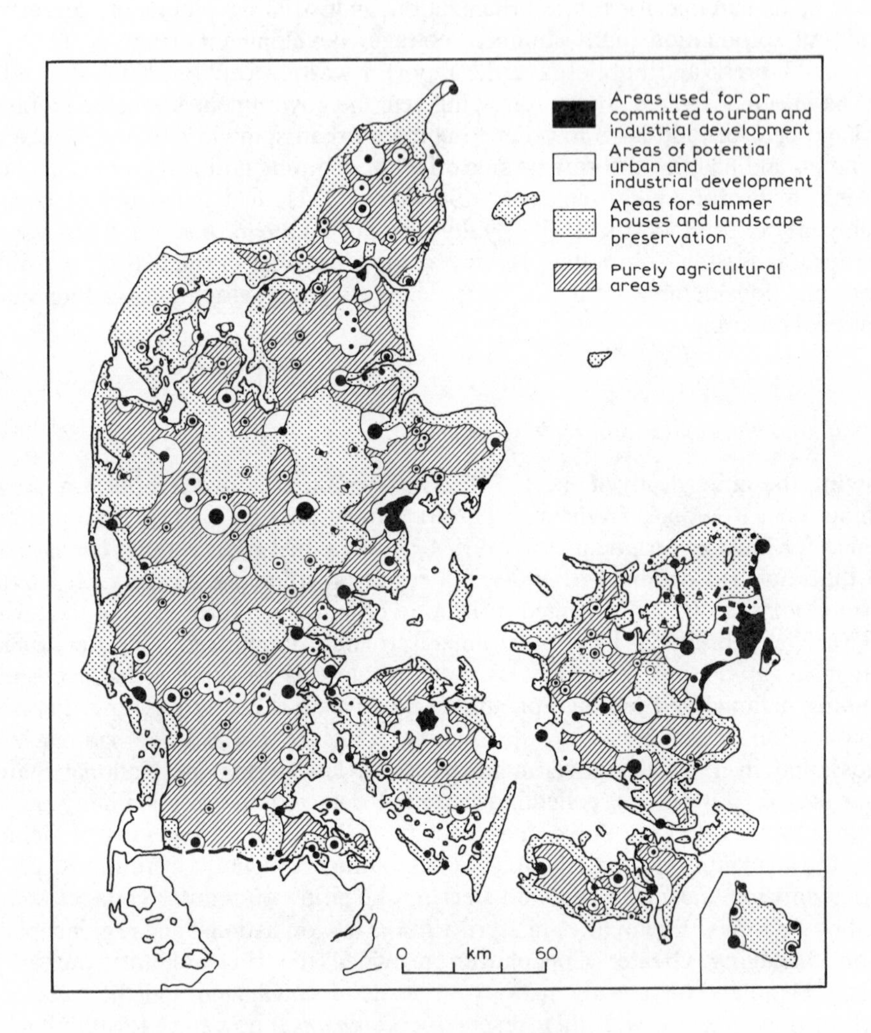

Figure 11.2. Denmark: Zone Plan 1962.

relatively easy in the small country of Denmark, which is not much greater than some of the individual regions of the larger countries of Western Europe.

Planning for the Future in a European Context

Changes in the distribution of population and manufacturing activities in Denmark cannot be assessed adequately without some consideration of actual and proposed changes in her transport infrastructure. With a major portion of the most developed part of the country consisting of an archipelago in a 'stepping stone' position between Continental Europe and the remainder of Scandinavia, the improvement of transportation in Denmark is both vital and expensive. Agreements were signed in 1973 between Denmark and Sweden regarding two projects that are still at the planning stage. The first involves the joint venture to construct the 'H–H' rail tunnel between Helsingør in northern Zealand and Hälsingborg in Sweden. Secondly, Sweden agreed to a joint venture whereby Denmark would develop a new airport on Saltholm island to relieve pressure on Kastrup airport outside Copenhagen and also to link Saltholm to the island of Zealand. Sweden was to construct a tunnel between Saltholm and her own coast. Furthermore, Denmark agreed to build a permanent connection between Zealand and Funen, but the final form and route have not yet been decided. Late in 1974 complications over interpreting the terms of the contract arose, but if the schemes are implemented the repercussions on the future geography of Denmark will be immense. Copenhagen's influence over the rest of the country will be increased enormously and the strong competition from Hamburg that is very noticeable in southern Jutland may be counteracted.

With such perspectives looming in the future, the need for long-term planning has become apparent among politicians and planners alike. Hence in 1968, a long-term government committee for economic planning was established. Fifteen-year 'perspective plans' will define the various choices facing political authorities in selecting priorities for public investment and economic development. The first plan, covering 1970–85, appeared in 1971 and stressed the merits of planning sectors of the economy. Such an exercise, in conjunction with physical planning, will be particularly useful. However, certain major assumptions about future growth rates have to be made in long-term planning. The scale of difficulty involved in this respect is exemplified by the fact that the first perspective plan was unable to envisage either the oil crisis of 1973 or the economic recession of 1974. Its deliberations were carried out on the basis of the blissful assumption that growth rates would continue only slightly below those of the late 1960s. In fact, a second 'perspective plan' covering the period 1972–87 was published in December 1973. Account was taken both of entry into the E.E.C. and of the oil crisis. Furthermore, investigations were broadened to include both the private and the public sectors of the economy.

The Impact of Planning on the Geography of Denmark

Thus it was not until the early 1970s that effective coordinated planning began to take effect in Denmark. That is not to imply that planning did not operate before that time, indeed some might argue that in some areas there was already too much planning. But in many instances it was uncoordinated and non-mandatory. It is therefore difficult to identify how great was the impact of the planning machine during the 1960s. Some

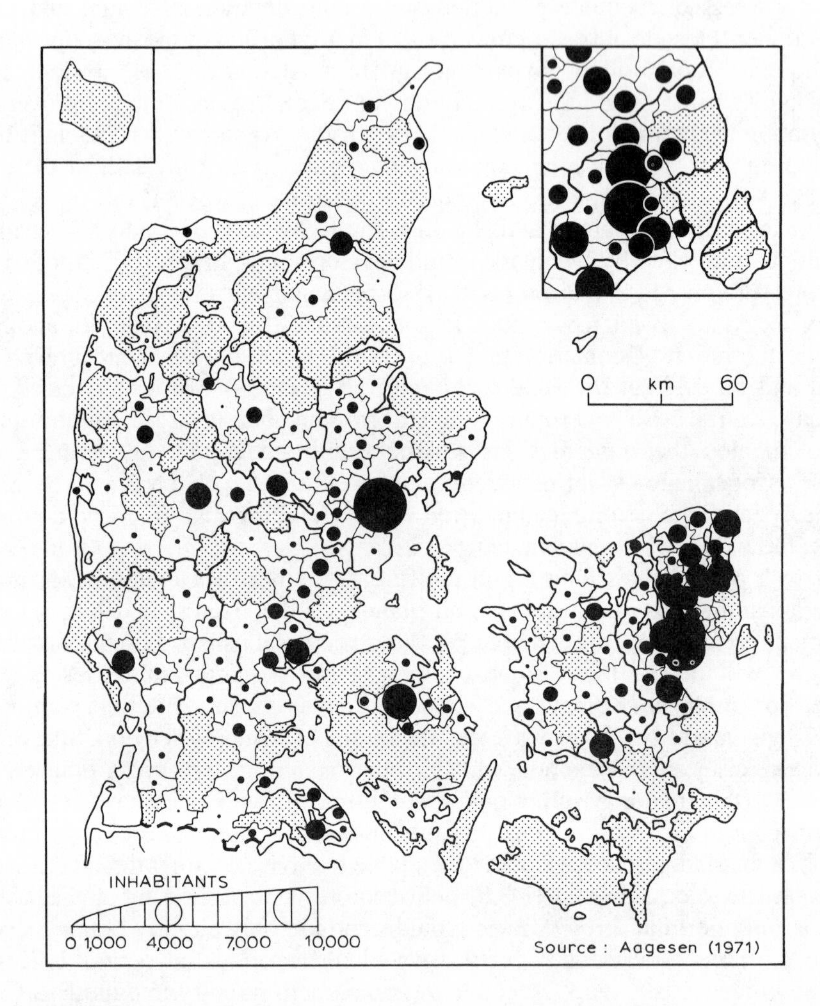

Figure 11.3. Denmark: population increase 1965–70. Areas losing population are stippled, those with no change are left blank. (Reproduced by permission of A. Aagesen and *Geografisk Tidsskrift*.)

Table 11.3. Change in Workforce, Manufacturing Industry, 1955–65

Towns plus suburbs	1955–61		1961–65	
	Additional workforce	Percentage growth	Additional workforce	Percentage growth
Copenhagen	19,000	12·0	−8,300	−4·7
Aalborg + Aarhus + Odense	12,800	28·5	1,500	2·6
Towns of 20–100,000 population	17,100	34·4	8,200	11·3
Towns of 10–20,000 population	5,300	28·3	2,500	10·6
Towns of 3–10,000 population	3,900	35·3	4,300	29·3
Other towns + rural areas	15,700	22·7	17,400	34·2

processes at work, such as urbanization and migration of population, might well have followed the paths they did even without any planning intervention. In the realm of manufacturing employment Table 11.3 reflects an appreciable westward shift of industrial job provision between 1955 and 1965. This was partly accounted for by the fact that many farmworkers left the land and provided a large local labour force that might be absorbed in Western Denmark by labour-intensive, low-wage industries, such as manufacture of footwear, textiles, clothing and wood products. In addition, Copenhagen and the other major cities were very rapidly expanding their tertiary, rather than their manufacturing functions. The pattern of population migration between 1965 and 1970, shown in Figure 11.3, certainly reflects a westward shift in the distribution of population but emphasizes the continuing dominance of Copenhagen in national life (Aagesen, 1972; Jensen, 1972). Attempts to develop outlying regions, such as northwestern Jutland and Lolland–Falster, are recognized as being less effective than had been hoped, although moderate success was achieved in northeastern, central and southern Jutland.

In the 1950s Denmark's schemes for planning and regional development had been too few and uncoordinated to be effective. And the administration and control of the planning process was not adequate for the task in hand. These problems were overcome gradually during the 1960s and early 1970s as sophisticated planning methods were being learned theoretically and applied in practice. Denmark now has a coordinated planning machine, but such machines do not operate in a vacuum and are subject to outside interference that may not be controlled, or only very marginally so. Hence the critical conditions of 1973 and 1974 will undoubtedly have far-reaching effects on the Danish people and the landscapes and regions that they have fashioned. With unemployment levels running at 9 per cent and inflation at 15 per cent in late 1974, Denmark will have to be more vigilant and well-equipped than ever before if she is to keep the various parts of her own house in order.

References

Aagesen, A., 1972, The population of Denmark, 1965–1970, *Geografisk Tidsskrift,* **70**, 1-21.
Betts, D. C., 1966, Jutland's new town: Hanstholm, *Geography,* **51**, 138–140.

Bunting, B. T., 1958, Recent trends in land use in Denmark, *Geography,* **43**, 53–54.
Bunting, B. T., 1968, The present reorganization of agriculture in Denmark, *Geography,* **53**, 157–162.
Cabouret, M., 1974, Quelques traits de l'évolution récente de l'économie agricole danoise, *Annales de géographie,* **83**, 684–714.
Diem, A., 1973, The growth and planning of Copenhagen, *Revue de Géographie de Montréal,* **27**, 41–51.
Hall, P. G., 1967, Planning for urban growth: metropolitan area plans, *Regional Studies,* **1**, 101–134.
Holmes, M., 1973, Slices of Danish land, *Geographical Magazine,* **45**, 772–773.
Jensen, R. H., 1972, Changes in the geographical distribution of Danish industry, *Geografisk Tidsskrift,* **70**, 41–58.

12

Northern Europe

Kjell Stenstadvold

Introduction

This chapter examines problems of regional development in three of the five Norden countries (Denmark, Finland, Iceland, Norway and Sweden). The group is defined mainly in terms of cultural heritage, although Finland, Norway and Sweden are also related physically. With respect to cultural conditions, similarity of language is the most striking aspect, with Danes, Swedes and Norwegians being able to understand each other without much difficulty. During the Viking period and late middle ages each of the five countries were at some time united to or conquered by some of the others. The latter phenomenon led not only to extensive Swedish settlement in Finland but also to Iceland being settled by Vikings mainly from Norway. The island remained a Danish colony until 1943.

Now the five countries have joined together to form the Nordic Council which was established in 1953. It is an advisory body and makes recommendations to member parliaments and governments on matters of common interest. The most striking results have been the abolition of passport controls and the creation of a free labour market and a system of common postage on intra-Nordic mail.

The Council has not suggested common rules or measures with respect to regional policy, but in 1966 it recommended that a liaison committee be set up for the exchange of information between policy-making bodies and for assisting cooperation in research. From 1970 onwards the NordREFO group (Nordiska arbetsgruppen för regionalpolitisk forskning) has published a quarterly journal which contains short notes and summaries about research projects, policies and meetings relating to regional development.

In the light of such close links between the countries of Northern Europe it might seem sensible to examine their regional development collectively. However, there are two good reasons why this will not be done. First, there are so many differences in detail that uniform presentation would not be possible. Second, the really interesting differences in policy emerge only if one studies the distinctive political scene of each country. Whilst the major directions of policy tend to be much the same, it is with respect to detail that each country has its distinctiveness. The present chapter will be restricted to Finland, Norway and Sweden. Iceland is unique among the Norden countries with respect to its location, size of population and economy (Kristinsson, 1973). Denmark is also omitted because of its atypical size and geological structure when compared with the three other countries (Table 12.1). As elsewhere in Norden, regional problems in Denmark result from a marked reduction in the agricultural labour force and from considerable urban concentration. But Denmark does not have the same problems that typify the northernmost countries which cross the Arctic Circle: namely, sparsity of population, great distances between settlements, isolation and harsh physical conditions that restrict primary activities. Denmark is a more homogeneous, densely-populated country and its regional problems might be compared more appropriately with those of the Netherlands or of lowland England.

Table 12.1. Basic Data on Northern Europe

	Population, 1970	Urban percentage[a]	Population density (per km²)	Land area (km²)	Arable percentage	Forest	Maximum length (km)
Denmark	4,938,000	80	116·5	42,370	62·6	11·1	370
Finland	4,598,000	64	15·0	305,475	8·9	63·7	1200
Norway	3,874,000	66	12·6	307,920	2·6	27·1	1800
Sweden	8,077,000	81	19·6	411,480	7·4	55·2	1600

[a] Defined as agglomerations of at least 200 inhabitants.

Nevertheless, the three remaining countries display certain marked differences. Norway is quite distinctive in physical terms because of its pronounced topographic diversity, while both Finland and Sweden are rather closer to the ideal 'homogeneous' surface. Historic and cultural links between Sweden and Finland have been strong and this fact has had considerable influence on the development of Finnish regional policies over the last ten years. At a very general level one may recognize two main lines of policy, with the Norwegian brand on one hand and the Swedish (plus Finnish) version on the other. The present chapter will concentrate therefore on Norway and Sweden, with Finland being included to further exemplify the types of policy implemented in Sweden or to point out where the Finns have created their own particular types of response to problems. Finally, this account reflects the author's evaluation of important and interesting themes and therefore pays attention to Norwegian experience.

Types of Regional Problem

Problems of regional disparity are not of recent origin in Norway, Sweden and Finland but it was only after the Second World War that they were identified as such and regional policies were developed. The underlying reasons for such disparities are obviously very complex in character but there are certain main issues that are more important than others: decline of primary activities; pattern of industrial development; provision of public services; and the problems of the northern parts of each country.

The first, and probably the most important, is the general problem of declining labour requirements in farming, forestry and fishing. Although these three activities have been of differing importance in the three countries, their decline since the Second World War has been both continuous and great (Table 12.2). However, it should be noted that policies of land clearance have operated until recent years, especially in the northern regions, and that such programmes have influenced the process of agricultural contraction. Sweden discontinued its land clearance policy in the late 1950s and in 1967 Finland turned away from a programme of interior colonization to one of reducing its agricultural surface because of serious problems of over-production at that time. By

Table 12.2. Economically Active Population 1950 and 1970, by sector (%)

	Finland		Norway		Sweden	
	1950	1970	1950	1970	1950	1970
Primary	45·8	20·2	25·9	11·6	20·2	8·1
Agriculture	39·5	17·2	18·0	8·9	17·4	6·4
Forestry	6·0	2·9	2·9	0·8	2·5	1·6
Fishing	0·3	0·1	5·0	1·9	0·4	0·1
Secondary (+mining)	27·1	34·2	36·5	37·2	40·6	40·3
Manufacturing	21·2	24·7	25·8	26·8	32·7	29·2
Tertiary	25·7	44·2	37·2	50·9	38·1	51·3
Total (in thousands)	1984	2118	1388	1462	3105	3413

contrast, Norway continues to support clearance activities, mainly for the addition of extra land to existing farms.

The second problem stems from the pattern of industrialization in Scandinavia. Northern Europe's industrial revolution occurred from the mid-nineteenth century to 1914, with developments in Finland being delayed by roughly a quarter century. A large number of new factories were located at isolated, waterfall sites to make use of indigenous sources of energy both before and especially during this period of industrial change. The waterfalls were harnessed to actually drive machinery. At first it was the textile mills that were located at the largest waterfalls not too far from the main cities. These factories were followed by pulp and paper mills, and finally by electro-chemical and electro-metallurgical plants. The latter were usually located in the remote and often sparsely populated areas which contained the largest waterfalls, and thus company towns had to be constructed for industrial workers in such areas. In Sweden this type of settlement was started in the Bergslagen mining and ironworking region but then spread northwards as sawmills and pulp and paper factories were established from the mid-nineteenth century onwards. Such settlements in Norway are based on each of these types of industry, with electricity-consuming industries being the most important and most recent (post-1905) in this respect. Company towns alongside industrial plants often experience considerable isolation and cannot function adequately as central places since they lack populated hinterlands. Alternative jobs within reach of daily commuting are rarely available and serious problems arise if the 'mother' factory closes. Both Norway and Sweden have considerable problems in this respect, but so far it has been Sweden that has seen most examples of factory closure, with a consequent abandonment of company towns in many cases.

The third type of problem is more modern in origin. It derives from the deliberate policies of the social democracies of Scandinavia to try to extend all the benefits of modern society to every family in the land, regardless of where they live. For instance, electricity supply is now available virtually throughout Norway. At the beginning of 1973, only 392 households (830 persons) were without supply, and two-fifths of these were in North Norway. Such a programme has been completed without modifying the settlement pattern, but, by contrast, the school reforms of the 1950s and 1960s have led to considerable internal migration and have exacerbated problems in very thinly-populated regions. The new centralized schools teach children up to the age of 16 and were designed to offer a wide range of facilities and to support specialist teachers. In order to achieve such aims it is necessary to attract a large number of children which involves pupils travelling by bus for up to 2 hours each way in northern Norway and northern Sweden. Swedish schools for children aged 16–19 years have also been centralized, but this reform is still at the planning stage in Norway. Similar problems have developed with respect to other aspects of infrastructure, such as hospitals, post offices and railways, and Norway experiences special difficulties associated with a reduced reliance on sea-transport and increased use of land routes. Problems arising from the application of national standards to remotely-settled areas have been acknowledged only recently and are now receiving increasing attention. However, the proposed solutions for Norway and Sweden are somewhat different.

Table 12.3. The Marginal Regions of Northern Europe

	Area (km^2)	Population 1970	Population density (per km^2)	Maximum length (km)
North Finland	151,000	602,000	4·0	750
North Norway	108,000	454,000	4·2	1000
North Sweden	272,000	1,460,000	5·4	1050
Crofting counties of Scotland (Highlands and Islands)	37,000	285,000	7·7	650[a]

[a] including Shetland.
Finland: Laapin/Lapplands, Oulu/Üleaborgs *lääni/län.*
Norway: Finnmark, Troms, Nordland *fylker.*
Sweden: Norrbottens, Västerbottens, Västernorrlands, Jämtlands, Gävleborgs, Kopparbergs *län.*
Scotland: Shetland, Orkney, Caithness, Sutherland, Ross and Cromarty, Inverness, Argyll administrative counties.

The fourth type of regional problem in Scandinavia involves the northern regions themselves in the three countries that cross the Arctic Circle (Table 12.3). The first three problems are, of course, present in these regions, but they are more extreme and together add to serious locational disadvantages both in the national and international contexts. The great size of these northern regions (Figure 12.1) makes even a growth-centre policy relatively expensive to implement since functions and services must be provided in quite a large number of settlements. For example, in 1969 the county administration of Norwegian Finnmark presented a scheme for increasing investment in fifteen fishing ports along the 500 km of outer coastline. This rugged strip of land facing the Arctic Ocean contains only 30,000 people but the scheme was criticized in many quarters for concentrating investment at too *few* settlements.

The Political Basis for Regional Policies

While it is possible to identify common problems and problem regions in the three countries there are certain important differences in policies that have been adopted. This is partly due to differences in physical geography but more probably due to the

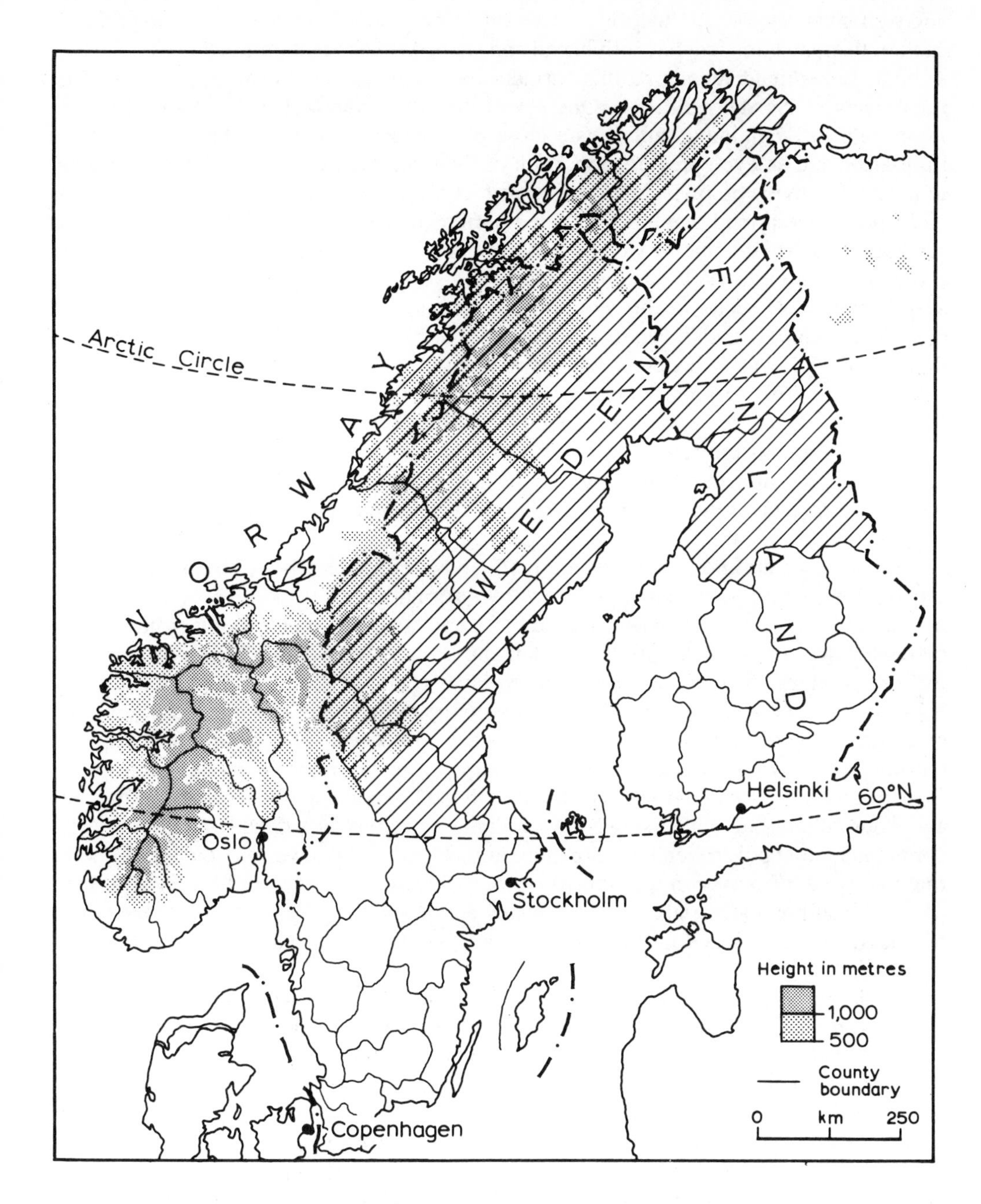

Figure 12.1 Northern Europe: its northern regions.

history of modern economic and political development in Norway, Sweden and Finland. It is right to acknowledge that all the Nordic countries are social democracies and welfare states, but for historic reasons differing groups have played varying roles in each of them and this fact has influenced and shaped subsequent regional policies.

First, one should recognize that farmers have a good 'public image' among most inhabitants of Northern Europe. One reason for this is the fact that urbanization was rather late in Scandinavia and thus a large proportion of town dwellers were born in rural areas and an even greater proportion of their parents were. It is still very common to have a relative who owns and runs a farm. A second reason involves the policies for clearance of fresh land that have operated until recently. This has given farmers a good reputation as frontiersmen extending the boundaries of settlement, and is especially pronounced in Finland. It has also given many people the impression that farming is still an expanding sector of the economy, particularly in the peripheral regions. A third reason in Norway and Sweden (and in Denmark) has been that farming has been dominated by owner–occupiers and therefore did not bear the taint of feudalism or landlordism. With varying degrees of emphasis these aspects largely explain why the farming population has had, and generally still has, considerable political power even though it represents only a small proportion of the population (Table 12.2).

The second major theme is what may be called the 'centre–periphery dimension' in the politics of each of the countries. This is clearly related to the political strength of the primary industries, but also has more general origins. To a very considerable extent variations in general policies between the three countries may be related to this spatial aspect of political power. There are now three formal levels of political administration in the Scandinavian countries (including Denmark): the central government; the counties (*fylke* in Norway, *län* in Sweden and Finland, and *amt* in Denmark); and the local communes. Most of the political power is found at the central and local levels because only the national councils and the commune councils are elected directly by proportional votes though this varies between countries. The communes acquired considerable powers through legislation dating from the mid-nineteenth century. At first the most important activities of the commune councils involved schools, churches and poor relief, but with the passage of time activities broadened into the present communal monopoly over land-use planning. Communes have two potentially very important sets of power in relation to regional development: the right to control land use (for example, industrial sites and housing areas) and the possibility to engage in any business, provide services to industry, or give loans or guarantees on loans to industralists.

Conditions in Norway will be examined to exemplify the working of the political system and then comparisons will be made with the other Scandinavian countries. The Norwegian *fylke* were originally established as administrative units to control and partly implement central government functions. In 1837 county councils were established with mainly an advisory role, but they have acquired more real power as time has passed, particularly during the last ten to fifteen years. The county councils are composed of members selected not by direct election but by nomination from commune councils and the allegiance of such members is directly to their own communes. The laws governing elections to county councils tend to make small, usually peripheral, communes relatively much stronger than their population size would suggest, because each commune is guaranteed one seat on the council but large communes cannot always obtain their proportional number of seats. The result of this system is to push power at the county level out towards the periphery. Outlying settlements therefore have the political power to engineer that finance is invested in infrastructures out on the smallest islands or in the most remote mountain valleys.

The political strength of outlying areas is also found at the highest level. As the result of an unwritten, but clearly discernible rule of political life, there should be a regional balance in the composition of members of the Norwegian government. With respect to communications, the investment programme for the national trunk road system is discussed by the county councils and priorities are announced to the Ministry which then present them in marginally modified form to the Parliament for the final decision. At the intermediate level, the counties are now primarily responsible for hospitals, secondary roads and secondary schools. Here again separate commune councils discuss the plans and instruct their representatives on the positions they are to adopt in county council meetings. In a classic case in Sogn-og-Fjordane this has led to a stalemate situation for nearly ten years over the problem of where to locate a new main hospital for the whole county. Once again the overall result is for matters to be decided in such a way as to benefit settlements lower in the hierarchy than the formal administrative structure may suggest. This clear tendency for decisions to be made at relatively low levels in the politico-administrative structure may, to a considerable extent, be explained by the relative political strength of the primary sector of the economy but also by the physical geography of Norway, which up to the last war made it necessary for local communities to survive in isolation because of poor and difficult communications. As a result, the Norwegian brand of regional policies is more favourable toward really peripheral communities than those operated in most other countries.

By contrast with conditions in the other Scandinavian countries, Finland developed more of a feudal rural society in the period from 1760 until 1917. Since then, land clearance schemes have given the farmers a favourable image, but the heritage of feudalism and the need for strong government after the civil war and during the Second World War made the balance of political power tilt in favour of central government rather than the communes and the periphery. The construction of the trunk road network over the last twenty to thirty years in Finland is a visual indication of this. These roads are inter-city links, running through the countryside without much regard to the smaller towns and settlements lying between the main nodes.

In Sweden as well it seems that the balance of power has been more toward the central side than in Norway. A good indication of this may be found in the history of commune boundary reforms over the last twenty five years. A Swedish commission in 1946 reported that modern conditions required larger communes and suggested that they should be reduced in number from 2500 to 1000. This reform was carried throughout the nation on 1 January 1952, but only a few years later it became clear that changing economic conditions required even larger administrative units. A commission was appointed in 1959 and when it reported in 1963 it advocated amalgamation to create 333 new units, a figure which was eventually reduced to 282 in 1964. This time the communes were given freedom to choose their own pace of amalgamation, but by January 1969 only 76 new communes had been formed. The main obstacle proved to be the necessity in the 1964 regulations for communes to be unanimous in their decision to amalgamate, but many communes were not willing to lose their identity. At the proposal of the government, the parliament set a time limit by which all amalgamations should be completed. The date was 1 January 1974, and now Sweden has 272 new administrative communes.

In Norway a similar commission was appointed in 1946, being partly inspired by the Swedish report of the same year, and presented its final proposals in 1962. Norway did not contain such small units as Sweden before the reform, having 744 before the first changes in 1958. The other contrast with Sweden involved the fact that in Norway many boundaries were completely rearranged. This was particularly apparent in coastal areas and was mainly in response to the changing provision of transportation by land rather than by sea. In the past, a commune had often included land on both sides of a fjord, but after the reform it was limited to one side and was linked by road to other parts of the country. There are two other features of the Norwegian reform which distinguish it from that in Sweden. First, the commission worked its way around Norway county by county and consulted local county representatives. As a result, they often arrived at different conclusions for what might appear superficially as rather similar cases. Also the time factor was significant, since proposals in the later reports differed from those in the earlier ones. Second, individual proposals were submitted to Parliament and thus the strength of local feeling and its representation in Parliament might influence the decisions. This method proved extremely slow and costly, and a provisional law was passed to speed up the reform after a few test cases had been scrutinized in detail in Parliament. By 1965 the number of communes had been reduced to 466 and by January 1974 there were only 443. The strength of political power on the periphery emerges again from later developments. Some of the amalgamations had clearly been forced on to the communes and unwilling partners demanded separation. These demands were so persistent that a government commission was appointed in 1972 to reexamine 21 amalgamations. When it reported in 1974, it suggested that three communes should be split again, and the minority report added a further eight to that list.

In both Sweden and Norway the main reason for the reforms was to improve efficiency in local administration and the provision of public services. In Sweden arguments for increasing efficiency seem to have won the day but these were not sufficient in Norway. Such arguments were often attacked by local politicians and members of Parliament in the following way. Local government should bring local influence to bear on decision making, and in particular spatial inequalities in economic viability should be reduced by grants from the central government. Several forms of equalization grant operate in both countries but unconditional grants are more extensive in Norway than in Sweden (Figure 12.2).

A word of caution is necessary before this survey of political conditions is brought to an end. Invididual communes or groups of peripheral communes do not always win the fight against central government in Norway, nor do they always lose in Sweden and Finland. Even if the communes have considerable nominal powers vested in them, the possibility of pursuing an active and independent local policy is usually constrained severely by lack of finance. Perhaps 90 per cent of expenditure is tied to national policies or to necessities such as schools or social services. Most of the remainder is usually destined for earlier projects started years ago. These constraints have increased almost everywhere over the last ten to fifteen years. Most of the differences in administration between Norway and Sweden are therefore due to political attitudes rather than to the immediate economic and legal situations.

A further point should be noted about Scandinavian local administration, and that is

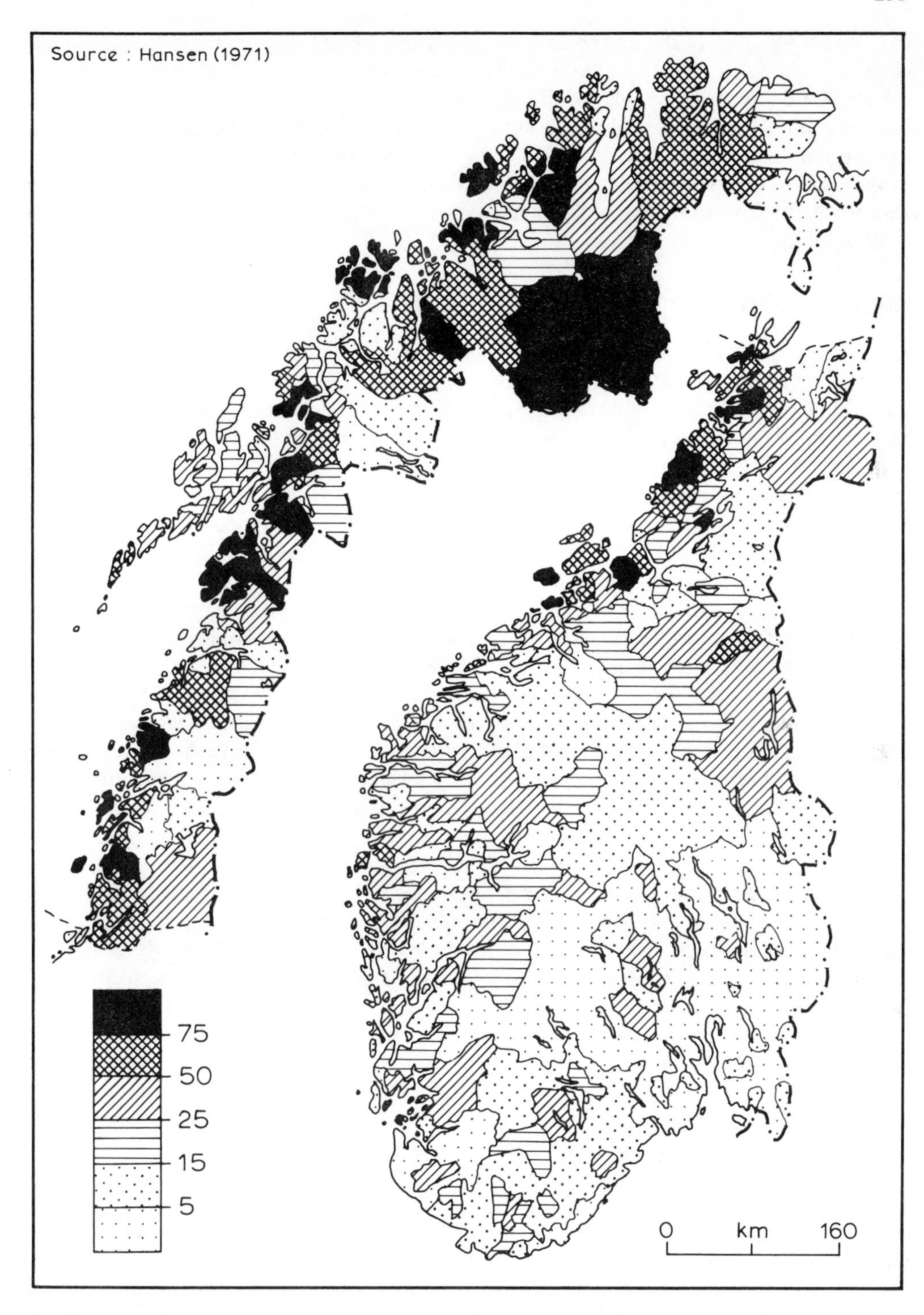

Figure 12.2. Norway: unconditional tax equalization grants, 1970, as a percentage of Communal Income Tax, 1968. (Reproduced by permission of *Norsk Geografisk Tidsskrift* and *Universitetsforlaget*.)

the unitary function of the communes. Virtually everything that relates to the local area is administered or at least cleared through the commune council. There are no separate school boards or water-supply bodies that report to outside agencies. However, more functions have been referred to the counties recently, but the communes still have a fair say through the county councils. This is especially so in Norway. At present each of the Scandinavian countries is discussing a reform of the counties to make them more democratic by instituting directly-elected assemblies that will assume responsibility for most activities at the intermediate level.

Objectives of Regional Policy

As in many other European countries, regional development in Norway, Sweden and Finland originally represented just one component in the national objective of promoting post-war economic growth; but since the mid-1960s regional objectives have come more prominently to the fore. The main objectives of Scandinavia's first comprehensive regional programme, the North Norway Plan (1952–60), were to increase incomes, provide new jobs in growth industries, and promote greater efficiency in the utilization of resources throughout the region (Wood, 1965). In 1956 a number of marginal areas in the northwestern part of South Norway were included for special assistance and in 1961 a national Regional Development Fund (R.D.F.) was established 'to support projects which will give increased, stable and profitable jobs in areas with particular employment problems or a poorly-developed industrial structure'.

The most important changes in the discussion of policy objectives in Norway occurred in 1965. For some time the Labour government had considered the possibility of introducing growth-centre policies and in the spring of 1965 they designated a few 'test centres' for a five-year period to try out the idea of growth centres under Norwegian conditions. The growth-centre programme was changed after the parliamentary election of September 1965 when a non-socialist coalition government took office. The main reason seems to have been opposition from outlying areas to a programme which discriminated in favour of growth centres, whereby investments in infrastructure and possibly even loans from the R.D.F. would be concentrated in the centres and not spread evenly over the commune or region. An alternative programme was launched in November 1965 when the government designated six 'development regions', which were growth *areas* as opposed to growth *points*. Areas comprising from one to six communes were designated, in which fishing or farming were dominant. The idea was not only to strengthen employment at specific centres but also to tackle the problems of wider areas as part of comprehensive development plans for entire regions. The growth centres were tactfully called 'test centres' in order to avoid any suggestion that localized growth might be at the expense of the surrounding region. At the time, it seemed to be mainly a difference of opinion over semantics. Now it is clear that there were contrasting views as to how the settlement pattern should evolve in Norway. Choice of a growth-centre policy implies change, while the discussion since 1965 has demonstrated that the approach that is favoured politically has been to preserve or to stabilize the existing dispersed pattern. As a consequence, regional policies have tended to be opposed to growth centres, and have often been anti-urban as well. The main concern of policy objectives has been to preserve the scattered settlements on the islands, along the coast, and in remote valleys in the mountains. In principle, each farm or fisherman's cottage should continue to be inhabited, or at least to be habitable, for as long as anyone wishes to live there. As a consequence of such discussion, priorities relating to economic growth (plus rising incomes) and the settlement pattern have undergone important change. A White Paper from the (non-socialist) government in February 1973 stated that 'it must be emphasized in future planning that the industrial structure must be adjusted to meet the goals set for the settlement pattern' (St. meld. nr. 50, 1972–3).

Norwegian regional policies now contain three main objectives: to stabilize the pattern of settlement as far as possible; to give fair opportunities for provision of jobs and services in all regions; and to improve the use of resources, with strong emphasis being placed on environmental protection. The idea of preserving the settlement pattern has not been adhered to in all cases, and there is reluctant acceptance that some change is inevitable. Nevertheless, the general attitude remains, and emphasizes the strength of peripheral areas in Norwegian political life. The farming and fishing lobbies are also making increasing use of the preservation-of-settlement argument to secure or increase the level of support and subsidy.

Reactions have been different in Sweden. It was not until 1963 that the 'Active Location Policy' report moved both government and Parliament to start a comprehensive programme of regional development planning. Prior to this the efficient and powerful Labour Market Board had run a 'private' regional policy as part of their work to keep the Swedish labour market in balance. Hundreds of millions of *kroner* were used to move thousands of workers from sparsely populated regions with inadequate employment opportunities to city-regions where industry needed more labour.

Although there were possibilities of moving jobs to the workers such facilities were rarely used. On the basis of the 1963 report, Parliament decided to launch a five-year preliminary development programme running from 1965 to 1970, which has now become a regular on-going scheme. From the very beginning, general policy goals were linked to economic and social policies in a way that paid little attention to the details of the settlement pattern. In fact, it was a stated objective that people should be encouraged to move to towns and service centres since they alone could provide the kinds of jobs and facilities that were considered necessary in the long run. The Swedes also accepted the growth-centre concept, with few of the Norwegians' qualms, and over the years made specific proposals for designating a complete national system of central places.

The first complete system was presented in 1970 and was concerned particularly about the so-called 'big city alternatives' (*storstads alternativ*), while it stopped at a regional, or supra-communal, level at the lower end of the hierarchy. The strong metropolitan emphasis of these proposals created considerable opposition, particularly from peripheral areas. To a limited extent this criticism was influenced by debates in Norway and there were even cases in the late 1960s of a couple of Swedish communes on the Norwegian border that applied for a transfer. The result was that the final version, which was approved by Parliament in December 1972, had simplified the system into three levels of centre below the three large city regions: the primary centres (incorporating the 'big city alternatives'); the regional centres; and the municipal centres. This meant that each enlarged commune contained a centre, even if at the lowest level of the hierarchy. The mounting criticism over centralizing policies has only slightly influenced these objectives. Unlike Norway, where the aim is to promote fair opportunities for *all* types of settlement or *area*, Swedish policy is distinctly orientated toward growth *points*.

Finnish experience has clear parallels with developments in Sweden. A regional programme was started in Finland in 1966 and this broadened into a more comprehensive policy for the period 1970–75. As part of the programme a national system of central places was identified and centres lying in Development Areas were designated as growth points for the purpose of investing public infrastructure funds. Such designation has also led private firms to concentrate their investments at such points.

An important question arises out of this difference in attitudes to growth-centre policies. Does long-term development necessarily require such discrimination between places for public investment and involvement? Because of economies of scale and the existence of threshold levels for many activities, the costs of focused investments may often be lower than for dispersed projects. Concentration may also assist external economies. But if society is willing to take on the immediate extra financial burden, the theory and literature relating to growth centres do not prove that concentration and spatial discrimination are necessary for further economic growth in an already well-developed economy.

It has been suggested that the Swedish policy is more long-term in character than the Norwegian, since the Swedes are planning for changes that seem to occur with varying speed, irrespective of what is done. Schemes in Norway are trying desperately to support the settlement pattern by giving grants and loans to firms that are willing to locate in remote areas. However there seem to be very few examples of the R.D.F. allocating financial help to firms in order to keep them going long after they have ceased to be viable operations. The two largest losses have been on a mining firm and a textile company located in a regional centre. So far, the Norwegian policies have been relatively non-discriminating and definitely less so than the Swedish. Attitudes against spatial discrimination for a long time steered the Norwegian Parliament away from offering a precise definition of Development Areas or from giving an exact criterion for the R.D.F. to use, but since 1969 Parliament has had to define boundaries for some of the new measures that have been introduced.

Differences between the approaches employed in the Scandinavian countries have been stressed in the preceding discussion, but it should be remembered that there are also many similarities. Concern about raising people's incomes and improving other factors behind their standard of living are common, and these are the objectives that welfare-state governments seem to be working towards. It is among the 'other factors' that the differences are found, most especially with respect to the spatial aspects of the general goal.

Measures for Regional Development

The North European countries have continuously broadened their range of measures for regional development over the last ten years. Instead of attempting to give a complete picture of the situation, the following paragraphs will concentrate on selected dominant features and, in particular, will relate the various measures in operation to the centre–periphery aspect of national politics. As a result of their close links the four North European countries (including Denmark) have tended to introduce more or less the same measures. Although Norway was the first to establish formal regional development, with the North Norway Plan of 1952, Sweden and Norway have shared the lead since the mid-1960s for introducing schemes for financial assistance. Such schemes have generally developed in the following fashion. At first tax concessions have been offered on funds for investment in development areas. These have been followed by loan guarantees and loans at reasonable interest rates from special funds or banks; then relocation grants to cover the transfer and training of labour; then investment grants; and, finally, direct wage support. So far only Sweden has used this last measure, and over a three-year period for new jobs in the Inner Aid Area.

A common feature has been an almost complete reliance on positive measures to support enterprises in designated regions, or to encourage them to expand, rather than direct controls such as the Industrial Development Certificates in Great Britain. On the other hand, both Norway and Sweden have a system of building licences to control the level of demand in the construction sector, but this has functioned mainly as a queuing device rather than as an instrument to present extra-metropolitan locations as alternatives to central city sites. Controls in Norway now cover only the major city regions and therefore have a 'regional' effect, but the general impact is rather small.

This is demonstrated by the measures introduced in Norway and Sweden in 1970 and 1971 which required all projected schemes for expansion (above 300 m^2 and 500 m^2 respectively), for new buildings and for conversions in the main city regions to be referred to an Advisory Board which may then 'delay' the projects for a period of time whilst 'force-feeding' the firms concerned with information on alternative locations. In autumn 1974 both countries were seriously contemplating an I.D.C. system comprising powers to veto new projects in locations deemed to be unsuitable by the authorities namely in the main city regions and in some coastal areas. Finland is also planning a similar system and, in 1973–74, introduced an investment tax on buildings which accommodate leisure and service activities in the three major city regions.

The countries have avoided giving direct investment grants for as long as possible. Sweden was first in 1965 but the grants were on a limited scale when compared with the amount devoted to low-interest loans. Investment grants were introduced in Norway in 1971 and in Finland two years later. It is noteworthy that respondents to a survey of relocated firms in Finland in 1969 were 'by and large unanimous' in their opinion that investment grants would not be a useful measure (E.R.I.F.I. 1969, p. 131). By contrast, various forms of conditional exemption from taxes have been used extensively. The greater part of investment in industry during the North Norway programme of the 1950s came from such a source, as did the first phase of regional development in Sweden (1965–70).

During the last few years increasing awareness of the complexity of regional development has led to two important patterns of events. One is increased emphasis on closer coordination and integration of the various public sectors so that they do not counteract each other either at the regional or the local levels. There had been a tendency for regional development to become just another sector of government, together with communications, agriculture, education and many others. Now steps have been taken to improve top level coordination at the early stages of planning in the appropriate ministries.

The second pattern of change is the introduction of more comprehensive planning at intermediate levels between the commune and the central government. It must be remembered that the communes have full responsibility for physical (land-use) planning in the northern countries. As a consequence, more extensive planning in larger areas would have to rely on cooperation between communes or, alternatively, might have to concentrate on planning aspects without direct spatial connections, such as population targets, jobs, or communication networks.

Norway generally attempted the first course of action through a Planning Act (1965) which introduced 'regional plans' relating to sub-regions of the *fylke,* usually comprising 3–5 communes. The plans should be outline or 'strategic' land-use plans, with each commune being responsible for local, detailed planning within its boundaries. As a result, individual communes might effectively block whole regional plans. In fact this has often happened. As late as April 1974, only the plans for *two* of the 80–90 regions had been approved by the Ministry of the Environment. Even the number of regions is still undecided, since in some cases it has proved impossible even to draw regional boundaries, let alone to make the communes cooperate in preparing plans. This failure may also to some extent be due to the lack of planners at the local, regional and national levels in Norway, since there are no training courses geared specifically to meet such needs. In Sweden and Finland regional (or sub-county) planning has proved more successful than in Norway but it has fallen short of expectations. One important shortcoming has been the problem of ensuring coordination between the contents of these plans and the objectives of central government organizations.

In this way planning at the county level was introduced in Scandinavia. Central government operated its regional administration at this level and the communes elected representatives to a council at the same level. In addition, preparation of regional (or sub-county) plans created problems and, therefore, integration at the county level seemed to be a good compromise. Sweden was first to come up with this 'new deal' in planning in 1967 (*Länsplanering* 67) and it then spread to the other countries, including Denmark. By 1974 it was Norway that had done relatively little at the county level of planning, and this may partially be explained by reference to the centre–periphery dimension in politics. In the other countries, comprehensive county planning was simply decided by Parliament and has been carried out mainly by the government side of the county administration. Central authorities in Norway are more sensitive to the attitudes of the communes and their defence of their traditional monopoly of planning. Nevertheless, the necessary laws for instituting county planning were enacted by the Norwegian Parliament in June 1973 and have prepared the way for a very comprehensive system of county planning. This was possible because new legislation was to follow in 1974 to enable direct elections to take place, of which the first will be in September 1975. County plans will be only at a preliminary stage before the new political assemblies come into existence and take responsibility for strategic planning. The new county authorities will obviously reduce the present power of the individual communes. The other northern countries also have proposals for giving broader powers, especially with respect to planning, to the county councils. The difference is therefore one of degree rather than kind; and this example supports previous suggestions about inter-Nordic differences in the centre–periphery dimension in politics.

The significance of this dimension can be seen clearly in the spatial aspects of some measures for regional development. A good example is the designation of growth centres and the consequent measures used to secure growth. Designated centres in Norway or, more precisely, their mother communes drew up new land-use plans and also invested quite heavily in new infrastructure, using borrowed funds. However, the expected growth often failed to materialize, partly due to a lack of positive means to guide activities to the centres, but also in some cases because other communes in the appropriate county united against the unit containing the designated centre and influenced the county council to oppose installing new activities there. It was hinted that growth centres were designated by the central government (in fact, after proposals received from the counties) and that the government should therefore take responsibility for them. The proper duty of the county councils, it was argued, was to look after the needs of less 'fortunate' communes. In Sweden and Finland, national settlement systems were defined with associated priority centres in the Development Areas which were supported by finance from central government to improve infrastructure. In Sweden firms were advised to locate at these centres, not least by the threat that there would probably be no long-term public investments in non-designated settlements.

Another discriminatory measure was a Scandinavian version of the British industrial estates system. It was first used in Norway where a central government corporation (S.I.V.A.) was established in 1968, and from the start an estate was designated in each of the five regions of the nation. Today seven estates are in operation, of which four employ more than 250 apiece, and a further five were designated in 1973–74, on which construction work has started. In 1966–67 it had been proposed that a S.I.V.A. estate should provide 1000–2000 jobs, occasionally providing a minimum of 500 in areas of very slight population. Over the years, however, the target size has been played down, as the attitude to growth-centre policies might suggest. Few of the twelve estates will provide more than 1000 jobs apiece and many will probably offer fewer than 500, particularly the recently-designated estates, of which one (Evje) has only 5000–6000 inhabitants in a radius of 30–40 km.

The idea of such industrial estates has also been adopted recently in Sweden and Finland. The first five estates in the two countries are now being completed. In Sweden, as in Norway, they are run by a government corporation and seem to be concerned with helping industrial expansion in some of the most peripheral of the designated priority centres (for example, Lycksele and Strömsund). In Finland they are run by separate semi-private corporations to which the government will lend money on favourable terms. They are relatively central in their location, with the minimum population base being in the order of 15–20,000. The figures are lower in Sweden (6–12,000), which emphasizes their role as a last resort in which private enterprise has not been willing to invest, notwithstanding offers of grants and loans. It must, however, be added that in each northern country the communes have, and often take, the opportunity to construct factories to rent at subsidized rates to new firms. In Norway the R.D.F. has assisted technically and financially with 240 such buildings since 1962, comprising 400,000 m^2 of floorspace. By comparison, the S.I.V.A. had built only 83,000 m^2 at the end of 1973. The start of Swedish regional policies in 1963 was mainly concerned with investment in industrial buildings to be rented out by the communes.

Development Areas have been delimited and graded for degrees of financial assistance in somewhat differing ways in the three countries. Sweden and Finland have relatively simple systems and have divided their development areas into two zones (Figures 12.3 and 12.4). Conditions in Norway are more varied, though North Norway has tended to have a special status with respect to most measures. Thus a two-zone system resulted, but following the introduction of investment grants in 1971 four zones were defined. In all countries grant-allocating bodies have been instructed not always

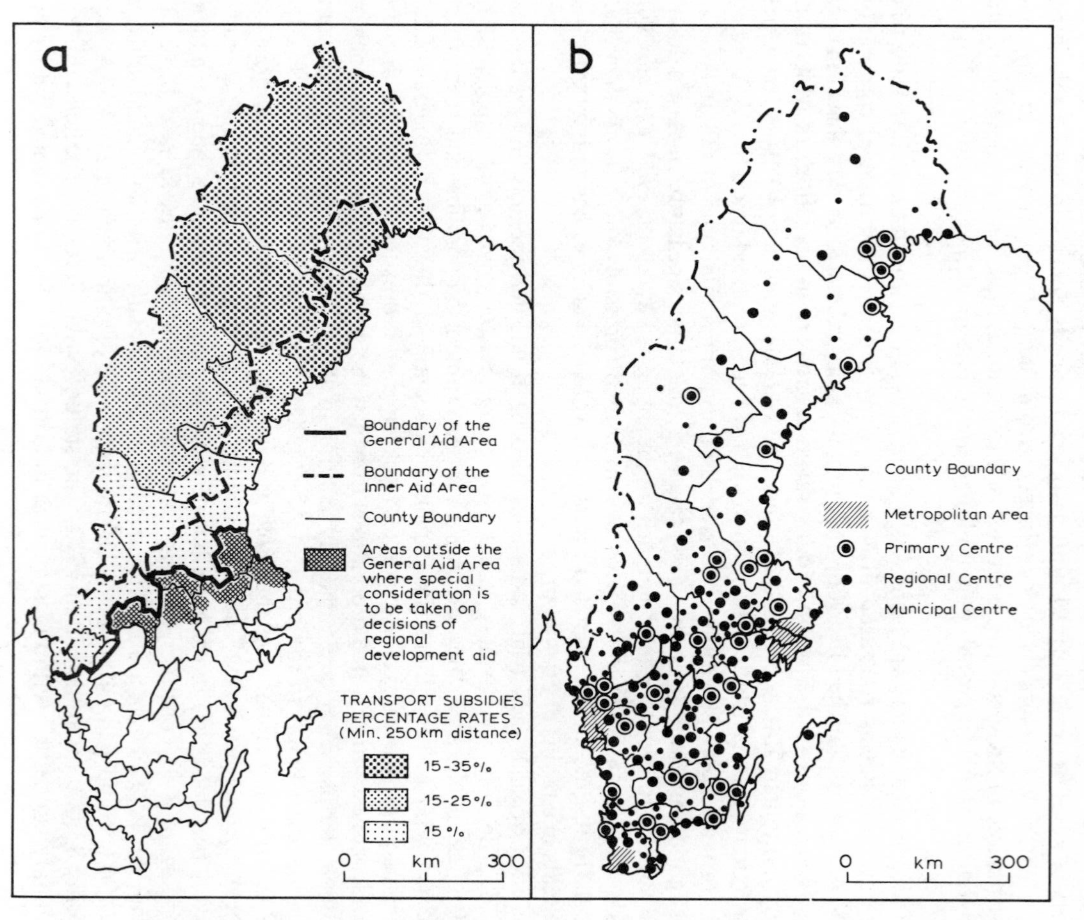

Figure 12.3. Sweden: (a) Development Areas; (b) urban centres.

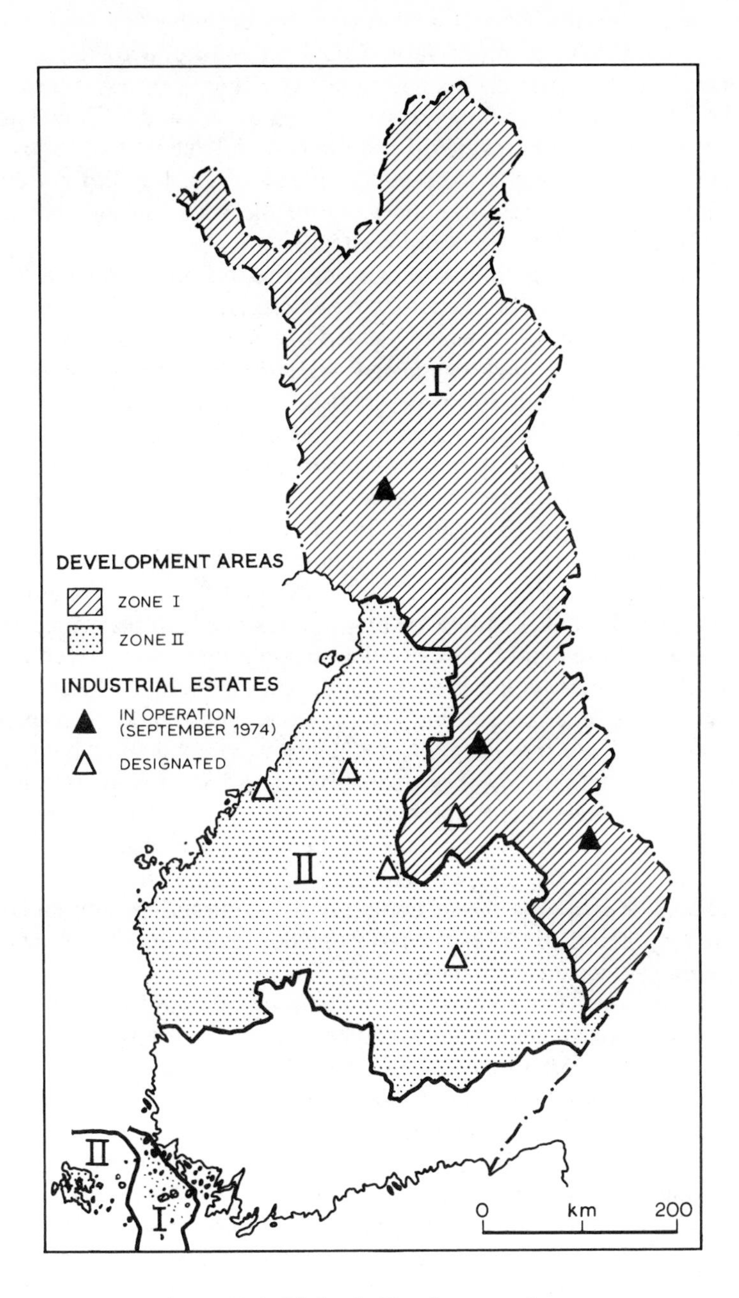

Figure 12.4. Finland: Development Areas.

to give the maximum degree of assistance but to work towards a fairly continuous spatial grading.

The most comprehensive background survey for defining development areas was carried out by the Finns. A multi-variate analysis was undertaken using eighteen socio-economic variables for each commune. The three poorest zones were given 'development area' status, with the two worst being designated as Zone I and the other as Zone II. Boundary changes resulted from the political debate that followed, and the Swedish-speaking Åland Islands were added even though they did not come off too badly in the analysis. This was obviously due to the Åland Islands having a special status in Finland because of their cultural heritage.

In Norway the process of delimitation was less rigorous in technical terms. Initially, the North Norway Plan was defined to cover northern areas. The second stage of delimitation took place in 1961 when the R.D.F. started work. Six variables were selected which experience had suggested were important indicators of regional problems. But these were not followed slavishly. Generally this produced a 'fluid' border within which loans and guarantees from the R.D.F. might be granted. Thus until 1969 there was no sharp definition of development areas in Norway, with the exception of North Norway which had special status. The R.D.F. presented its opinion of the 'poorly developed areas' of the country in its annual reports to Parliament. As in Finland, political bargaining was of crucial importance when rigid boundaries had to be defined in 1969 for allocating tax-free funds and in 1971 for providing investment grants. The boundaries that resulted display interesting features which relate to the anti-growth-centre attitude and to the political bargaining power of certain peripheral areas (Figure 12.5). For example, the whole of Sogn-og-Fjordane county was included in the maximum grant area. It had long been considered as the least developed county in South Norway and, on this account, the decision seemed fair enough. But when compared with delimitations in other parts of western Norway it is rather odd. The boundary for tax-free investments coincides with the boundary between 15 and 25 per cent investment grants. Communes containing heavy electro-chemical or metallurgical plants have been put on the 15 per cent side of this line, despite the increasingly difficult problems faced by these communes in attracting alternative jobs. In Sogn-og-Fjordane this distinction was not made, even though Norway's largest aluminium smelter is located there (at Årdal).

All three countries are working on the problem of decentralizing offices of central government. Sweden has achieved most and Norway least in this respect. The Swedes published a research report in 1970 and one year later Parliament decided that 30 administrative bodies, employing about 6000 people, should be relocated to 13 named towns. A further ten bodies with 4000 workers were added to the list in 1973. In Finland, reports were published in 1973 and 1974 which proposed decentralizing a total of 26 administrations (plus parts of four others) with 8000 workers to 19 towns. A dozen of these settlements are located in development areas. A commission was appointed in Norway in 1961 to review the possibility of decentralizing government offices. Its report was published in 1964 and debated by Parliament in 1966. Virtually nothing happened, but interest was renewed in 1969 and a committee looked into the possibility of locating *new* national agencies outside Oslo. Most notable has been the creation of the Oil Directorate at Stavanger in May 1973. A further committee was appointed in that year and will report on the issue of decentralization in 1975 and 1976. It is difficult to know why Norway has lagged behind but it may partly be due to the lack of a policy for an urban system. By contrast, most primary centres designated in Sweden are to receive relocated government offices. In Norway, policies have been more intimately concerned with the problems of rural areas where office relocation would be inappropriate.

Results of Regional Development Policies

It is difficult to evaluate the results of regional policies for two main reasons. First, such results are produced not only by regional development measures but also by the general economic trends and by other policies pursued by the government. Second, there is a serious lack of feedback studies on the results. In Norway aid-giving authorities have not been required to provide such information, and aid has not generally been linked to specific employment targets. Another problem is the choice of criteria to evaluate the success of regional policies. Normally the number of jobs created is the accepted criterion, but it might be just as relevant to use some index of change in the distribution of population, particularly in Norway.

Statistics on industrial expansion in Scandinavian development areas are often inadequate to allow a distinction to be drawn between moves, expansion of local factories, and completely new factories. In Swedish development areas 13,500 new jobs were created between mid-1965 and the end of 1971 at a cost of SKr 1,570,000,000. One-fifth of the funds was in the form of grants and the remainder comprised low-interest loans. Figures for Norway are less accurate. The R.D.F. has made rough estimates of increases in employment since firms first received loans or grants from the R.D.F. In 1972 such firms employed about 80,000 and the increase was put at 30,000 jobs. A year later the increase was estimated at 34,600. Total employment in aided firms in Sweden was 60,000 at the end of 1971.

Regional aid represented 53 per cent of total investment costs in Swedish industrial development schemes, running at SKr 117,000 per job. In Norway about NKr 50,000 of public funds were allocated for each new job (1 SKr = 1·40 NKr before 1972). There are two main reasons for this difference. First, the Norwegian programme has run for a 20-year period compared with only 6 for the Swedish scheme; thus both inflation rates and the period of growth since the first financial input have been different. Second, there are differences in the industrial structure of development areas. Sweden has a particularly large part of her heavy pulp and paper industry in these areas, while much of the industry in Norwegian development areas consists of fish-processing and other light industries. The heavy electricity-consuming plants have usually been financed by tax-free corporate funds and not by loans from the R.D.F. Finally, it must be added that the Norwegian figures include substantial loans to non-manufacturing sectors such as tourism and fisheries. The R.D.F. has given some indication of the relative insignificance of moves in the present policies. Between 1966 and 1972 191 manufacturing firms moved or set up branch factories with financial help from the R.D.F. In comparison, the R.D.F. was helping to finance over 2200 manufacturing firms at the end of 1972, of which more than 1200 were new engagements in the preceding seven years.

Change in population distribution is another criterion for measuring the success of regional development policies. Stabilization of the settlement pattern was the major objective in Norway, but this was not particularly important in the other two countries. Table 12.4 indicates demographic trends in Northern Europe for 1969. An annual

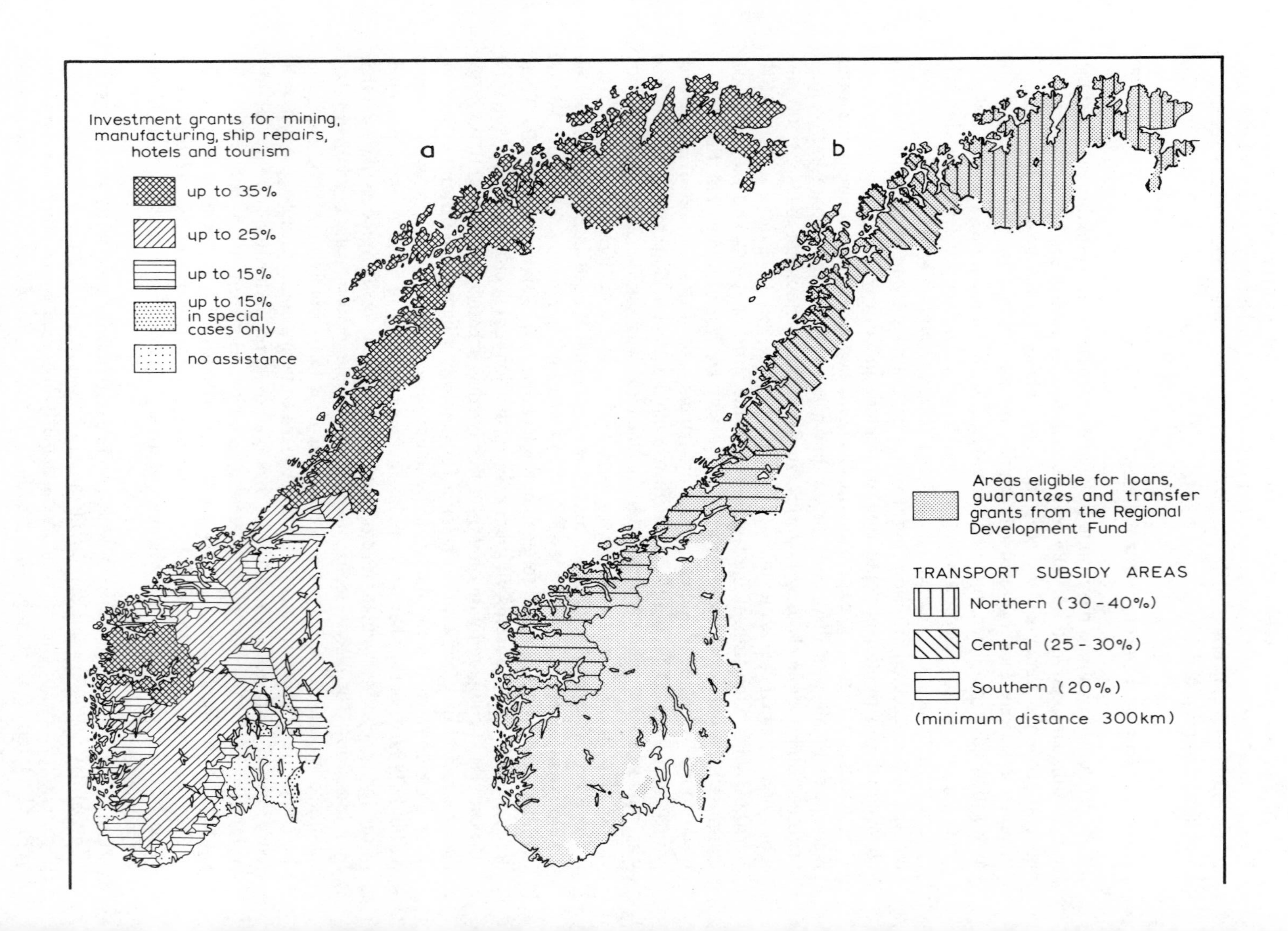

Investment grants for mining, manufacturing, ship repairs, hotels and tourism
up to 35%
up to 25%
up to 15%
up to 15% in special cases only
no assistance
a
b
Areas eligible for loans, guarantees and transfer grants from the Regional Development Fund
TRANSPORT SUBSIDY AREAS
Northern (30-40%)
Central (25-30%)
Southern (20%)
(minimum distance 300km)

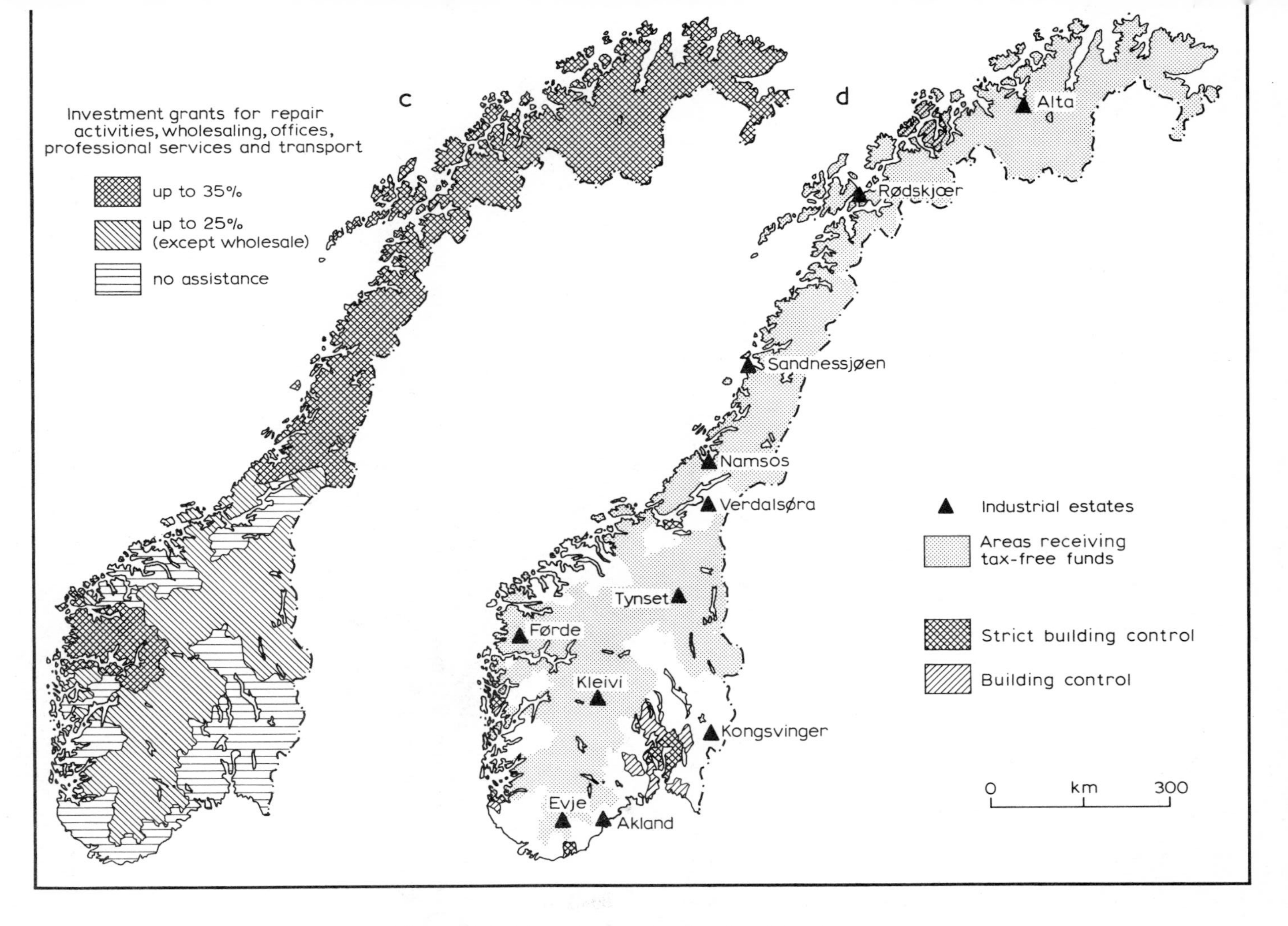

Figure 12.5. Norway: aspects of regional development.

Table 12.4. Comparative Demographic Data for Communes in Finland, Sweden and Norway, 1969

		Percentage of communes with			
		Population decrease			
	Number of communes	total	more than 1·5 per cent	More deaths than births	Net out-migration
Finland	521	75	50	38	75
Sweden	848	52	26	49	61
Norway	451	49	17	11	65

Source: J. C. Hansen, 1972a, The problem of marginal areas in Norway: population trends and prospects, *Meddelelser fra Geografisk Institutt ved NHH og UIB, 19.*

decline of more than 1·5 per cent has been called 'critical depletion' by Hansen (1972a), and it appears that Norway as a whole is doing very much better than her neighbours. However, it must be added that conditions in northern parts of each country fall well below the average. In northern Sweden four communes out of every five lost population, compared with two out of five in the South; and the critical annual loss of 1·5 per cent affected 55 per cent of the communes in northern Sweden, compared with 35 per cent in North Norway. The year 1969 was a period of heavy migration from Finland to Sweden, involving some 15,000 people, with important moves from northern Finland to central and southern Sweden. In the same year, North Norway experienced the highest out-migration ever recorded, involving about 5300 compared with 2500 in 1971 and 120 in 1972. (Admittedly the 1970 figure was higher, but that was partly due to the census revealing unregistered moves.)

This evidence indicates that Norwegian policies have been fairly successful, but it should not be forgotten that it is a political objective that not only each commune but every community in each commune should survive. Hansen's (1972a) provisional analysis of data for enumeration districts shows relatively less success at that level, in North Norway at least. Some 414 of the 586 enumeration districts in North Norway lost population, with 209 having lost more than 16 per cent of their November 1960 total by January 1970. Districts within commuting range of settlements of 1000 inhabitants or over experienced less severe losses than districts in more remote locations. Only half of the 190 enumeration districts within commuting range lost population and of these a mere 34 declined by more than 16 per cent. By contrast, 175 of the 396 peripheral enumeration districts experienced losses in excess of 16 per cent. But it must be remembered that the peripheral districts contain relatively few people. In 1970 two-thirds of the inhabitants lived within the commuting hinterlands defined by a 30-minute journey to work. Even so, 150,000 people lived beyond these hinterlands. If the minimum settlement size is increased from 1000 to 2000 only 50 per cent of the inhabitants of North Norway live within 30 minutes' travelling time from such employment centres.

The Swedes and Finns do not seem to be preoccupied in the same way with their smallest communities and perhaps that is a major reason behind their greater difficulties in keeping the total population in their peripheral areas fairly stable. Experience would suggest that policies for retaining population over a fairly wide area may be effective in saving individual *communes*, but if one attempts to save only the central settlement of a commune, as in Sweden, it is likely that the loss from the surrounding rural area may be so great that the viability of the central place is lost and *its* population declines in turn.

It is also necessary to stress differences in economic structure in the present discussion, as well as policies and attitudes. Agriculture, forestry and fishing are each undergoing rapid rationalization. The resource base for agriculture and forestry is relatively limited and thus a quadrupling of productivity in forestry (especially logging) over the last 15–20 years has involved a commensurate decline in demand for labour since cutting could be increased by only 20–40 per cent.

The Norwegian fisheries have also improved productivity by a factor of 3 or 4 since the mid-1950s. Catches have roughly doubled, and have been rather more impressive in North Norway, so that the decline in labour demand has not been quite so sharp. In addition, larger catches and a change from simple salting and drying to more advanced freezing and filleting made it possible to increase onshore employment in the fish-processing industry by 50 per cent from 1960 to 1970. Pulp and paper manufacturing in Sweden and Finland, by contrast, reduced its labour demands in the 1960s. Nevertheless, there are still differences that may be attributed to deliberate policies. In North Norway there are many fisherman–farmers (*fiskerbonde*) who are supported both by regulations and finances to assist fishing and by agricultural subsidies which offer advantages to smallholders. (There are also considerable regional variations in certain agricultural subsidies, particularly for milk.) Finland also included a similar bias in favour of smallholders in its subsidy structure but had to discontinue some measures in 1968–69 because of over-production of certain commodities. Agricultural policy in Sweden has been guided mainly by questions of productivity since 1947 and especially after 1967. Its regional implications have been predominantly negative for the sparsely-populated, marginal areas of the North.

It is argued with increasing vigour in Norway that agricultural subsidies should be considered as an addition to regional policies because they tend to maintain the local population and thus a demand for local services. Surplus farm labour is also available for deployment in other jobs. Indeed, there have been some quite successful examples of job-sharing between agriculture and manufacturing industry. Two men may sign on together for a single full-time job in a factory. Each works there for half of every week and then spends the rest of his time on his smallholding. Productivity was reported to be very good from one firm running such a scheme, particularly because of a reduction in absenteeism.

The implications of agricultural policies on regional development bring the discussion back to the problem of policy coordination. It may be argued that regional development is the result of three sets of policies: the direct (e.g. grants); the indirect (e.g. growth centres, general planning); and the unconscious (e.g. railway closures, centralized government purchases, agricultural policies). The last of these groups is the joker in the pack. It is probable that the most important factor influencing population change at the local and commune level in Norway and Sweden during the 1950s and 1960s was the new school laws. Centralization of schools into larger units to improve specialist teaching facilities was very dramatic. The number of primary schools in Norway fell from 5200 in 1955 to 3000 in 1971. The idea was to give people equal opportunities but paid no real attention to spatial differences. Children probably now have better opportunities to choose freely between jobs or sectors of national society, but they have reduced freedom of choice between places to live since they receive education for jobs which only city regions can provide.

Scandinavian policies have dealt almost exclusively with sparsely-populated regions of primary production. The problem of decaying industrial regions is relatively unknown. Regional development funds have been used in certain small industrial areas in southern Sweden, particularly in the Boräs and Norrköping textile zones. Some attention has been given to the leather- and shoe-making town of Halden in southeastern Norway, near the Swedish border, and also to some smaller settlements around the Oslo-fjord which were established about a hundred years ago during the first phase of industrialization. Recently, however, the many one-company towns throughout Norway have given rise to concern. Their lack of alternative employment and complete dependence on the market for individual products show that problems can stem from this kind of industrial development. Communes in which these factories are located have started to show the same kind of general decline in population and increase in the proportion of older people as is found in surrounding rural communes. In two recent cases, whole communities were threatened because the resource base of a mine was depleted or because the technology used for nitrogen production was no longer competitive. The mining community of 300 people was abandoned, but the other plant employed 1200 and was the basis of survival for the town of Rjukan with 6–7000 inhabitants. In another case, an aluminium smelter announced plans for halving its labour force of 2000 over a period of five years unless more electricity could be supplied. This is difficult since most of the cheap hydro-electric power has already been harnessed and nature conservation is claiming many sites that might otherwise be used.

On the whole, may one conclude that regional policies have been successful in Scandinavia? It was suggested that Norwegian policies had been quite successful because population decline in peripheral communes was relatively smaller than in comparable areas in Sweden and Finland. However, this does not mean that these countries have 'failed' because they have consciously adopted a policy of 'decentralized concentration' or growth centres. This inevitably involves population loss from areas between the centres. According to projections made in 1969, only 6 out 43 groups of communes in the four northernmost counties of Sweden were expected to record population increases between 1965 and 1980, while 27 were expected to lose more than 2 per cent each year over that period. The authorities could not acquiesce to such a decline and planning proposals have been made to stabilize population numbers whilst accepting some intra-county movement. At a county level conditions seem to have remained stable up to the end of 1973 at least. This may have been due to regional measures introduced after 1965, but it is still difficult to say whether or not Swedish regional policies have been a success.

Policies may be evaluated at the other end of the scale by comparing the growth rates of city regions of different sizes, assuming that relatively low growth rates are desirable for the largest cities at least (Table 12.5). Finland scores least well when such a comparison is made, with a tendency towards excessive concentration. During the 1960s the Helsinki region received 93 per cent of the national population increase,

Table 12.5. Growth of City Regions,[a] 1950–70 (1950 = 100)

	Capital region1	Level 2	Level 3	Level 4	Total urban population	Total national population
Finland	168	142	132	121	138	115
Norway	130	133	128	124	129	119
Sweden	134	143	145	147	143	115
Size range (in thousands) and number of city regions at each level.						
Finland	800	234–239	101–143	30–88		
	(1)	(2)	(6)	(19)		
Norway	745	145–228	79–113	31–63		
	(1)	(3)	(5)	(11)		
Sweden	1276	389–603	100–178	30–95		
	(1)	(2)	(12)	(47)		

[a] With more than 30,000 inhabitants in 1970.
Source: E.F.T.A., 1973, *Regional Policy in E.F.T.A.: national settlement strategies a framework for development*, E.F.T.A. Secretariat, Geneva. Reproduced by permission of The European Free Trade Association.

which means that the urban hierarchy has tended to develop into a primate-city system. After 1970 the Finns took steps to control this tendency. Norway displays the smallest difference between total population increase and metropolitan increase which indicates that there has been some success in stabilizing the distribution of population.

Conclusion

It is very much a matter of personal opinion whether or not the three North European countries have developed their own particular versions of policies for regional development. Certainly there are four vital issues that one should consider when examining what has been achieved.

The first point to note is the nature of the regional problems. The three countries contain problem regions which are the most extensive and, in absolute terms, probably the most marginal of all in Western Europe. At the same time, however, even the economic cores of these countries may be recognized as marginal to the bulk of Western Europe. This underlines the importance of relativity in considering marginal regions. For example, services and infrastructure that are considered to be 'necessities' in problem areas in Great Britain or the Netherlands may not even be thought of in northern Scandinavia. In similar fashion, policies that are effective in Great Britain or Western Europe may at best be irrelevant or at worst be detrimental to meaningful development in the far North.

The second point to stress is the political and administrative system of Scandinavia, in particular the role of the communes. This has had a considerable impact on the nature and results of regional development since political power is focusing and channelling all activities through the commune councils.

A third point is an increasing awareness of the regional dimension in all national planning. Starting with social goals, and passing through optimal use of national resources, growing emphasis is now placed on the distribution of population. In Norway this means trying to stabilize as much as possible of a scattered and often very rural settlement pattern. In Sweden and Finland it means creating an urban system that gives people maximum provision of services and choice of jobs. In all countries it now means reducing the growth of the largest city regions. Regional dimensions are now included in the main national 'welfare function' and more or less override the earlier goal of economic growth.

The fourth point to note relates to a greater awareness of which policies actually promote regional changes. In the early phases the so-called 'regional' policy measures were the only ones considered, but experience has shown that regional development may be influenced to an even greater degree by general sectoral policies. Expansion in the national economy may create more jobs in peripheral regions than a doubling of regional aid during a 'stop' phase of national economic policy. Specifically it has led to a greater emphasis on coordination of sectoral plans and to a strengthening of an intermediate planning level attached to the counties. The plans of national corporations, particularly in the communications sector, have recently been studied closely for their regional effects and to see if they conformed with regional plans.

It has not proved possible to conclude whether the three countries' regional policies have been particularly successful when compared with each other or with other West European countries. There is a lack of feedback studies on the effects of regional policy measures. In addition, and in the light of the increasing importance being attached to non-economic factors in evaluating living conditions, it has become more difficult to find relevant indicators. For example, should one examine service provision or the value of open space? To use the Norwegian case, when personal incomes are examined it would appear that Norway's policies have been successful since regional differences have been narrowed. But there still seems to be dissatisfaction with the lack of success in stopping the urbanization process and internal migration, despite Norway's success compared with Finland or Sweden.

References

Berg, P. O., 1973, Norway, in M. Broady (Ed.), *Marginal Regions: essays on social planning,* Bedford Square Press, London, 47–56.

E.F.T.A., 1973, *Regional Policy in E.F.T.A.: national settlement strategies, a framework for development,* E.F.T.A. Secretariat, Geneva.

Enequist, G., 1968, Agricultural holdings in Sweden 1951–66 and 1980, *Acta Geographica,* **20**, 75–89.

Economic Research Institute of Finnish Industry, 1969, A study of the factors influencing the choice of location of manufacturing establishments in Finland, *E.R.I.F.I. series B,* 16.

Gilpin, M. C., 1968, More and Romsdal, Norway: a study of changes in the rural transport system of a coastal area, *Geography,* **53**, 146–156.

Hansen, J. C., 1971, En Geografisk kritikk av de norske beforlkningsprognoser, *Norsk Geografisk Tidsskrift,* **25**, 175–188.

Hansen, J. C., 1972a, The problem of marginal areas in Norway: population trends and prospects, *Meddelelser fra Geografisk Institutt ved NHH og UIB,* 19. (A shortened version appears in L. Kosinski and R. M. Prothero (Eds.), 1975, *People on the Move,* Methuen, London, 255–276.

Hansen, J. C., 1972b, Regional disparities in Norway with reference to marginality, *Transactions Institute of British Geographers,* **57**, 15–30.

Hustisch, I., 1968, Finland: a developed and undeveloped country, *Acta Geographica,* **20**, 155–175.

Kiiskinen, A., 1965, Regional problems and policies in Finland, *Papers and Proceedings of the Regional Science Association,* **14**, 91–106.

Knox, P., 1973, Norway: regional problems and prosperity, *Scottish Geographical Magazine,* **89**, 180–195.
Kristinsson, V., 1973, Population distribution and standard of living in Iceland, *Geoforum,* **13**, 53–62.
Mead, W. R., 1974, *The Scandinavian Northlands,* Oxford University Press.
Ministry of Labour and Housing, 1973, *Planning Sweden: regional development planning and management of land and water resources,* Allmänna Förlaget, Stockholm.
Odmann, E. and G. B. Dahlberg, 1970, *Urbanization in Sweden,* National Institute of Building and Planning Research, Stockholm.
Pred, A. R., 1973, Urbanization, domestic planning problems and Swedish geographical research, *Progress in Geography,* **5**, 1–60.
Smeds, H., 1960, Post-war land clearance and pioneering activities in Finland, *Fennia,* **83**, 1–31.
Sommers, L. M. and O. Gade, 1971, The spatial impact of government decisions on postwar economic change in North Norway, *Annals of the Association of American Geographers,* **61**, 522–537.
St. meld nr. 50, 1972–3, *Tillegg til St. meld. nr.27 for 1971-2 Om regionalpolitikken og lands-og landsdelplanleggingen,* Miljøverndepartementet, Oslo.
Törnqvist, G., 1970, Contact systems and regional development, *Lund Studies in Geography, Series B Human Geography,* 35.
Varjo, U., 1971, Development of human ecology in Lapland, Finland, after World War II, *Geoforum,* **5**, 47–74.
Wood, J. S., 1965, *The North Norway Plan: a study in regional economic development,* Chr. Michelsens Institutt, Bergen.

13

Austria

Fritz Schadlbauer

The National Context

Austria is a landlocked nation set between two major territorial blocs with markedly different political, social and economic attitudes. The Austro-Hungarian Empire was finally dismembered in 1919, when Czechoslovakia, Yugoslávia and other new states were created. South Tirol was lost to Italy and the Italian frontier extended to the Brenner Pass. The port of Trieste, which had functioned as Austria's Adriatic and Mediterranean outlet, was also handed over to Italy. Finally, Burgenland was transferred to Austria from Hungary. A generation later Austria was divided between the Allied Forces after liberation in 1945, having been forcibly incorporated in the German Reich in 1938. Vienna was split into four separate occupation zones. A provisional government was supervised by a four-power control council for ten years until independence was regained in 1955 and the federal republic of Austria was established. Permanent neutrality was written into its constitution, with the U.S.S.R. being a guarantor of independence. Much of its originality and many of the problems of spatial management stem from these facts. Austria borders on seven other countries and one-third of its 2640 km of external boundaries are with Czechoslovakia and Hungary. This gives rise to considerable economic problems in the border zones.

Although only a small country, Austria has a diverse physical base which, in turn, contributes to a range of management issues. Five geographical regions are usually recognized: the Alps; the Granite and Gneiss Uplands; the Alpine and Carpathian Foreland; the Eastern Foreland; and the Vienna Basin (Figure 13.1).

The Alps are young mountains separated by a series of structural longitudinal valleys that have been reshaped by glacial action. Two great troughs separate the Central Alps from the Northern and Southern Limestone Alps and are the main arteries for trade and commerce in an east–west direction. The northern trough runs from Vorarlberg through the Inn, Salzach, Enns and Muerz valleys into the Vienna Basin. The southern trough runs along the Puster and Drau valleys into the Klagenfurt basin. Population is concentrated in the larger valleys and basins which support farming and manufacturing. The highly developed textile industry in Vorarlberg is important, as are the old mining and iron and steel industries in the Mur–Muerz valley. Many H.E.P. plants are located in this zone.

The Granite and Gneiss Uplands north of the Danube are flat-topped plateaux with a raw climate and rather infertile soils. They are thinly populated and are still forest covered. This rather poor economic situation became even worse after the Second World War because the border with Czechoslovakia was closed. The Alpine and Carpathian Foreland is an undulating region. The Alpine section stretches from Bavaria to the Vienna Woods and is excellent farming country interspersed with gravel 'islands'

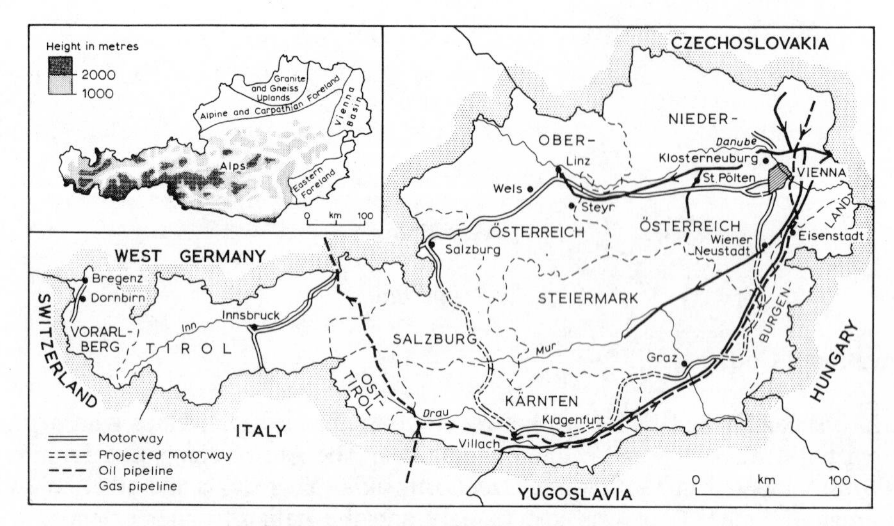

Figure 13.1. Austria: the national context.

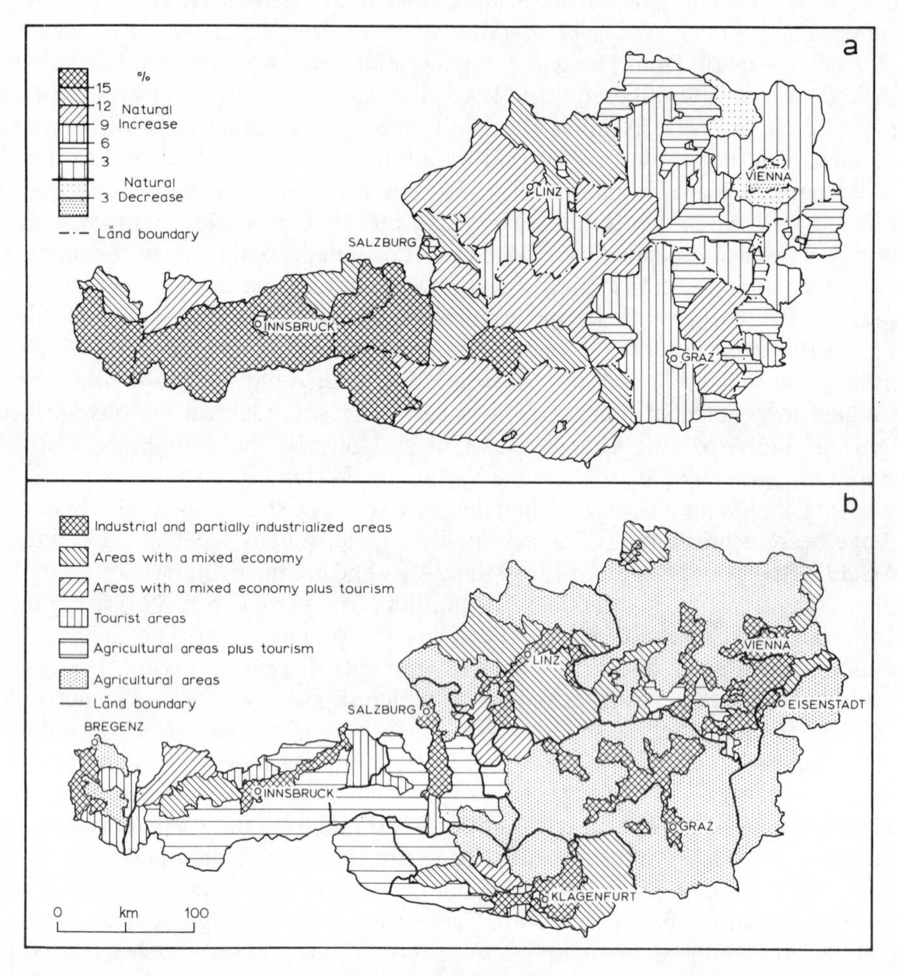

Figure 13.2. Austria: (a) natural change 1961–71 (excess of births over deaths as percentage of the 1961 population); (b) structure of the Austrian economy.

covered by forests. The Danube, which broadly speaking marks the northern boundary of the Alpine Foreland, cuts into parts of the Granite Upland so that its valley is characterized by an alternation of gorges and broad basins. H.E.P. stations have been built at the lower ends of some of these gorges to make use of the constant flow of the Danube, and during the Second World War a large industrial complex was developed at Linz. The Carpathian Foreland, lying between the Granite and Gneiss Uplands and the Vienna Basin, is a very fertile area producing sugar beet, barley, vines and dairy products.

The Eastern Foreland borders on Yugoslavia and Hungary. This farming area is also fertile, but as a result of fragmented holdings is in a rather poor state economically. Manufacturing industries are concentrated around Graz and include engineering and production of cars and cycles. Finally, the Vienna Basin stretches from Neunkirchen to the Czech border. Its northern part is very productive and comprises rich farmland. Southwards from Vienna it includes the largest industrial area in Austria.

The national population has increased almost continuously since the first census (1869) with the exception of the two World Wars. Austria's total rose from 4,500,000 in 1869 to 6,530,000 in 1923 and 7,460,000 in 1971. There have, however, been important regional variations in population change since the First World War and the break up of the Austro-Hungarian Empire. Vienna was no longer the capital of a state of 53,000,000 inhabitants. The city experienced a reduction in economic attractiveness and population declined in Vienna and in neighbouring *Länder* of Nieder Osterreich and Burgenland. By contrast, numbers increased in the rest of the country, with *Länder* totals increasing by more than 50 per cent in Salzburg, Tirol and Vorarlberg. As a result of Austria's relief features the distribution of population displays clear contrasts. Densely populated zones are below 600 m above sea level and are generally outside the Alpine ranges. Some 27·5 per cent of Austrians live below 200 m altitude, 29·7 per cent between 200 m and 400 m, and a further 27·1 per cent between 400 m and 600 m. This tendency for population to concentrate in just a few areas is aggravated by the enormous difficulties of small farms in mountainous areas which, nevertheless, support Austria's highest rates of natural increase (Figure 13.2). Some 43 per cent of her population live in 44 towns and cities with more than 10,000 inhabitants apiece. The same centres accommodate 250,000 foreign workers, mostly from Yugoslavia and Turkey, who make up about one tenth of Austria's total labour force. But in some areas like Vorarlberg, Austria's smallest but fastest growing *Land*, the proportion is as high as 25 per cent.

The Economy

About 426,500 people (14 per cent of the workforce) were engaged in farming and forestry in 1971, but they contributed only 6 per cent of the G.N.P. Since the Second World War agricultural production has improved greatly in response to specialization, mechanization and rationalization. Now it accounts for 85 per cent of Austria's food requirements. The economic situation of agriculture is characterized not only by extensive mountainous areas but also by the small size and scattered nature of many farms, although much has been done to make them more viable by plot consolidation. Main foci for arable production and dairy farming are situated outside the Alpine region. Pastoral farming is characteristic of higher areas. Vines covered 47,690 ha in eastern Austria in 1972 in the Weinviertel, along the Danube, the southwestern part of the Vienna Basin, and in the southern and eastern Tirol.

Some 43 per cent of Austria's land surface is occupied by forests. In addition to their economic importance they fulfil a social role for the whole nation providing opportunities for outdoor recreation and preventing soil erosion, landslips and avalanches in mountain areas. In order to prevent the occurrence of avalanches 11 per cent of Austria's forests are protected and may not be used for economic purposes. Overcutting was significant in previous centuries and avalanches threaten those areas where this happened. Two-thirds of the avalanches in Tirol come from areas that are physically capable of supporting timber (Eckmüller, 1970). As well as these protected forests, a further 10 per cent of Austria's forests are in high mountain areas and may not be used economically because of low timber quality or high costs of transport. The most densely forested areas are in Steiermark, Karnten and Nieder Osterreich, with one fifth of forests being owned by the state and public bodies and the remainder being in private hands. Almost three-quarters of private woodland is owned by farmers with less than 5 ha of timber apiece.

Table 13.1. Austria: Mines and Mining Output, 1937–72

	1937		1965		1972	
	Mines	Output (tons)	Mines	Output	Mines	Output
Coal	8	230,220	2	58,611	0	0
Lignite	41	3,241,770	16	5,450,356	13	3,755,510
Iron ore	3	1,884,694	3	3,536,300	3	4,132,200
Lead and zinc ore	1	116,238	1	196,964	1	359,635
Copper ore	2	7,221	2	121,201	2	196,167
Graphite	7	18,158	8	85,755	3	18,777
Magnesite	10	397,838	9	1,815,608	6	1,429,414
Crude oil		32,933		2,854,544		2,477,862
Natural gas (in m^3)		0	7	1,723,714	5	1,963,130

Mining has a long tradition in Austria and is still important with respect to oil and natural gas, magnesite, lignite and metal ores (Table 13.1). The main oil fields are in the north-east of the Vienna basin, with minor ones in the Alpine Foreland of Ober Osterreich. Domestic oil output satisfies only one third of demand. For this reason the Adriatic–Vienna Pipeline was built as an offshoot from the Trans-Alpine Pipeline and serves a refinery at Schwechat near Vienna. Natural gas is also produced mainly in the northeastern parts of the Vienna basin and is piped to Vienna and industrial centres in Steiermark, Nieder and Ober Osterreich. Since 1968 Austria's pipelines have been connected to the pipeline from the Ukraine at Bratislava. Magnesite is mined at Radenthein, Veitsch and Launersbach and is an important earner of foreign exchange. Production of hard coal declined markedly during the post-war period and the last colliery was closed in 1968. Lignite is mined in Steiermark and in Ober Osterreich, but at a considerable economic loss. For example, Fohnsdorf in Steiermark is the deepest lignite mine (1200 m) in the world. Iron and manganese ores are mined in Karnten and Steiermark. The Erzberg (Ore Mountain) has been exploited since prehistoric times and formed the main initial factor for the growth of the iron and steel industry in the

Mur–Muerz valley. Lead, zinc and copper ores are also mined. Some of these satisfied Austria's needs in earlier times but now many minerals have to be imported.

Unlike other resources that are in relatively short supply, Austria has no problem of electricity production. Rivers that flow steadily through the year (such as the Danube, Enns, Inn, Salzach and Drau) and meltwater form the bases for many important H.E.P. schemes. Electricity generation increased tenfold between 1939 and 1972. But although many H.E.P. plants were built during this period the share of Austria's electricity output accounted for by H.E.P. fell from 83 per cent to 59 per cent. Construction of reservoirs and power stations in the mountains sparked off a booming tourist industry. One reason was that efficient transport facilities had been installed for construction purposes and were later modified for transporting passengers. Thermal power stations are located near lignite mines and near larger cities. A nuclear power station is under construction at Zwertendorf in Nieder Osterreich and another is planned at St. Pantaleon.

Manufacturing industry accounts for 36 per cent of Austria's G.N.P. The main industrial area has long been the Vienna basin which contains almost all branches of manufacturing. The traditional focus of the iron and steel industry is the Mur–Muerz valley, and other old centres include Steyr and St. Poelten. Large paper mills are situated in valleys on the edge of the wooded Alpine regions. The location of both of these industries results from initial dependence on local raw materials. An important textile industry has developed in the Rhine valley in Vorarlberg and in the upper Inn valley in Tirol. Large new industrial sites were built during the Second World War in Linz (iron and steel, chemicals), Ranshofen (aluminium), and Lenzing (cellulose, synthetic fibres). Non-ferrous metals are refined at Gailitz-Arnoldstein (Karnten), Brixlegg (Tirol) and Lend (Salzburg). Construction of a new copper refinery at Mitterberg (Salzburg) where ore is mined is under discussion. As in agriculture and forestry smallish production units characterize Austria's manufacturing industry. Thirty per cent of the industrial workforce is employed by firms with under 20 workers apiece and a further 28 per cent in firms with 21 to 50 employees. Only 9 per cent work for firms with over 250 employees. This generally reduced scale of manufacturing industry is a disadvantage with respect to international competition in mass production. Austria's marginal position to the principal world markets raises transport costs and this also impedes competitiveness. For years the country has successfully orientated its manufacturing to high-grade products.

Tourism has been of major importance in Austria's post-war economic revival. Foreign exchange earnings from tourism compensate for 90 per cent of the country's trade deficits. Tourists spent almost 93,000,000 nights in Austria during 1972, with two-thirds of these in commercial accommodation (Table 13.2). Foreigners dominate the tourist scene, accounting for 72,000,000 overnight stays. The western *Länder* receive the largest number, with some 77 per cent of all Austria's foreign visitors coming from West Germany, 6 per cent from the Netherlands and 5 per cent from the United Kingdom.

Austria's largest trading partner is West Germany, which supplied 42 per cent of her imported goods in 1972 and to which 22 per cent of exports were sold. The other traditional commercial link, namely the eastern European countries, is declining in importance. Trade with members of Comecon now accounts for only 12 per cent of Austria's exports compared with 16 per cent in 1967. In spite of growing links with West Germany, exports to the Six fell from 50 per cent in 1963 to 39 per cent in 1972.

Table 13.2. Nights Spent by Tourists in Austria, 1972

Land	Total nights (in thousands)	per cent	Nights spent by foreigners (thousands)	per cent
Burgenland	1,039	1·1	489	47·0
Karnten	14,822	15·9	12,594	85·0
Nieder Österreich	5,308	5·7	1,209	22·7
Salzburg	16,330	17·6	13,316	81·5
Steiermark	7,947	8·6	2,859	36·9
Tirol	29,953	32·2	28,374	94·7
Ober Österreich	7,829	8·4	4,604	58·8
Vorarlberg	6,135	6·6	5,686	92·7
Wien	3,575	3·9	3,111	87·0
Austria	92,938	100·0	72,242	77·7

But Austria's imports from the Six have been rising and this is reflected in dramatic increases in her trading deficit. With the enlargement of the E.E.C. Austria must export more to the Nine unless she is to rely even more heavily on its earnings from predominantly West German tourists. Under an agreement of 1972 an industrial free-trade zone was established between Austria and the Nine, with tariff barriers to be removed in five stages. However, so-called 'sensitive' goods (such as aluminium, alloy and carbon steels, and agricultural produce) have been excluded. The Austrian government views its trading future with the E.E.C. optimistically, since the country did very well in the E.F.T.A. in competition with such industrial states as Sweden, Switzerland and the United Kingdom.

The Spatial Structure of Austria's Economy

The spatial distribution of sectors of the national economy may be summarized by a six-fold classification proposed by Bobek (1970) (Figure 13.2). Industrial and partially industrialized areas cover 14 per cent of Austria's territory but contain 57 per cent of the population and 61 per cent of the labour force. Each *Land* except Burgenland contains part of at least one of these areas which include each of the provincial capitals. The leading industrial area is Vienna plus its southern extension in the Vienna Basin being followed by Linz–Wels–Steyr in Ober Osterreich. The main centre in Steiermark covers the Mur–Muerz valley and the Eisenerz area. Partially industrialized areas are found around Graz and in Karnten, Salzburg and Tirol *Länder*.

Areas with a mixed economy (agriculture and industry) border the industrial and partially industrialized zones that have been identified already. Areas with a mixed economy plus tourism (having more than 1000 overnight visitors per 100 inhabitants) involve the Tirol (Ausserfern) and parts of the Salzkammergut region and Karnten. Tourist areas are found mainly in western parts of Austria, extending as far east as the Carinthian lakes and the Gastein Valley. Only two tourist areas have developed in the East (Mariazell and Semmering-Weschel). Agricultural areas plus tourism (with more than 1000 overnight visitors per 100 inhabitants) cover most of the higher parts of western Austria and there are two such zones in the East (Guttenstein, Tuernitz). Finally, truly agricultural areas are situated east of a line between Salzburg city and Spittal-an-der-Drau. They cover most parts of the Alps and form a broad belt along Austria's southern boundary.

Regional Development Problems and Policies[1]

As in other West European countries, economic and social problems in Austria are found in relatively backward rural areas, in declining mining or manufacturing communities, and in congested urban zones. But such issues are of a smaller dimension than in most of the other countries examined in this volume. In particular, concentration of population and economic activity has not yet assumed really alarming proportions. Admittedly, a quarter of the national population lives in the Vienna agglomeration (*c.* 1,850,000) but rates of increase are low, being accounted for by in-migration since more deaths than births are recorded in the city (Figure 13.2). Limited rates of natural increase in the eastern part of the country contrast with very pronounced growth to the west. Nevertheless, Austria's predominantly Alpine environment imposes formidable physical constraints on the distribution of population and patterns of agriculture and communications. The rural exodus continues to contribute an important source of supply of additional manpower for urban industries and tertiary activities. Rural areas require remodelling in response to these changes and towns need planning for increased population numbers. A special problem is caused by Austria's 'dead frontiers' with her communist neighbours, since it is in these parts of the country that industrial employment is weak and out-migration is intensified by the lack of cross-border interaction. When income levels and unemployment rates are analysed it becomes clear that the most important problem areas are in the eastern border zones. It is there too that problems of countryside management are most severe and mountain farming is undergoing the most serious decline in the eastern Alps. By contrast, mountain agriculture in western Austria is bolstered by incomes derived from tourism and, in addition, rural residents take advantage of commuting across the border to industrial work in Switzerland and Bavaria.

The main objective of regional management in Austria is to encourage economic development in all parts of the country in such ways as to improve the living standards of the population and to contribute to the growth of the national economy in the most efficient manner. Income levels need to be raised in lagging areas and transport facilities strengthened to link remote areas to foci of economic development. Improvements in energy supply, housing and education are also vital in Austria's regional management. The federal government has prepared a general regional policy plan which contains guidelines for national ministries and provincial administrations. Detailed maps on communications, farming, tourism, population, other socio-economic indicators and links between town and country were presented in a series of documents in 1969 and 1970 (Sieper, 1972). The concept of regional policy has only recently come to the fore in Austria, with responsibility falling to the *Länder* and local communities rather than to the federal government. Hence the purpose of the general policy plan is to help coordinate legal measures and administrative action. At the lowest level, independent local authorities are required to prepare land-use plans which are submitted to the government of the appropriate *Land* in order to ensure that they conform with the general objectives of spatial management. The *Länder* are also largely responsible for implementing regional policy measures and meeting their costs. A key objective is to ensure that new industries are not allowed to develop in a random or scattered fashion but are clustered at industrial development nuclei where infrastructural improvements are concentrated. Such schemes are financed by subsidies, loans and tax remissions from provincial budgets which, in turn, receive substantial inputs from the federal budget in the context of numerous programmes for industrial or agricultural support and the installation of new infrastructures. Individual communities may attract new firms by providing building sites at low charges, and pegging costs of road construction and water or electricity supply. A special credit organization helps to finance industrial installation in development zones, especially in eastern Austria.

Regional management in Austria therefore requires detailed collaboration between the various levels of government in this federal state. Planners are very much aware of the special role of tourism in the national economy and they realize that it is a particularly important source of income in predominantly rural areas with attractive landscapes, where industrial development is largely impossible or is feasible to only a very limited extent. Industrialization programmes must therefore take tourism fully into account and must ensure that recreational environments are not polluted or otherwise degraded. A compromise between taking jobs to the workers and workers to the jobs is being implemented. Industrial installation is being favoured in areas and towns where it is considered to have a chance of being viable, but in parts of the country with little potential for factory development (but perhaps great potential for tourism) commuting to jobs at industrial nuclei is being encouraged through transport improvements. In addition, subsidies are paid to mountain farmers with the partial objective of keeping them on the land, for without their presence the infrastructure and visual attractions of Austria's agricultural landscapes would diminish rapidly.

References

Bobek, H., 1970, Ausgliederung der Strukturgebiete der österreichischen Wirtschaft, in *Strukturanalyse des österreichischen Bundesgebietes,* Vienna.

Eckmüller, O., 1970, Struktur und Problematik der österreichischen Forstwirtschaft, in *Strukturanalyse des österreichischen Bundesgebietes,* Vienna.

Kopetz, H., 1972, Economic problems of mountain farming in Austria, *O.E.C.D. Agricultural Review,* **19**, 14–15.

Krebs, N., 1961, *Die Ostalpen und das heutige Osterreich,* 2 vols., Darmstadt.

Lichtenberger, E., 1975, *The Eastern Alps,* Oxford University Press.

O.E.C.D., 1970, *The Regional Factor in Economic Development,* Paris.

O.E.C.D. Economic and Development Review Committee, 1972, How Austria has avoided stagnation, *O.E.C.D. Observer,* **60**, 4–6.

O.E.C.D., 1974, *Salient Features of Regional Development Policy in Austria,* Paris.

Osterreichische Akademie des Wissenschaften, 1961, *Atlas der Republik Osterreich,* Vienna.

Scheidl, L. and H. Lechleitner, 1972, *Osterreich: Land, Volk, Wirtschaft,* Vienna.

Sieper, M., 1971, Migrations alternantes et centres d'attraction au Tyrol, in P. Brunet (Ed.), *Exemples des sociétés rurales européennes en transformation,* Caen, 217–311.

Sieper, M., 1972, Chronique autrichienne 1970, *Revue de Géographie de l'Est,* **12,** 421–444.

Note

1. By Hugh D. Clout.

14

Switzerland

Frédéric Chiffelle

Limited Space for Economic Activities

Physical Factors

The spatial organization of Switzerland is characterized by two major features: its relatively small area for potential economic use and its regional disparities. Both of these issues are major preoccupations of the authorities. Various physical, demographic and economic factors limit the availability of land for economic use. Switzerland's accidented relief considerably reduces the proportion of the national land surface (41,300 km^2) that may be used for farming, industry, transport and settlement. Over a quarter of the country is either unusable or no longer usable for day-to-day activities by the Swiss population. This includes 20 per cent occupied by glaciers, rock outcrops and lakes plus a further 9 per cent covered by buildings. Tourism and H.E.P. production are virtually the only activities that make profitable use of Switzerland's mountains. Further use of existing built-up areas presupposes raising the densities in residential zones.

Some 21 per cent of Switzerland is occupied by mountain pastures of limited productivity. The agricultural value of 10,000 km^2 of Alpine pastures is roughly equivalent to only 500 km^2 of lowland grazing. High pastures are no longer a source of great profit for Swiss farmers and are being used in an increasingly extensive way because of the decline of the agricultural labour force in the mountains. Only tourist activities compete with farming for use of land in higher parts of Switzerland, for example for developing resorts or building ski lifts; lowland areas are infinitely more favoured for locating manufacturing activities.

Forests cover 24 per cent of the country and this area may not be used for accommodating future needs of housing, industry and services since a federal law of 11 October 1902 stipulated that 'the woodland area of Switzerland must not be reduced'. This restrictive measure stemmed from growing recognition of the protective role that forests play against such natural disasters as soil erosion, landslides and avalanches. The law applies not only to 6800 km^2 of public forests but also to 2800 km^2 in private hands. (In the first category 6300 km^2 are owned by local communities (*communes*) and 500 km^2 by the Confederation). Clearance of woodland requires permission of State Forestry authorities which demand that an equal surface should be replanted in compensation.

It follows that the most important aspects of the Swiss economy are concentrated on the remaining 26 per cent of the land—the agricultural surface. Unless the density of non-agricultural activities can be raised in existing urbanized areas, future expansion of settlement, industry and tertiary functions will have to occur at the expense of the best agricultural land sited on the Swiss Plain (between the Alps and the Jura) and in the valleys. Ultimately, Switzerland's problems of land-use competition involve only 11,000 km² out of the national total of 41,300 km².

Demographic and Economic Factors

In spite of the fact that Switzerland contains so much land that is relatively useless for agriculture and industry, the national population density is quite high, reaching 152/km² in 1970 (Table 14.1). When glaciers, rock outcrops and lakes are subtracted from the total land surface the corrected average density is raised to 199/km². Relatively rapid population growth since the Second World War has not been due to high rates of natural increase, since Switzerland has experienced demographic changes similar to those in other industrial countries in Western Europe. Birth and death rates have fallen and generate only a modest rate of natural increase (6·7 per thousand in 1970, Table 14.1). The declining birth rate is linked to psychological factors, the widespread desire to produce two-child families, the smallness of many housing units, and the gradual ageing of the national population. The over-65 age group represented only 6 per cent of the total at the beginning of the century but now makes up 11 per cent.

Immigration has been much more important in contributing to Switzerland's population growth. Before its industries developed, the country had been primarily a departure zone but now it has become a significant reception area. In 1970 only 160,000 Swiss lived abroad, compared with almost 350,000 in 1930. By contrast, the number of foreigners resident in Switzerland increased rapidly from a mere 285,000 (6 per cent of the population) in 1950 to 1,000,000 in 1970 (16 per cent). When 150,000 foreign seasonal workers, who enter the country for 8–9 months each year, are added, it is clear that Switzerland has one of the highest proportions of foreigners inside its borders of all the European countries. These 1,150,000 foreigners represent 18 per cent of the 6,300,000 people living in Switzerland. Most are of Mediterranean origin and in 1970 53·6 per cent were Italians, 11·8 per cent Germans, 10·4 per cent Spaniards, 5·2 per cent French, 4·4 per cent Austrians, 2·3 per cent Yugoslavs, and 1·1 per cent Turks, with the remaining 11·1 per cent being composed of smaller groups. Immigration has been vital in sustaining population growth in Switzerland, not only by transferring people from one country to another but also by inflating rates of natural increase. Since 1968 the volume of natural increase among the foreigners living in Switzerland has been greater than that among the indigenous Swiss population (Table 14.2). Such large flows of foreigners, mainly Roman Catholics from southern Europe have been viewed as a threat to Swiss identity. Public opinion, expressed in recent referenda on 'overpopulation by foreigners', has caused the authorities to restrict immigration for the future.

The population of Switzerland is far from evenly distributed, with relief playing a leading role in explaining the various densities. These fall to less than 5/km² in the high Alps, reach the Swiss average (c. 150/km²) in the Jura and much of the Swiss Plain, and exceed 300/km² in five *cantons* (Argovie, 310; Basel rural, 480; Zurich, 640; Geneva city 1175; Basel city, 6300). Four-fifths of the population lives below 600 m altitude, with 18 per cent living between 600 m and 1200 m, and only 2 per cent above that altitude. Admittedly most Swiss live in the lower parts of the country but, contrary to popular opinion, the mountain zones are far from deserted. For while the farming population of the Swiss mountains has declined greatly this loss has been more than

Table 14.1. Switzerland: Demographic Data 1880–1970

Year	Total population	Urban* population	Density (per km²)		Crude birth rate (‰)	Crude death rate (‰)	Natural increase (‰)
			Crude	Corrected†			
1880	2,832,000	378,000	69	90	29·6	21·9	7·7
1900	3,315,000	728,000	80	105	28·6	19·3	9·3
1950	4,715,000	1,720,000	114	149	18·1	10·1	8·0
1960	5,429,000	2,280,000	131	172	17·6	9·7	7·9
1970	6,270,000	2,843,000	152	199	15·8	9·1	6·7

* Towns defined as having more than 10,000 inhabitants.
† Calculated after the removal of glaciers, rock surfaces and lakes from the total national territory.

Table 14.2. Switzerland: Volume of Natural Increase, 1960 and 1970

	1960			1970		
	Births	Deaths	Natural increase	Births	Deaths	Natural increase
Foreigners in Switzerland	11,400	3,400	8,000	29,700	4,000	25,700
Swiss-born population	83,000	48,700	34,300	69,500	53,000	16,500

compensated for by population attracted by industrial development and especially by tourism. For this latter reason, areas of Switzerland above 1500 m now contain more inhabitants than at the beginning of the century. In 1960 35,000 lived above that altitude, compared with 23,000 in 1900.

Concentration of population and movement down-slope in Switzerland has not occurred in such a sharp and dramatic fashion as in other countries. Several reasons explain this, of which the federal structure of the country is the most important. The 25 *cantons*, which act almost as autonomous states as far as taxation and education are concerned, have been able to maintain and create employment opportunities on their respective territories and have kept down the volume of migration between *cantons*. A number of medium-sized towns function as regional growth centres because they are the administrative centres of their *cantons*. Sion in Valais *canton*, Chur in Graubünden *canton*, and Neuchâtel in the *canton* of the same name are good examples.

In spite of only a moderate decline in the agricultural population, concentration of people in urban settlements is very evident. A century ago only one-tenth of the Swiss lived in 'towns' of more than 10,000 inhabitants. Now the proportion is nearer to one-half, involving 2,843,000 people. Small *communes*, with fewer than 2000 residents, house only 30 per cent of the population, compared with 80 per cent in 1850. National statistics define 'towns' as settlements with at least 10,000 residents and on this basis Switzerland contained 92 towns in 1970, having had only 8 in the middle of the nineteenth century.

The Mittelland may be likened to a great 'suburb', using the term to designate a state of advanced but quite diffuse urban development. Nevertheless, there are a number of large urban centres. Two 'megalopolitan' zones, albeit quite small in area, accommodate an appreciable proportion of the national population. The first is located at the foot of the Jura and stretches in an uninterrupted chain of towns from Winterthur to Biel and Berne, passing through greater Zurich which is the largest individual component with 675,000 inhabitants in 1970 (Figure 14.1). The second is the Léman megalopolis, which extends along the northern shore of Lac Léman (Lake Geneva) from greater Geneva (321,000) to Montreux, and includes greater Lausanne (219,000). Greater Basel, with 373,000 inhabitants, forms a third urban zone and is the second largest conurbation in the country. Since Basel is only 80 km from Zurich it is easy to appreciate that the two cities represent a favoured area for economic growth. The megalopolis of German-speaking Switzerland thus comprises three main components

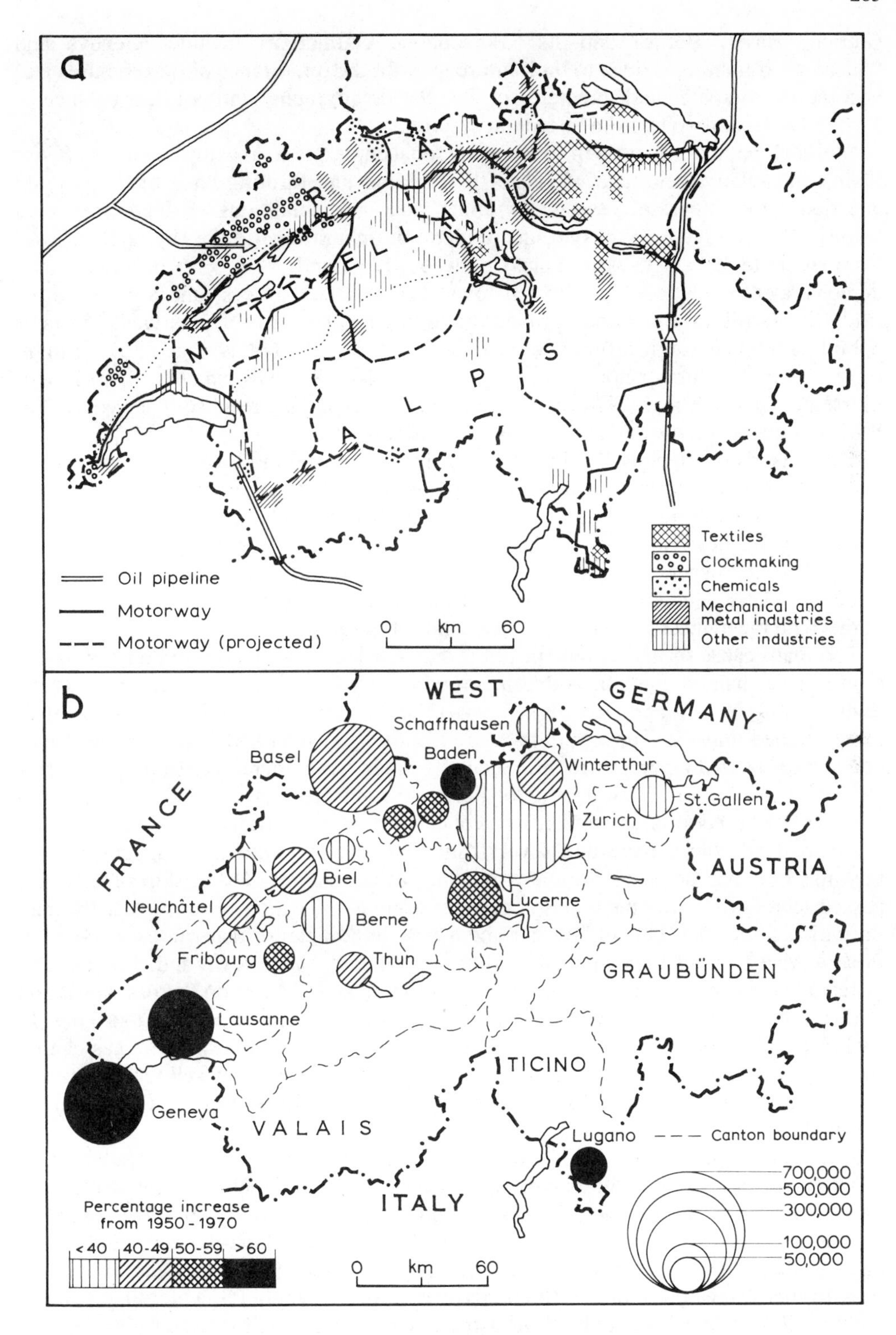

Figure 14.1. Switzerland: (a) economic development; (b) growth of urban population, 1950–70.

(Zurich, Basel, Berne) and its Francophone counterpart includes Geneva and Lausanne. If nothing is done to limit future growth, the importance of these centres will become even greater and threatens to alter the demographic and political balance of Switzerland that derives from its federal system.

National population growth, increasing concentration in the towns of the Swiss Plain, and relative shortage of space for building and farming have made the land question very important, especially under post-war conditions of increasing land values. The average price of one square metre of unbuilt land in the city of Basel rose from 30 F to 250 F (Swiss francs) between 1945 and 1962. Urban studies have demonstrated a relationship between town size and land prices, and a comparative study of Marseilles and Zurich proved the high cost of land in Switzerland. Prices of unbuilt land in central Marseilles were *c.* 800 F/m^2 in 1962 but averaged 2670 F/m^2 in central Zurich. Furthermore, greater Zurich (600,000) was smaller than greater Marseilles (800,000). In 1974 prices of unbuilt suburban land reached or exceeded 100 F/m^2 around most large and medium-sized Swiss towns.

The reasons for increased prices are well known. Some are linked to the supply of land and others to demand for it. Among supply factors one must remember the limited amount of agricultural land (11,000 km^2) that is being nibbled away at the rate of 20 km^2 a year for new buildings and highways. Prices have also been inflated by the fact that many landowners have hung on to their property, hoping to receive higher sale prices at a later date. In general terms, rising land prices have spread outwards from points with the highest values, rather like drops of oil on water.

The main cause of higher land prices is the very localized pattern of demand in and around five major cities. In addition, growing affluence has increased the average Swiss family's demand for living space. Four-roomed dwellings, rather than three-roomed units, are now built most frequently in Switzerland. The average rate of occupation is 0·86 persons/room. Second homes have increased from just a few thousands in 1945 to 130,000 in 1970. Finally, in an inflationary period, investment in land has been used to safeguard against reduced purchasing power of the Swiss franc.

Rising land prices have had several unfavourable consequences. Speculation on building land has occurred around expanding urban centres. Ownership of land has passed increasingly into the hands of private corporations which currently own the land on which more than half of Switzerland's new buildings are constructed each year. Finally, rising prices have created land-use conflicts between town and country. The difference between agricultural land values (running at 2–3 F/m^2 on the best soils) and the price of building land (expressed in tens or hundreds of F/m^2) is most striking. In such a situation of conflict, farmers on the margin of expanding urban areas cannot compete with other prospective buyers. Many give up farming and sell their property; some with particularly well located holdings become millionaires (in Swiss francs) overnight.

A number of laws and administrative rules attempt to render land use rather more coherent. They operate at the *commune, canton* and federal levels of decision making. The *commune* authorities held, and still hold, some of the most effective means of managing land development. They are required to establish land-use zoning plans for their respective territories and to ensure that they are adhered to. *Communes* are able to steer further extension of their settlements by granting or withholding building permits. Finally, they enjoy a monopoly of infrastructure provision and have the right to refuse connection of 'undesirable' buildings to electricity-supply grids, water mains, gas pipelines, sewers, and road systems. In other words, land-development control is essentially at the *commune* level in Switzerland.

Further up the administrative hierarchy, *canton* governments exercise some power. They maintain the right to overview communal activities and have established a series of laws which, in the absence of land-development plans at the *canton* level, are ecologically orientated. There are laws on new construction, protection of landscapes and monuments, and conservation of water and flora. The *cantons* will probably strengthen their role in land development in the future.

However it is the federal government that has strengthened its position most effectively with regard to land planning. Until 1970 it had only indirect legal backing, in the form of federal laws on forest management (1902), maintenance of rural landed property (1951), protection of water against pollution (1955), and protection of nature and the landscape (1960). But it was only after 1969 that the Swiss people accorded planning rights to the federal government by adding a new article to the federal constitution: 'The Confederation will establish legal principles to be applied to land-development plans that the *cantons* will be called upon to draw up to ensure judicious use of land and rational occupation of territory.' A federal law on spatial planning was thus enacted to embody that article. This was of vital significance not only as the first *federal* planning law but also because it represented something of a 'revolution' in such a decentralized country as Switzerland, where citizens jealously guard their autonomy at *commune* and *canton* levels. Intervention of the Confederation in spatial planning corresponded exactly with the time when regional disparities were increasing and tensions threatened between economically dynamic *cantons* and those that were stagnating or were on the decline.

Regional Disparities

Agriculture for Food Production and for Conserving the Landscape

As has been shown, variations in population density represent one of the most basic forms of spatial disparity. Concentrations of population are most pronounced in urban areas and in lower parts of the country. These demographic variations reflect the unequal distribution of employment in the various sectors of the economy. Growing opinion is assigning two roles to Swiss agriculture for the future: one is traditional and involves providing the population with food; the second has been voiced increasingly since 1960 and involves conserving the landscape. Lowland farming responds particularly well to the first objective, and mountain farming to the latter.

'Lowland' farming in Switzerland is highly productive but does not enjoy the best of conditions. Indeed, it suffers from a number of handicaps. The Mittelland is also referred to as the Swiss Plain or Plateau but in no way resembles the lowlands of the North European Plain. Completely flat surfaces are rare. Farming on the 'Swiss Plain' involves hilly terrain, dissected by valleys, with a minimum altitude of 300 m and elevation rising progressively towards the Alps. Factors of production are scarcely more favourable than physical geography. Average farm size on the Swiss Plain reached 15 ha in 1970, by comparison with 8 ha for the whole of Switzerland. This size of holding is only just above the threshold of viability, below which the farmer would be unable to meet financial obligations without help from sources outside agriculture, such as from industrial or tertiary work. As elsewhere in Europe, average farm sizes are increasing but at the slow rate of one additional ha during the decade 1955–65. Reasons for this slowness must be sought in the very high prices of land which prevent farmers enlarging their properties as they may wish. In addition, the fact that too many farmers continue to work the land in some parts of the country also retards property enlargement.

Extreme fragmentation of holdings into a large number of plots seriously hinders mechanized farming. In 1939 the average holding in Switzerland comprised 10 plots. That number has now been reduced to 7. The average size of plot increased from 0·5 ha to 1·0 ha over the same period. Structural change is scarcely less slow in areas where formal plot-consolidation schemes have operated. Such slowness had direct implications for farm enlargement. Some holdings have been enlarged by acquisition of dismembered parts of farms that have disappeared; however, they experience an increase in their number of plots unless old and new parcels are directly juxtaposed.

Excessive fragmentation of holdings and small farm size rarely make mechanization viable. Swiss farming is generally over-mechanized. Each farm has one or two tractors, and the survival of a mixed 'polyculture plus stock-rearing' farm system necessitates purchasing a wide range of machines. Thus, machines for planting, weeding, and harvesting potatoes may be used over only small areas and for short periods of time. The same is true for hay harvesters, machines for spreading manure and chemical fertilizers, and combine harvesters. High prices encourage farmers to try to cooperate with neighbours in buying machinery; climatic conditions are variable, however, and harvesting, for example, can be carried out under optimal conditions for only short periods each year. As a result, a single machine (for harvesting hay or grain, for example) may not be shared efficiently by more than 3 or 4 farmers.

Shortage of agricultural labour is the principal reason for over-mechanization. Virtually all paid labour has been drawn away from the land by higher wages in manufacturing and tertiary jobs, so that the typical Swiss farm is a family holding that employs just one man. Only large farms of more than 30 ha, or those that concentrate on such specialized crops as the vine, tree crops and market-garden products have been able to recruit seasonal farmworkers from the Mediterranean lands by offering the same range of wages as those available in factory work.

Finally, two-thirds of farmland in Switzerland is held in owner-occupation. The remainder is leased. Owner-occupation presents some advantages, but also the inconvenience of immobilizing large quantities of capital. The fact that Swiss agriculture is more in debt than ever before partly relates to the desire of many farmers to buy the land that they work, and to their often doing so at very high prices. In 1974 the price of Swiss farmland averaged 2–3 F/m^2. Leased farms are important in the French-speaking part of the country, where only about 50 per cent are in owner-occupation. Moreover, the role of leasing is increasing throughout Switzerland, rising from 25 per cent of holdings in 1955 to 35 per cent ten years later.

These factors (size of holdings, fragmentation, over-mechanization, lack of labour) all influence systems of production. Although they rarely reach the stage of monoculture, many Swiss farms have moved from a subsistence-type of polyculture combined with stock rearing. There are, of course, striking regional specialisms. Generally speaking, arable farming declines and stock farming increases in importance from west to east in the Mittelland. Thus there are many cereal farms near Geneva that support no livestock, whilst arable crops are often absent from stock farms in northeastern Switzerland. Nevertheless, livestock continues to provide the main source of revenue on most farms. Animals and animal products accounted for 77 per cent of the Swiss agricultural income in 1969, with milk representing 32 per cent, beef 20 per cent, pork 18 per cent and poultry 5 per cent. Crops accounted for only 23 per cent of farm income, with tree and cereal crops (6 per cent each) in the leading position, followed by vines (4 per cent) and market-garden crops and potatoes (3 per cent each). Whilst livestock farming occurs throughout Switzerland, arable cultivation is becoming increasingly concentrated on the Mittelland and in valleys. These low-altitude agricultural areas account for almost all crops grown in Switzerland, producing more than 90 per cent of cereals, vegetables and potatoes, and all marketable tree crops and grapes.

High agricultural productivity and protective tariffs allow 60 per cent of the calorie requirements of the Swiss to be met from home produced farm goods; and the proportion is on the increase. In 1970 Switzerland grew 55 per cent of the wheat that it consumed (compared with 30 per cent in 1930), produced more potatoes than it needed for its own requirements (120 per cent, compared with 79 per cent in 1930), and 20 per cent of its sugar requirements (6 per cent in 1930). Little has changed with respect to livestock farming since Switzerland produced all the milk it needed (as it had done in 1930), 97 per cent of the pork it consumed, and 50 per cent more cheese than it needed. Considerable quantities of potatoes and cheese are exported.

Mountain farming covers two-thirds of the agricultural surface but is quite different from what has been discussed so far. It remains viable in middle-mountain areas, such as the Jura and the Pre-Alps, as a result of farm enlargement. Farms in the 30 ha range are necessary if grassland farming is to be profitable. But it must be admitted frankly that most holdings in the High Alps no longer provide their occupiers with acceptable incomes. It is in the Alpine *cantons* of Valais, Ticino, and Graubünden that one finds not only the smallest (less than 10 ha and often less than 5 ha) but also the most fragmented farms (with 20–50 plots per holding). It is not surprising that these regions provide only a small proportion of total farm output. Crop production is virtually nil in the mountain zone. Admittedly upland farming is mainly orientated to stock rearing, but even that sector is weaker than in lowland areas. Mountain farms contain only 25 per cent of Swiss dairy cows and 20 per cent of pigs, but account for the majority of sheep and goats. Here again is evidence of their marginality.

Many mountain farms survive only with the help of external finance, derived from supplementary jobs in manufacturing or tertiary activities, or from federal subsidies. Farmers in the *cantons* of Valais, Ticino and Graubünden earn money from jobs in factories situated in the valleys, from temporary work on ski-lifts and in other branches of tourism, and from jobs on the roads or at H.E.P. installations. Financial aid from the Confederation provides another source of revenue. This is designed to maintain a farming population in the mountains by providing decent incomes. It therefore aims to avoid total desertion of higher areas in the country and to guarantee that mountain landscapes are maintained. One might ask if the mountain farmer of the future will have become a State employee whose task it is to maintain the mountains for tourists and to help prevent avalanches. (It is well known that snow slides more easily over land from which the hay has not been harvested.) Numerous federal laws also offer price supports for cereals and potatoes grown over 800 m. In short, farming in the Swiss mountains is probably one of the most subsidized in Western Europe. This is the ransom that Switzerland has agreed to pay to ensure that, in spite of enormous differences in incomes between mountain and plain, the Jura and the Alps remain among the most populated mountain zones in Europe.

Regions and Nodes of Industrial Growth

Switzerland was not predestined to be an industrial nation. Mineral raw materials, coal and oil are non-existent. The country's continental location and distance from maritime ports are additional handicaps. Only Basel, at the terminus of the Rhine navigation system, has fluvial access to the sea. Nevertheless, Switzerland is among the most industrialized countries of Europe. Some 48 per cent of its workforce was engaged in manufacturing in 1970, with a further 45 per cent in services, and only 7 per cent in farming.

Although Switzerland does not possess any fossil fuels, it does benefit from considerable H.E.P. resources which have formed the starting point for its subsequent industrialization. H.E.P. is produced in two types of area: the Alps, which contain over 300 Swiss H.E.P. plants and reservoirs; and the valleys of the Rhône and, especially, the Aar and the Rhine. Cheap H.E.P. has stimulated industries such as chemical production at Monthey and Viège (Valais *canton*) and Ems (near Chur, Graubünden *canton*), and aluminium at Chippis near Sierre (Valais *canton*). Now it is admitted that almost all Switzerland's viable H.E.P. resources have been harnessed. Hence an effort has been made to diversify energy sources by establishing thermal power stations. Two have been opened in the proximity of oil refineries, one at Aigle, at the extreme eastern end of Lake Geneva, and the other at Cressier between the lakes of Neuchâtel and Biel (Figure 14.1b). Since the Swiss government wishes to reduce the country's dependence on oil, two atomic power stations have also been constructed, one on the river Aar at Mühleberg (between Berne and the lake of Biel) and another at Beznau close to the Aar–Rhine confluence. Half a dozen more will be built, provided that increasingly strong public opinion does not prevent it on the grounds of safety and environmental protection.

Switzerland has long been one of the most important banking nations of Europe. She owes this position to four main factors. First, free movement of capital is guaranteed by the absence of threats of nationalization or embargo. Second, security of investments is ensured by monetary and political stability and by national neutrality which strengthen Switzerland's role as a safe deposit. Third, taxation on large incomes and fortunes is low; since precise taxation levels are fixed by *canton* and *commune* authorities, some localities are highly favoured in this respect. Finally, secrecy in banking and the system of numbered accounts, which may not be revealed to taxation authorities in Switzerland or abroad, are other attractions. All these factors help explain how the Swiss economy, and in particular its manufacturing industry, have been able to benefit from bank loans under very favourable conditions.

Swiss industry has overcome the handicap of distance from the sea and the necessity of importing mineral raw materials by specializing in light, high quality goods with a large proportion of value being added by labour inputs. Clock- and watch-making is a very successful example of this kind of activity, since labour accounts for over three-quarters of the cost of a watch or clock. The pattern of industrial development reemphasizes the preeminence of the Mittelland, and especially its northeastern section in the Olten–Schaffhausen–Winterthur–Zurich quadrilateral. This is the prime industrial region of Switzerland which continues to increase its dominance. Zurich is the leading city in the country and has a wide range of high-level service facilities and equipment. It is the only town in the country to have both an international airport and two institutions of higher learning (the University and the Federal Polytechnic). The other cities are less well equipped. Lausanne and Berne lack airports, and the one at Basel is small. Zurich is also at the heart of Switzerland's road and rail networks. In addition, the city supports not only a wide range of industries, which affords a measure of security in times of crisis, but also machine manufacture which is undergoing important expansion. This industrial sector is by far the most important in the city and in its satellite towns of Winterthur, Schaffhausen, Lucerne, Baden, Aarau, Olten and Zoug.

Table 14.3. Switzerland: Factories and Industrial Employment (Main Sectors), 1966 and 1970

	Factories		Employment	
	1966	1970	1966	1970
Food processing	605	589	44,800	46,000
Textiles	850	727	68,400	60,000
Clothing	1478	1225	72,000	62,800
Wood	1682	1438	44,600	42,500
Printing	979	982	49,100	52,300
Chemicals	434	433	54,700	64,700
Metallurgy	1838	1692	120,900	120,900
Machinery	2650	2195	263,900	267,400
Clocks and Watches	1278	1177	72,600	72,800

The Basel region is another of Switzerland's industrial strongpoints, being founded on an expanding chemical industry which offers the highest salaries of any industrial sector (Table 14.3). The industrial area around Saint Gall is still predominantly orientated to textile production (embroidery, silk, cotton, synthetic fibres, clothing) but attempts at industrial diversification in the form of machine manufacturing may help this area escape from the kind of crisis that threatens many textile areas.

The watch-making area of the Jura is perhaps more exposed to economic depression. It is paradoxical that this is Switzerland's best known industry abroad but is no longer the economic leader that it was immediately after the Second World War. It has been overtaken by the manufacture of machines and chemicals in terms of export earnings. (In 1970 machine production earned 8 billion (thousand million) francs, chemical products 5 billion francs, but watches and clocks only 3 billion francs). Similarly clock-making employs only 73,000 compared with 267,000 in mechanical engineering and 120,000 in metallurgy. At present the Swiss watch industry suffers serious competition from producers in the U.S.A. and, especially, Japan, where giant factories employing thousands of workers, can achieve economies of scale that are quite impossible in Swiss workshops, averaging 40 workers apiece. Competition is particularly serious, since 97 per cent of Swiss clocks and watches are exported, allowing Switzerland to capture about half of the world market. Reliance on overseas trade and the 'luxury' character of watches makes this industry particularly vulnerable to economic fluctuations.

Machine production, metallurgy, chemicals, textiles and watches are the leading sectors of Swiss manufacturing. Processing of food must be mentioned, as brand names like Suchard and Nestlé are known throughout the world; but the food industry (and woodworking and printing for that matter) is not concentrated in any one region. It is not a leading industrial activity, paying relatively low wages and employing mainly an unskilled labour force.

'Concentrated Decentralization'

In spite of federally inspired correctives, important regional imbalances are clearly in evidence in the agricultural, industrial, tertiary and quaternary sections of the Swiss economy which, of course, are developing at different rates. Whatever opinion one may hold, the motorway network currently under construction will enhance the advantages of existing dynamic regions. The authorities have sensed this danger of excessive concentration of power in the Zurich–Basel–Berne triangle and along the Geneva–Lausanne axis. For this reason, the federal planning service has adopted the concept of 'concentrated decentralization' which proposes that further growth of the five largest cities should be restricted (for example, by reducing the volume of public investment) and that development of a selected number of medium-sized towns should be encouraged to allow them to reach 'metropolitan status'. Hence the term 'concentrated decentralization' expresses a median position between extreme concentration and total dispersal. The solutions adopted will certainly be of vital importance in shaping the future human geography of Switzerland, but one wonders how the relationship between *cantons* and regions will fare, and whether Swiss federalism will emerge weakened or revitalized.

References

Barbier, J., J-L. Piveteau and M. Roten, 1973, *Géographie de la Suisse,* Presses Universitaires de France.

Burtenshaw, D., 1970, Switzerland's planning priorities, *Planning Outlook,* **8**, 55–68.

Chiffelle, F., 1968, *Le Bas-Pays neuchâtelois: étude de géographie rurale,* La Baconnière et Payot.

Chiffelle, F., 1973, Le remembrement parcellaire au service de l'aménagement régional: le cas de la Suisse, *Annales de Géographie,* **82**, 28–41.

Colin, G., 1970, *La Suisse,* Centre de Documentation Universitaire.

Diem, A., 1969, Planning in the Zurich region, *Canadian Geographer,* **13**, 150–162.

Gallusser, W., 1973, L'organisation suisse des routes nationales comme innovation de l'espace culturel, *Revue Géographique de l'Est,* **13**, 289–300.

Gretler, A. and P-E. Mandl, 1973, *Values, Trends and Alternatives in Swiss Society,* Praeger.

Gutersohn, H., 1958, *Geographie der Schweiz,* Kümmerly and Frey.

Kilchenmann, M. and coworkers, 1972, *Computer Atlas of Switzerland: Population, Housing, Occupation, Agriculture,* Kümmerly and Frey.

Piveteau, J-L., 1969, Les régions attractives de la Suisse, *Annales de Géographie,* **78**, 435–461.

Raffestin, C., 1971, Les mutations de l'industrie genevoise, *Revue de Géographie de Lyon,* **46**, 317–325.

Thomas, W. S. G., 1967, Remaniement parcellaire in Switzerland, *Geography,* **52**, 307–310.

15

Iberia

John Naylon

Regional Problems

Since the beginning of the 1960s Spain and Portugal have lifted themselves out of the ranks of the developing countries and transformed themselves from mainly agricultural into mainly industrial nations. Spain in particular has achieved an 'economic miracle' since the Stabilization Plan of 1959, her economic growth rate during the 1960s being exceeded only by Brazil and Japan. In 1973 Spain was the twelfth country in the world in the value of her G.N.P. (i.e. ahead of Sweden, the Netherlands, Belgium and Switzerland) and independent commentators such as the Chase Manhattan Bank and the Hudson Institute forecast rapid economic expansion well into the future. Portugal lags well behind her neighbour: during the first half of the 1960s she too enjoyed one of the highest rates of growth of G.N.P. in Western Europe, but since the middle of the decade the cost of her African struggles (rising to over 40 per cent of the national budget) has gradually slowed progress. Nevertheless, despite the contrasts in their stages of development, Spain and Portugal both now stand at an intermediate level between the most advanced and the emergent nations of the world. Both countries are experiencing the most radical social and economic changes they have known for centuries, and the study of these phenomena can offer valuable lessons to nations further down the development ladder—lessons in development problems and their analysis, in regional policies and their prospects of success.

Unfortunately, as nations develop economically, all their component regions do not benefit equally. On the contrary, the development process emphasizes the disparities between rich and poor regions, the absence of frontiers within a nation facilitating the free flow of capital, labour, goods and services from the backward agricultural areas to the dynamic, urbanized industrial ones. Internal differences of all kinds are generally greater, therefore, within semi-developed countries such as Spain and Portugal than within advanced modern societies; and in fact many Iberian regions show classic underdevelopment features.

In Spain rural poverty and regional underdevelopment have been recognized and prescribed for since the eighteenth century, yet in the 1960s they were still marked enough for Linz and de Miguel (1966) to write of 'The eight Spains'. In fact, despite its early unification in 1515 under a single monarchy, Spain with a population of 34,365,000 remains probably the most heterogeneous society in Europe. The 8,900,000 inhabitants of Portugal are a much more homogeneous poeple, yet within this small country levels of regional development are no less uneven. Regional differences in Iberia are of various kinds. To the strong contrasts in climate, relief and soils can be added differences in economic development; in social structure, property and land-tenure patterns, employment and education; in linguistic and cultural traditions; in political and religious attitudes; in social mobilization and participation; and in social values, norms and family patterns. Whichever criteria are used to measure economic differences, the pattern emerges of vast areas with little or no industrialization in the modern sense, and a marked concentration of wealth in the national capitals, in the industrial and commercial metropolis of Barcelona and its satellites, in the heavy industrial centre of Bilbao, in Guipũzcoa and Valencia with their small manufacturing towns, and to a lesser extent in Oporto and in the coal–steel complex of Asturias. These latter regions constitute 'bourgeois' Spain and Portugal, related to a modern industrial–commercial economy and enjoying an income per head and a buying power up to four or five times greater than those of the poorest provinces. At the other end of the scale come Iberia's underdeveloped regions; interior Galicia, the provinces of the Cordillera Ibérica, Extremadura, Andalusia and the South-East (large parts of Murcia and Albacete), and well-nigh the whole of interior Portugal. A particular problem, especially for Portugal, is the international frontier, which inhibits commercial exchanges and communications in a zone already meagrely endowed with resources. This *raya de Portugal* (Portuguese boundary) is one of Europe's most backward regions, as evinced by figures such as 51 per cent illiteracy (1970) in the Portuguese district of Portalegre and an infant mortality rate of 70·7 per thousand in Vila Real.

A more familiar case of resistance to development are the eight southern Spanish provinces of Andalusia and their continuation into the Alentejo of Portugal. As in Italy's Mezzogiorno, the essence of the problem here is not so much lack of physical resources or harshness of natural environment as human institutional factors. Andalusia and the Alentejo are the domain of the *latifundio*—the large estate—whose concomitants are low agricultural productivity, low wages, large pools of casual labour, poor living conditions, low levels of education and technical skill, very little vertical social mobility, a great gulf between the favoured few and the masses, little enterprise and little expectation of progress. The small but wealthy absentee landowner group lives a *rentier* life and has failed to provide opportunities for more varied and rewarding employment. Little or no industry, therefore, is to be found in the large southern cities, towns and 'villages' (the latter, in Andalusia, often having up to 30,000 inhabitants) other than isolated modern plants, built largely on the initiative of the central government or the big national banks and contrasting strongly with their rural surroundings.

In contrast to the *latifundismo* and the regressive society of the south, common problems throughout the countryside in the northern half of the peninsula are *minifundismo* and the fragmentation of landholdings, prime obstacles to the rationalization of Iberian farming. Popular response to both situations has been an intense internal and overseas migration, involving well over 5,000,000 Spaniards and at least 1,300,000 Portuguese during the decade of the 1960s (Figure 15.1). Two flows predominate in this diaspora—an uncontrolled flight from the land to the city, and an emphasizing of the familiar Iberian pattern of a densely populated periphery and a thinly populated interior (apart from Madrid and a few smaller oases such as Zaragoza, Valladolid and Burgos). Towns and provinces which have failed to industrialize are experiencing decay. Thus, the population of the southern Spanish Meseta fell by 5·47 per cent during the period 1960–70, the province of Badajoz showing the worst decline with only 9 out of 162 municipalities increasing their populations. On the Portuguese side of the frontier the abandonment of the countryside is much worse: during the same intercensal period the districts of Bragança and Beja respectively lost 23 per cent and

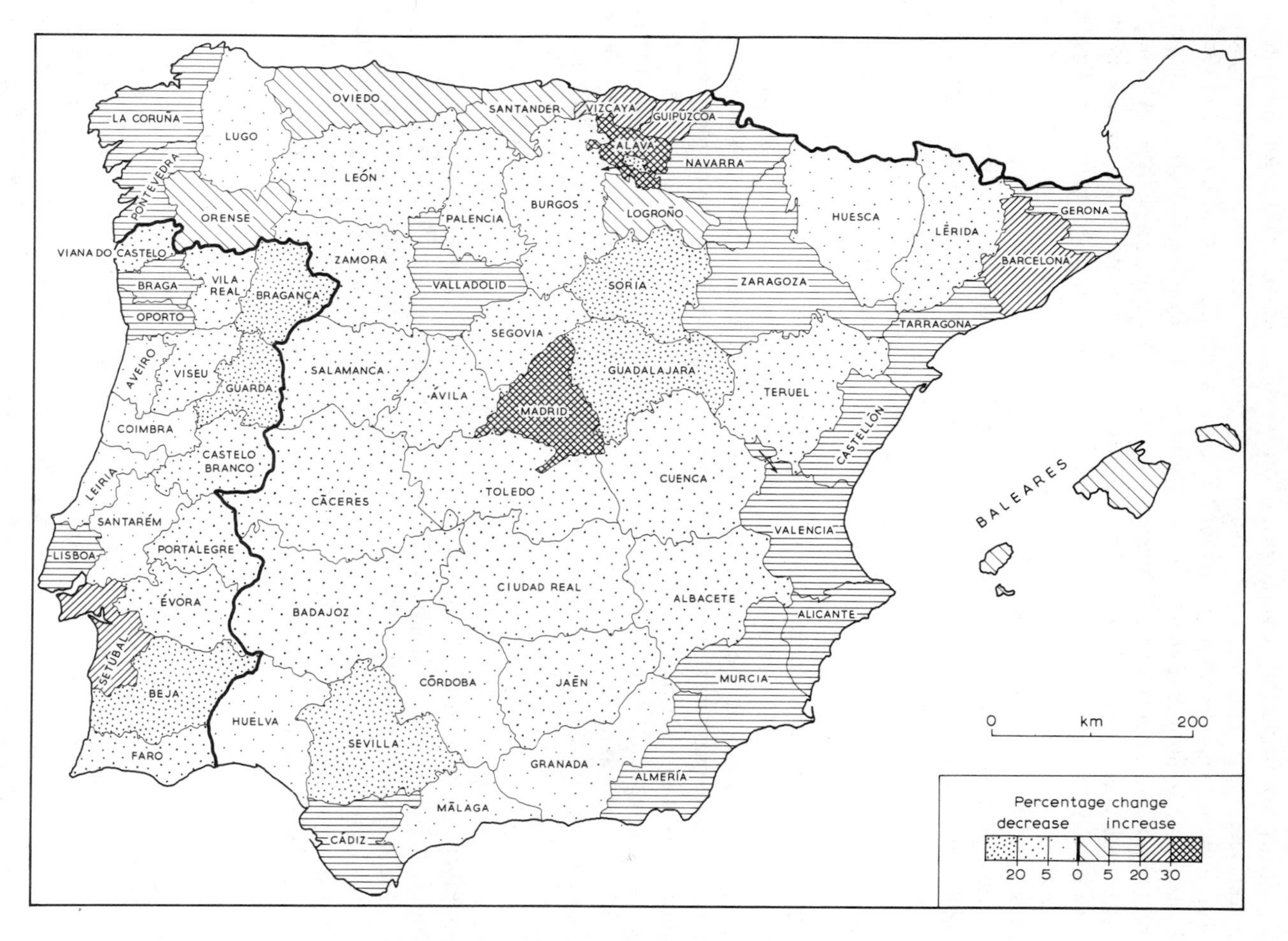

Figure 15.1. Iberia: population change, 1960–70.

25 per cent of their populations. In contrast, intense industrialization since the beginning of the 1960s, reinforced in some cases by the parallel tourist boom, accounts for urban population increases in Spain such as Barcelona 36·53 per cent, Madrid 39·21 per cent, Tarragona 79·78 per cent and Vitoria 85·71 per cent. These Spanish urban influxes are imposing severe strains on housing, health, education and other public services, and creating new social and political tensions; the Portuguese situation, however, is even more worrying, albeit in a different way. During the 1960s only Lisbon and Setúbal, and to a far lesser extent Oporto, Braga and Aveiro, managed to attract population; every other Portuguese district experienced a decline. Although perhaps as many as 100,000 in-migrants live in shanty towns and illegal dwellings on the outskirts of Lisbon, even the national capital only increased its population by 16 per cent. Faced not only with economic stagnation, unemployment and poor living conditions in the countryside, but also with prolonged compulsory military service in Africa and a growing gap in living standards between Portugal and the rest of Europe, one-eighth of the Portuguese population (well over 1,000,000 people) opted to leave their homeland altogether, either legally or clandestinely. As a result, Paris is now the third—or second—Portuguese city in Europe and Portugal has become the only European country to lose population, having 165,852 inhabitants fewer in 1970 than in 1960.

In summary, Spain and Portugal exhibit the characteristics common to countries in an intermediate stage of development: an intense urbanization, closely linked to the growth of industry and services but confined to certain limited regions; wide income differences; a wholesale rural exodus; and the decay of many small towns and villages, especially those with only administrative and commercial functions. This uneven social and economic development has produced an industrialized and advanced Iberia alongside a rural and backward one—what J. R. Lasuén (1962) has called 'a quasi-feudal world co-existing with a twentieth-century bourgeoisie'. In Spain, notoriously, the differential location of economic, intellectual and other centres of power has affected efforts at national integration, producing strong political and social tensions between the regions themselves and also with the central government. The comparatively early industrialization of Catalonia and the Basque Country, and their ensuing high living standards, have bolstered their consciousness of distinct ethnicity, language and history; the results have included separatist movements and resentments against the Castilian central administration—among the factors which led to the Civil War of 1936–39—as well as a disdain for the inhabitants of other Spanish regions such as Andalusia who now flock to their cities in search of a better life.

Regional Development Policies in Iberia

For backward areas to improve their relative positions generally calls for conscious government action in regional development; they must look to basic changes politically imposed rather than to the casual effects of spontaneous economic development. Unfortunately, the dominant characteristic of Portuguese and, especially, Spanish state regional development action has been a proliferation of disconnected projects. In part this is because it is not so easy—as it is in Portugal or Italy—to group Spain's backward areas. Whereas regional contrasts express themselves in a North–South dualism in Italy, and an interior–coastal dualism in Portugal, Spain's depressed regions are more widespread. It is true that northern Spain is generally wealthier than the South (although there are pockets of poverty in the northern half of the country, such as Soria, rural Leôn and Orense); but the overall Spanish pattern is that the main centres of economic activity are like widely separated islands, imperfectly linked by a sparse transport system, so that industry is orientated towards regional rather than national markets, the national economy is far from integrated, and there are marked differences in incomes, prices and the possession of consumer durables. This situation, in turn, is a reflection of Spain's physical geography: large cities and the more densely-populated basins and valleys are separated from each other by difficult mountains or dry empty countryside. There is a particular contrast in settlement and economic activity between the periphery and the interior.

Another factor to be borne in mind is the special political and administrative system in Spain and Portugal, with a general lack of coordination between government departments, sometimes complicated by political antipathies.

The Shortcomings of Regional Planning Organizations

There has been no lack of organizations theoretically dedicated to ameliorating the regional problems of Iberia. Several ministries in Madrid and Lisbon maintain provincial field services whose brief is the identification and solution of local problems. In theory, provincial and municipal authorities, who are generally lacking in planning skills, can call upon the expertise of the central government technical agencies to draw up and execute public works and services plans. In practice, however, these state field services are fragmented and are subject to little or no control or coordination at local level. The technical agencies pursue their own group interests and policies, even agencies dependent on the same ministry tending to go their separate ways. These attitudes stem from the highly centralized and hierarchically structured nature of the Spanish and Portuguese states and their tradition of discouraging local initiatives and participation. Each technical agency is answerable to Madrid or Lisbon rather than to the local authorities, with whom it is rarely disposed even to consult. The agencies jealously defend their autonomy and compete, rather than cooperate, in the services they provide. This competition extends to the taking over of public services normally provided by municipal and provincial authorities. There is no single authoritative agency capable of seeing each province's problems as a whole and allocating resources rationally. At the same time, local initiatives are stifled and local councils are characterized by inefficiency or even passive resistance to central government efforts to remedy deficiencies in such minimal services as roads, housing, education and health provision. The central government, by the same token, finds itself cut off from sources of information about local needs and resources.

These attitudes and antagonisms have frequently thwarted endeavours to promote regional development; these endeavours, moreover—as a review of Spanish efforts will suggest—have been very varied in their organization and administration (Figure 15.2). Sometimes plans have involved several ministries, as in the case of the Costa del Sol Planning and Development scheme. Sometimes several General Directorates of the same Ministry have been concerned, as with the Agrarian Extension Plan for La Coruña and the Agrarian Reappraisal Plan for the Amblés Valley (Avila). Occasionally the central government has launched independent regional development plans concurrently with its National Investment Planning Programmes and its four-year Economic and Social Development Plans; this is the case with the Tierra de Campos

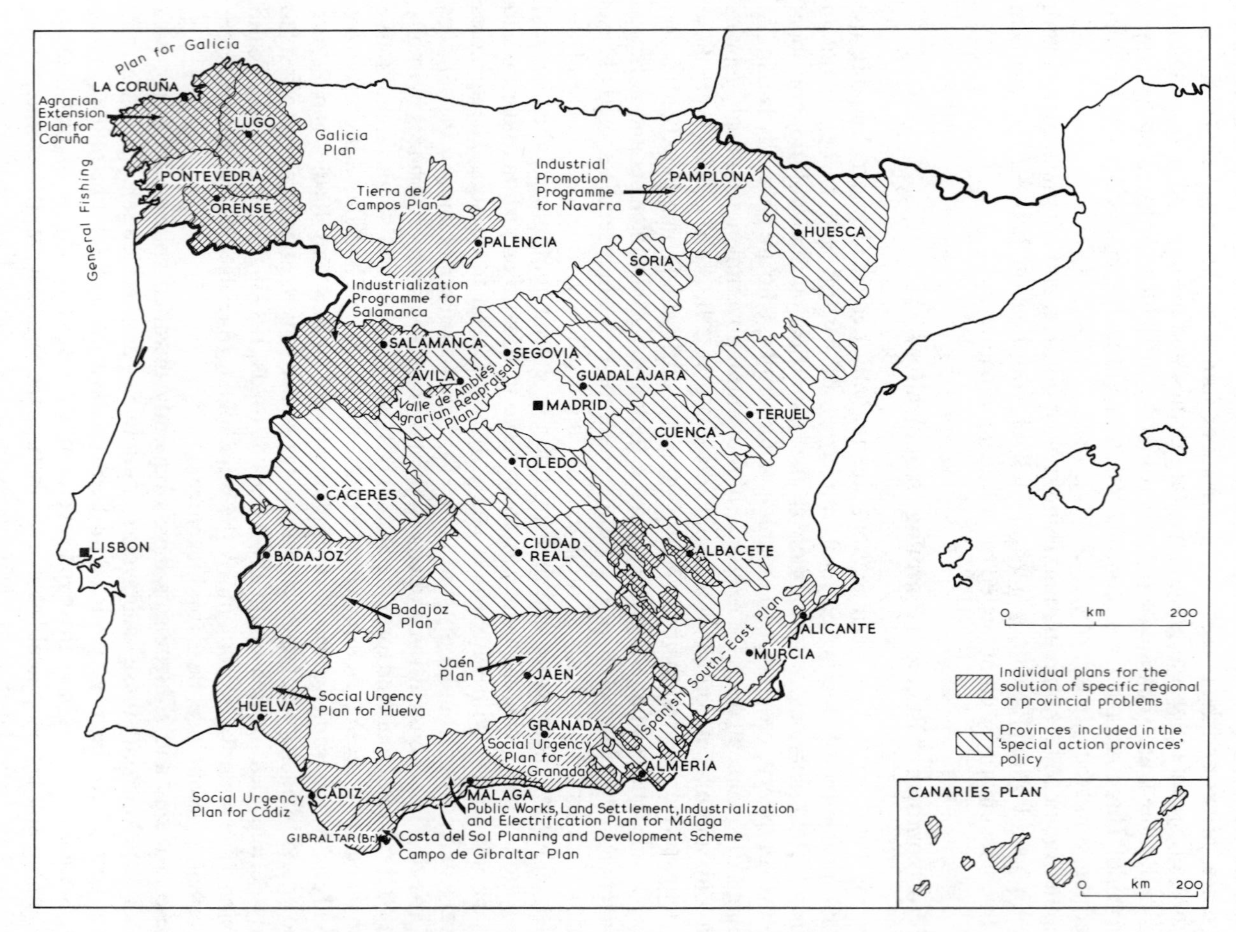

Figure 15.2 Spain: regional development plans.

and Campo de Gibraltar Plans. In 1946 Provincial Planning Boards were set up so that the state field services could assist provincial governors and mayors to adapt national policies to solve local problems, and between 1946 and 1958 over 50 provincial studies (the Planes de Ordenación Económico-Social de las Provincias Españolas) were carried out. Unfortunately these plans mainly remained on paper, not least because the money to translate them into reality would have to be assigned from the national budget; and while the Interior Ministry might empower provincial authorities to put the plans into action, the Finance Ministry might withhold the necessary funds. The Provincial Planning Board members, being appointed rather than elected, held little or no consultation with local councils and were, in effect, national officials out of touch with local realities. Perhaps for this reason, although in 1958 the responsibilities of the Boards were extended to the provision of amenities in rural villages and regional centres, and although over 24,800 items of work (roads, water and electricity supplies, drainage, telephones) were carried out under this policy between 1958 and 1970, these exercises proved incapable of stimulating the rural economy or abating rural migration.

Another series of academically valuable but largely unrealized Spanish plans has been formulated by the Technical Cabinet of the Syndical Organization, a group of economists, agronomists, statisticians, etc., set up in 1958 by the National Economic Council of Spain's official 'trade union' movement. In 1963 the Technical Cabinet completed the first comprehensive survey of the social and economic conditions and aspirations of each of Spain's 50 provinces. The speed of the surveys and the wealth of information they contain is explained by the position of the *sindicatos* at the heart of local economic activities. These surveys were much used by the World Bank Mission in its 1962 Report on the Economic Development of Spain and by the Madrid government in drawing up the country's first Economic and Social Development Plan of 1964–67. A second series of provincial studies is being carried out during the 1970s. In the mid-1960s the Technical Cabinet undertook a number of regional, as opposed to provincial, surveys and some of these, in turn, were taken up by the government and given concrete form: for example, the plans for the Tierra de Campos, the Campo de Gibraltar, and the South–East. However, although the Syndical Organization has played an important role in regional development planning by its numerous detailed research studies, it has not been allowed, for political reasons, to participate in the execution of its plans. Instead, this huge, elaborate and omnipresent Falangist organization has had the galling experience, once its surveys have been completed, of seeing new and autonomous bodies, often staffed by Opus Dei technocrats, specially created to carry out its recommendations.

Besides these centralist endeavours, from time to time provincial authorities and interest groups have contrived their own responses to local needs. Examples of these initiatives are the Technical Guidance and Assistance Institute of Southeast Spain, created by the provincial government of Murcia; the studies made by the provincial government of Cádiz; the Industrial Promotion Programme for Navarra; and the Industrialization Programme for Salamanca. The overall picture, however, is that although the provincial and regional Syndical Economic Councils in Spain have held many conferences on regional development themes, and though many provincial authorities have been encouraged to draw up plans for their areas, these have only come to anything when the central government has not only given them its approval but has also provided the financial means for their execution; and as we have already seen, a general tendency has been for the Ministry of Finance to starve local authorities of the means to help local development and to deprive them of the authority to raise revenues locally.

Not surprisingly, in view of this division of effort, a chronological outline of Spanish and Portuguese regional planning policies reveals a large number of unrelated schemes which have failed to go to the heart of the problem.

Hydraulic Works, Irrigation and Land Settlement

The regional planning approach with the longest parentage in Iberia is that involving the harnessing of hydraulic resources for hydro-electricity generation and river control, the irrigation of dry-farming land to achieve greater productivity, the settlement of the surplus agricultural population in new villages and farms, the creation of agro-industries to utilize the produce of the irrigated land, afforestation and the improvement of communications. This emphasis is understandable, given the dryness of much of the peninsula and the fact that as recently as the 1940s in Spain and 1950 in Portugal half the active population worked on the land. For more then 20 years after the advent to power of the present Spanish government in 1939, and more particularly after the passing of the 1949 Law on Colonization and Distribution of Property in Irrigable Zones, integrated hydraulic and land-settlement schemes were the official panacea for all regional ills, reaching maximum expression in the Badajoz (1952) and Jaén (1953) plans. Portuguese efforts, here as in other respects, have been less spectacular, albeit starting from an earlier date. Portugal began to evaluate her hydraulic resources in 1924, as compared with the first of the Spanish Hydrographic Confederations in 1926; but progress has been slow and the largest scheme, the plan to irrigate 170,000 ha of the Alentejo, only dates from 1959 (Figure 15.3).

Spain's irrigation and land-settlement schemes in particular are visually spectacular and technically of the highest order. Since 1939 Spain has completed dams at the rate of one per month. Of the world's 10,000 'large dams' (those over 15 m in height) 500 are in Spain, and of the further 1000 under construction, 80 are Spanish. Since 1939 the country has brought more land under irrigation than in the previous 2000 years. Forty per cent of Spain's agricultural production by value comes from the 12 per cent of the cultivated land which is irrigated (2,580,000 ha in 1971, with another 500,000 ha to be added by 1980). Nevertheless, such schemes have failed effectively to develop Spain's backward regions, while Portuguese policy has been so half-hearted that success could not be expected.

Irrigation and land settlement have proved to be very expensive ways of tackling regional problems, and too slow to cope with the pressures of a developing economy and society. Nor can the massive spending involved be said primarily to help the agricultural labourer and needy small farmer: over 60 per cent of the irrigated land remains in the hands of the original landowners while the state settlers installed alongside them cultivate *minifundios* of only 5–6 ha. The returns from some of these smallholdings do not even equal the national average wage, nor can the properties fully employ the families living on them, so that out-migration continues unabated. The most telling commentary on the Badajoz and Jaén Plans is that while the total employment created by the two schemes between the early 1950s and 1971 was 12,047 jobs in Badajoz and 11,156 in Jaén, net out-migration from these provinces in the intercensal period 1951–70 was 214,794 and 264,751 people respectively. It is important to realize that irrigation and land-settlement policies in Spain and Portugal

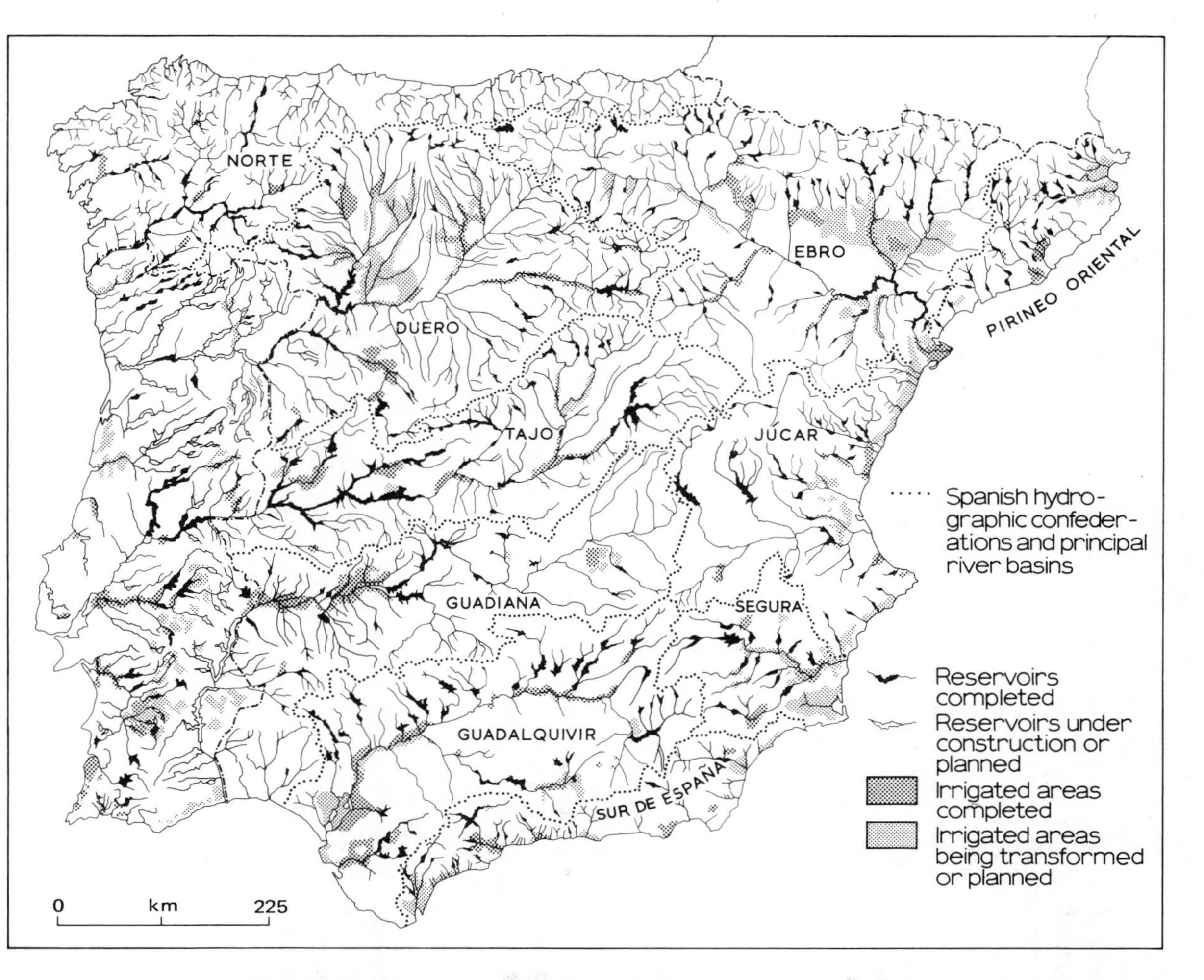

Figure 15.3. Iberia: irrigation and hydrographic confederations.

have not constituted a land reform. The Spanish Second Republic of 1931–36 sought to give the rural proletariat a real stake in their regions by a true land reform, namely, a redistribution of property. Their political affiliations have caused the Portuguese government since 1928 and the Spanish government since 1939 to turn away from the question of who should own the land to the less controversial task of making it yield more. Not surprisingly, then, the number of humble beneficiaries has been small—little more than 2000 families per year in Spain and incomparably fewer in Portugal. In the latter country, so effective has been the opposition from landowners and conservative politicians—and especially from the vertical agricultural syndicate known as the Cooperção de Havora, which is effectively an association of *latifundistas* and big tenant farmers—that the Junta de Colonização Interna, which is nominally charged with settling irrigated land, has not been able to function to date in the main irrigation zones; its activities have been confined to small colonization schemes on parcels of land donated by estate owners, to the rationalization of common lands in northern Portugal, and to the provision of credit and technical assistance in setting up cooperatives. Other general infrastructure investments by the Junta, such as roads and irrigation canals, benefit large landowners as much as anyone, since the provision of irrigation water has not been accompanied by a redistribution of land; Portugal's new irrigated areas are given over to monocultures of rice and tomatoes, and remain in the hands of *latifundistas.* On top of this vitiation of the irrigation and land-settlement legislation of 1959 by vested interests, the post-1961 guerrilla wars in Africa left the Portuguese government with even less time and money to devote to agrarian questions at home.

A major criticism of these costly schemes is that, although they have provided welcome social and economic overhead capital in the countryside, they have not been able to draw industry or commerce in their wakes. This is partly because the linkage effects of agriculture upon other producers are comparatively small and diffuse, and partly because the industrialization part of such plans is much less coordinated than the irrigation and land-settlement part: only rarely does the government involve itself in industrialization schemes, while the response from the private sector has been mediocre. While the Madrid government, unlike that of Lisbon, has provided its settlers with the essential support systems of technical training, equipment, fertilizers, pesticides, hybrid seeds and adequate credit, neither has paid enough attention to marketing networks, pricing policies and guarantees. In many cases hydraulic works have been technical exercises by engineers, with little concern for their economic viability; in Portugal, where the eye for the grandiose, and the haphazard empirical approach, have been most marked, dams have been completed before any thought has been given to soil conditions or to the crops that could be cultivated; and of course these projects pay no heed to the problems of dry-farming, livestock raising, forestry, education, housing, etc., in the countryside at large. Although in Spanish planning irrigation continues to get the lion's share of agricultural investment, its efficacy as a means of dispersing social equity and economic prosperity is seriously in doubt.

Land Consolidation and Rural Planning

The major physical obstacles to the rationalization of Iberian farming are *minifundismo* and property fragmentation. According to Siguán, writing in 1967, a 2-ha parcel of dry-farming land could not produce more than 250 pesetas (£1·50) per month; yet 52 per cent of all Spanish farmers possess less than half this amount of land. It is the sheer impossibility of obtaining a living wage from such atomized holdings, plus the endless litigation and frustration which are part of *minifundismo*, which are responsible for a large part of Iberia's rural exodus.

In contrast to hydraulic works, irrigation and land settlement, a happy feature of the land consolidation carried out in Portugal by the Junta de Colonização Interna and in Spain by the Servicio de Concentración Parcelaria (absorbed into the Instituto Nacional de Reforma y Desarrollo Agrario in 1971) is the high level of economic benefits achieved by a low-cost operation. Having said this, land consolidation is proceeding extremely slowly in Portugal's badly affected northern districts, while the much more efficient Spanish programme, although going on at the rate of 250,000 ha per year, still does not constitute an answer to the poor social conditions and low incomes of the majority of smallholders. There is no effective provision in the programme for enlarging holdings which are still tiny by modern standards even after consolidation, while the increased value of land after consolidation means that labourers have less prospect than ever of acquiring farms of their own. According to Xavier Flores it will take until 1995 to consolidate the 10,000,000 ha in Spain which need it; the rate quoted above would suggest that this is an underestimate; in the meantime, the rationalization of agriculture via land consolidation only serves to speed rural depopulation and even the abandonment of small municipalities. The removal of state subsidies for grain prices would undoubtedly cause the majority of Spanish dry-farming family holdings to fold up immediately.

José López de Sebastián has pointed out that, apart from the irrigation-based projects, which have proliferated in well-nigh every region with agricultural problems, the Spanish government's efforts to aid farming have favoured the Meseta provinces and especially Old Castile. Eighty per cent of land consolidation, for instance, is carried out in Old Castile, although *minifundismo* and fragmentation reach their most acute in Galicia and hosts of tiny smallholdings struggle for existence alongside the *latifundios* of Andalusia and Extremadura. This tends to be explained officially in terms of the impossibly high degree of fragmentation in Galicia, coupled with the difficulty of the terrain and the individualism of the Gallego farmer. Other writers such as de Miguel and Salcedo suggest that the real factors may be the paltry political weight of Galicia and Asturias and the association in the government's mind of the Castilian peasant with the quintessence of the Spanish national character. This preoccupation with the northern Meseta is also evident in the activities such as soil conservation, the promotion of livestock husbandry, and the formation of grain-farming groups and cooperatives, which are collectively known as 'Ordenación rural' (rural planning). Aimed at helping country folk to face up to the social and economic changes of the present day, none of this work seems able to offset the attractions of the city.

On the credit side, all these empirical efforts have added to the infrastructure and capital investment of agriculture in many Spanish provinces and a few Portuguese districts. In pursuing these policies, planners have gained experience of the problems inherent in regional development and of the difference between planning and reality. But, of course, Spanish and Portuguese country people have almost lost their traditional craving for a stake in the land; aware of the advantages of town life and of secure wage earning in factory, shop and office, they turn their backs on the prospect of a lifetime of drudgery on a smallholding and seek a future in industry and services.

National Economic Growth versus Regional Development

An important role can be played in regional development by state enterprises, with their large capital and technical stocks, in carrying out research, offering technical assistance, participating in financing and opening up markets for private firms in poorer regions. A prime example of such an enterprise is the Italian Istituto per la Ricostruzione Industriale, which has played an important part in the development of the Mezzogiorno. Portugal still does not possess any such instrument, although an Instituto Nacional de Promoção Industrial is mooted; but Spanish industry is dominated by the giant Instituto Nacional de Industria, set up in 1941 and indeed modelled on Mussolini's IRI. Unlike the Italian case, however, where the greater part of industrial activity in the Mezzogiorno is due to state enterprise, so that Italy 'is a country whose industry is fundamentally private in the north and public in the south' (Gustav Schachter), INI has never been required to locate in Spain's backward areas, nor has there been any INI policy of cooperating in other government development projects such as growth poles. Only in 1974 has the Institute been persuaded to participate in the creation of S.O.D.I.A.N.—a Society for the Industrial Development of Andalusia.

Lacking a lead from the state organization most fitted to provide it, the rapid industrialization which Spain has experienced since her Stabilization Plan of 1959 has conferred little direct benefit on her underdeveloped regions. Gradually opening the doors to foreign investment, know-how and competition, Spain achieved economic take-off during the early 1960s and has sustained a very high aggregate growth rate ever since. This has been achieved, however, at the cost of very wide regional income differentials. In the conflict between economic efficiency and social equity, the needs of the regions have been subordinated to the pursuit of national development. Responsibility for this attitude lies with the I.B.R.D.* mission which visited Spain in 1962 and whose Report of 1963 set out the criteria for the country's first four-year Economic and Social Development Plan 1964–67. The I.B.R.D. mission was doubtful that Spain could afford the massive expenditure necessary for an effective regional development policy; on the other hand there were productive investment opportunities in the already developed regions which would boost national economic growth. The guiding principle, therefore, was that the most efficient way of providing jobs and reducing regional disparities was by maximizing the overall growth rate; regional policies should only be pursued when they did not interfere with overall growth objectives. As far as Spain's poor regions were concerned, the I.B.R.D. Report considered that the worst of them should even have out-migration induced, with the concession that tolerable living standards should be maintained for those remaining behind by some degree of state social expenditure.

This was the thinking which lay behind the first and second Spanish Economic and Social Development Plans of 1964–67 and 1968–71. Although regional planning was nominally an integral part of these programmes—the text of the I Plan stating that the objective of regional planning was 'securing the balanced participation of all the geographic regions in the economic and social well-being of the country'—in practice Spain's vigorous industrialization has largely taken place in the already-industrialized regions. Rather than being directed by the national plans, Spanish economic policy during the 1960s responded to spontaneous market forces, making it difficult to implement a successful regional policy. The same philosophy can be seen in

*International Bank for Reconstruction and Development.

Portuguese planning up to and including the III and IV Planos de Fomento of 1968–73 and 1974–79. In the *Draft of the Third Development Plan for 1968–1973* the Minister of State in the Prime Minister's Department stated: 'The fundamental aims of the Third Plan . . . are the following: firstly, acceleration of the rate of growth of the national product; secondly, a more equitable distribution of income and revenue; thirdly, progressive correction of regional imbalances in development'. However, 'We believe that the first objective . . . should have preference over the other two . . . regional promotion . . . must be subordinated to the prime aim of the Plan, the acceleration of the growth of the national product'. The Portuguese IV Plano de Fomento follows the same line, aiming to increase economic development and lessen regional inbalances—but 'not in the sense of an even spread of employment but in terms of income per head'; in other words, via the mechanism of large-scale internal migration. Again, 'For the areas considered critical, where agriculture is the exclusive or predominant activity and without great potential for development or for stabilizing population, sectoral programmes will . . . be defined with a view to ameliorating the most urgent problems but, also, to guiding the excess population in the direction of locating itself in developing areas': in brief, planning for decline.

The Growth Pole Approach

If a regional policy was to be pursued, the 1962 I.B.R.D. mission to Spain urged that action be carefully selective and confined to centres with sound development potential, already possessing some degree of industrialization. This growth pole approach, reflecting the influence of French theorists such as Perroux, was adopted in the I, II and III Spanish, and the III and IV Portuguese, national plans (Figure 15.4). Investment has been concentrated in five types of locality: 'industrial promotion poles' in towns which did not have much industry but possessed the resources for industrialization if supported by state aid; 'development poles' in towns already having some industrial tradition; 'decongestion centres' whose function is to relieve pressure on Madrid, Lisbon and Oporto by receiving overspill industry and population; 'industrial estates' set up by local authorities, with or without central government assistance; and 'great industrial areas' based on heavy industry and deep-water sites.

In theory this growth pole strategy is well suited to Iberian conditions: it concentrates scarce resources on a limited number of centres which either enjoy locational advantages or offer a situation where economies of agglomeration can be built up fairly rapidly; and in countries like Spain and Portugal, where the main obstacles to economic development are social ones, the growth pole approach is one way of diffusing growth-mindedness. In practice, however, these efforts have had but a limited effect on regional inequalities and have exerted little influence further afield than their own municipal districts.

If one compares the location of the growth poles with maps showing various indicators of economic wealth and social provision, it is immediately apparent that not all the poles are sited in backward regions—not surprisingly, if the criteria guiding their choice are good situations in terms of markets, adequate urban infrastructures, fairly adequate industrial technical conditions, good labour supplies, location on important transport axes and, if possible, local natural resources. The fiscal incentives which the growth poles offer to industry are confusing, too low, and slow in their administration. They favour capital-intensive rather than labour-intensive industries and—a related factor—large rather than small or medium-sized firms. The benefit period is

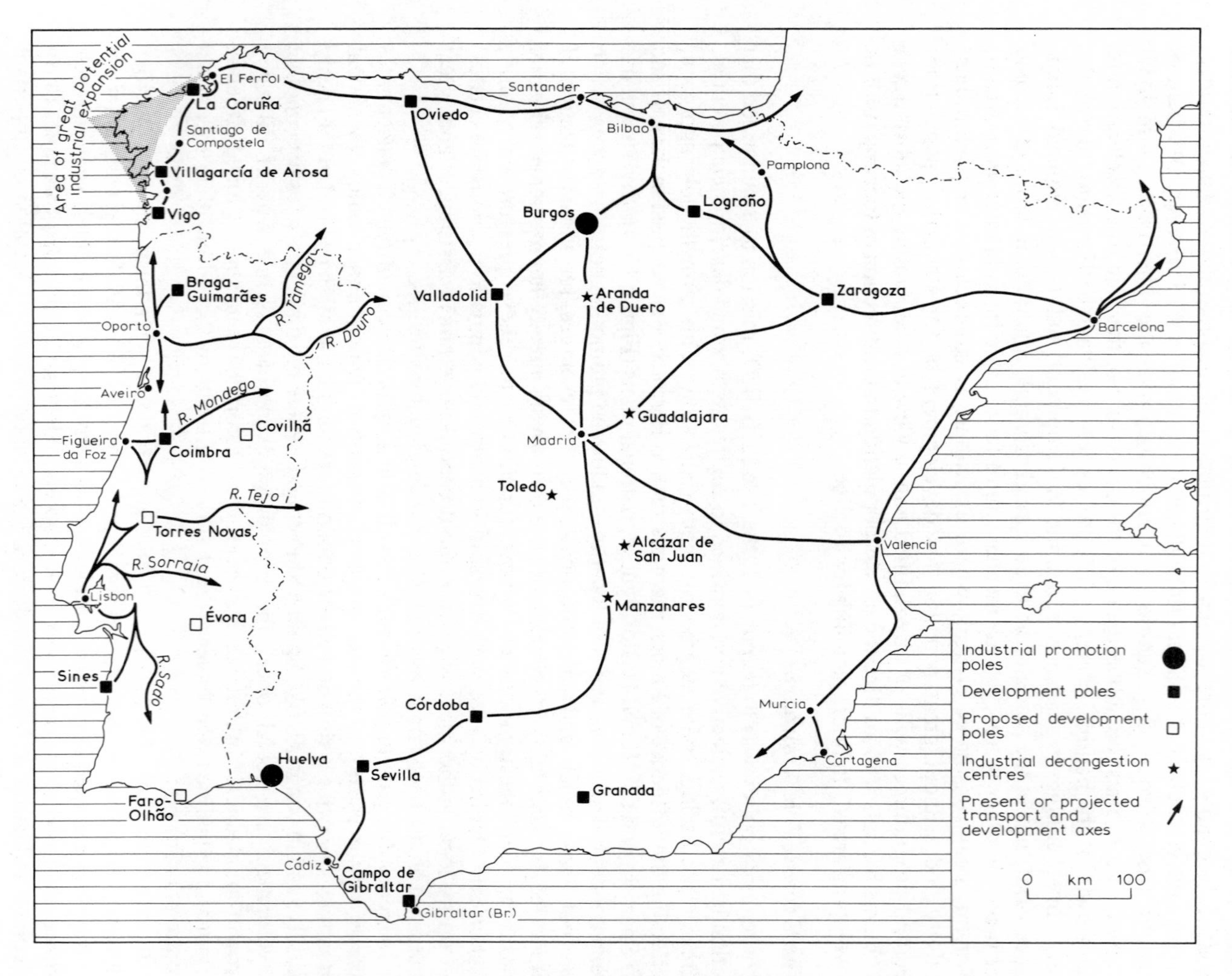

Figure 15.4. Iberia: aspects of industrial development planning.

unrealistically short, while the growth of some poles has been almost strangled by inadequate infrastructures. So far, the range of industries attracted to the poles has not been very wide, and the cost of creating jobs has been high. The central government departments have shown their customary lack of coordination and synchronization in providing investment, and especially have overlooked two elements essential for a successful growth pole strategy: technical education and cheap housing for immigrant workers. As for the industrial decongestion centres, it is doubtful whether they are economically necessary or perceptually attractive. Road traffic build-up and pollution in Madrid, Lisbon and Oporto can be combated in other ways than by decentralizing industry; all three centres have plentiful land for industry on their peripheries, ample labour from willing immigrants, and little congestion outside the city centres. It is likely to prove difficult to persuade population to move to the decongestion centres because of their lower level of amenities, lower wages, inadequate housing, and fewer work opportunities for those who are not heads of households, while the inducements offered to industry are not likely to offset the disadvantages of moving from the metropolitan centres.

Of course, it could hardly be expected that the Spanish and Portuguese growth poles would exercise an irradiating effect on their regions' economies from the very start. Their first effect has been to absorb the modest industrial activity already existing in nearby nuclei and actually exercise a negative influence on the latter; to date, 'backwash' effects have prevailed over 'spread' effects. Investment and industrial growth continue to concentrate in the already-industrialized regions and the growth poles have been able to do very little to vary this process. According to the Spanish Ministry of Industry, for instance, permission was granted in 1973 for a total industrial investment of 106,700,000,000 pesetas (£767,600,000); of this finance, no less than 51 per cent went to the four provinces of Tarragona, Barcelona, Valencia and Castellón. An analysis of the location of industries and jobs in the same year reveals a similar pattern: out of 10,711 new industries providing 72,513 jobs, the five provinces of Barcelona, Valencia, Madrid, Alicante and Tarragona took 42·6 per cent of the industries and 51·5 per cent of the jobs; out of 10,929 extensions to existing industries, providing 56,548 jobs, the four provinces of Barcelona, Valencia, Madrid and Alicante took 47·7 per cent of the industrial extensions and 58·7 per cent of the jobs. Alongside this polarization process the growth pole provinces make only a modest showing—in the first 11 months of 1973, for example, receiving only 9·2 per cent of Spanish investment in new industries. The drawbacks of 'indicative' planning are apparent. Industrial expansion in the growth poles is supposed to be achieved entirely by attracting private enterprise through state benefits; at the same time there are no negative controls—such as the British Industrial Development Certificate policy—on further industrial growth in the advanced regions and major cities, as a means of steering firms to the development areas.

It is perfectly understandable that private capital should flow to those areas where it can expect most returns. What is striking about the Spanish situation is that official credit, which ought to be mobilized to help the deprived regions, also flows in the same directions. The official Banco de Crédito Industrial is the biggest single credit organization in Spain, providing 28 per cent of all industrial credit; yet although Spain's backward areas ought to be able to look to such bodies for special aid, there is in fact no legislation such as that in Italy which singles out the poorer provinces as objects of preferential government interest and stipulates that a given percentage of official investment shall go there.

The Role of Tourism

The free play of market forces which characterizes the industrialization process is also a feature of the biggest Iberian industry of all—tourism. Since the devaluation of the peseta in 1959 Spain has virtually cornered the European mass tourist trade; in 1973 34,500,000 people visited Spain, representing an income of $3,091,160,000. To the latter figure can be added $3,216,100,000 spent by tourists on the purchase of goods and services. It was tourism which broke the vicious circle of Spanish underdevelopment and is the main financing force of the nation's development plans. Similarly, tourism has been Portugal's great growth industry and foreign exchange earner since 1964.

Tourism seems to be an eminently suitable machine for developing peripheral areas and, indeed, the industry is already crucial to the economies of some Iberian regions. It is sometimes asserted that tourism is fickle and may fall away as newer and cheaper competitors enter the field, or as a recession cuts the number of holiday makers; yet tourism is not more fickle than other sectors of world trade which are subject to the vagaries of import restrictions and embargoes. It is occasionally pointed out that several regions where tourism is a major activity (Málaga, Granada, Las Palmas, Santa Cruz de Tenerife, the Algarve) are down towards the bottom of the income per head table; but the more meaningful question is: where would these provinces be without tourism? Professor Alcaide Inchausti has argued that over the period 1962–71 provincial and *per capita* incomes have increased relatively more in Spain's seven main tourist areas than in the three industrial zones of Barcelona, Madrid and the Basque Country. This would seem to be a field in which government help might achieve a better distribution of income among Iberia's poorer but undeniably beautiful regions.

So far Spanish and Portuguese tourism has grown spontaneously and is heavily concentrated in a few, mainly coastal, areas. Little headway has been made in exploiting the attractions of the interior—the historic and artistic patrimony of small towns and villages; the numerous lakes and reservoirs which often stand in surroundings of rare natural beauty; the wild and empty landscapes which offer respite to the traveller from crowded Europe. To some extent this rural interior is already benefiting from the growth of domestic tourism. A combination of increasing car ownership and the metalling of local roads has brought life to *pueblos* which were moribund a few years ago. Like other Europeans, Spaniards and Portuguese are now investing in second homes distant from the cities in which they live and work. Central governments could disseminate tourism more widely by local road improvements; by encouraging local authorities to draw up inventories of local attractions and advising them how to promote them; and by offering official credits for water supplies, electricity, drainage, telecommunications, publicity, the building of adequate hotels and places of entertainment, the restoration of monuments, museums and art collections, and the professional training of personnel. Otherwise, the private capital which has promoted the tourist boom along the coasts will not be prepared to risk operations in the unspoilt but ill-equipped interior.

What is being urged here is a closer concern by central governments with the problems, aspirations and potentials of Iberia's provincial cities, towns and villages, and an active encouragement of local initiatives and enterprise. Such concern and encouragement have not hitherto been forthcoming from Madrid and Lisbon; but they have found some expression in the most recent Spanish and Portuguese development plans and could, with good will, finally reach the roots of regional backwardness in Iberia.

Policies for the Future: the Urban Hierarchy Approach

In the Spain and Portugal of the development plans the market mechanism has been considered the most efficient way of allocating resources—at the cost of emphasizing the inequalities between regions. Until recently, neither government has shown itself sensitive to complaints from the problem areas; the overriding aim has been national economic growth. Of late, however, the cumulative forces polarizing development have begun to threaten social, economic and political stability. These stresses can no longer be ignored, and a new emphasis on social justice and regional equity is marked in both the III and IV Spanish national plans and in statements by the Portuguese government since the coup of April 25 1974.

The new stratagems for promoting regional development are to take regional needs into account in deciding where to put public investments in irrigation, livestock raising, forestry, education, health, housing and communications; and to organize an articulated hierarchical system of metropolitan, urban and rural settlements. Both these approaches are copies of French theory and practice—the regionalization of investment dating from the IV and V National Plans to the case of France, and the urban hierarchy concept being an imitation of the French *aménagement du territoire*. Each element in the national urban system is defined and ranked, and provided with the infrastructures, finance and administrative machinery needed for it to provide the services required by its population. The hierarchy of cities, towns and villages then provides the channels by which the development process initiated in the industrial metropolises is diffused to the provinces (Figure 15.5). Official statements stress that indispensable factors in achieving this objective are the active participation of the populations of the backward regions, and the granting of greater autonomy and financial resources to local government.

Unfortunately, these notions will have to contend with certain adverse physical and ideological realities. A fundamental difficulty in the way of building up regional growth foci in Portugal is the non-existence, apart from Coimbra, of any cities with between 50,000 and 500,000 inhabitants. The small size of the Portuguese district capitals (Evora 25,000, Faro 19,000, Viseu 17,000, Beja 16,000, Castelo Branco 15,000, Vila Real 10,263, Bragança 8,075) and their lack of dynamic industrial, commercial, transport, educational or other service functions, makes them powerless to combat the flow of population from the interior to the coast, and what Herminio Martins calls 'the concentration and agglutination of superior assets' in Lisbon and Oporto.

Equally obstructive are the political attitudes characteristic of the Spanish—and up till recently the Portuguese—regimes. If the state were prepared to allow the emergence of independent-minded local authorities capable of commanding popular support, local development efforts might be forthcoming; but such a step implies more open local elections and the abandonment of state control over key local political appointments. These reforms are likely to come about in Portugal after the 1974 coup. In Spain they

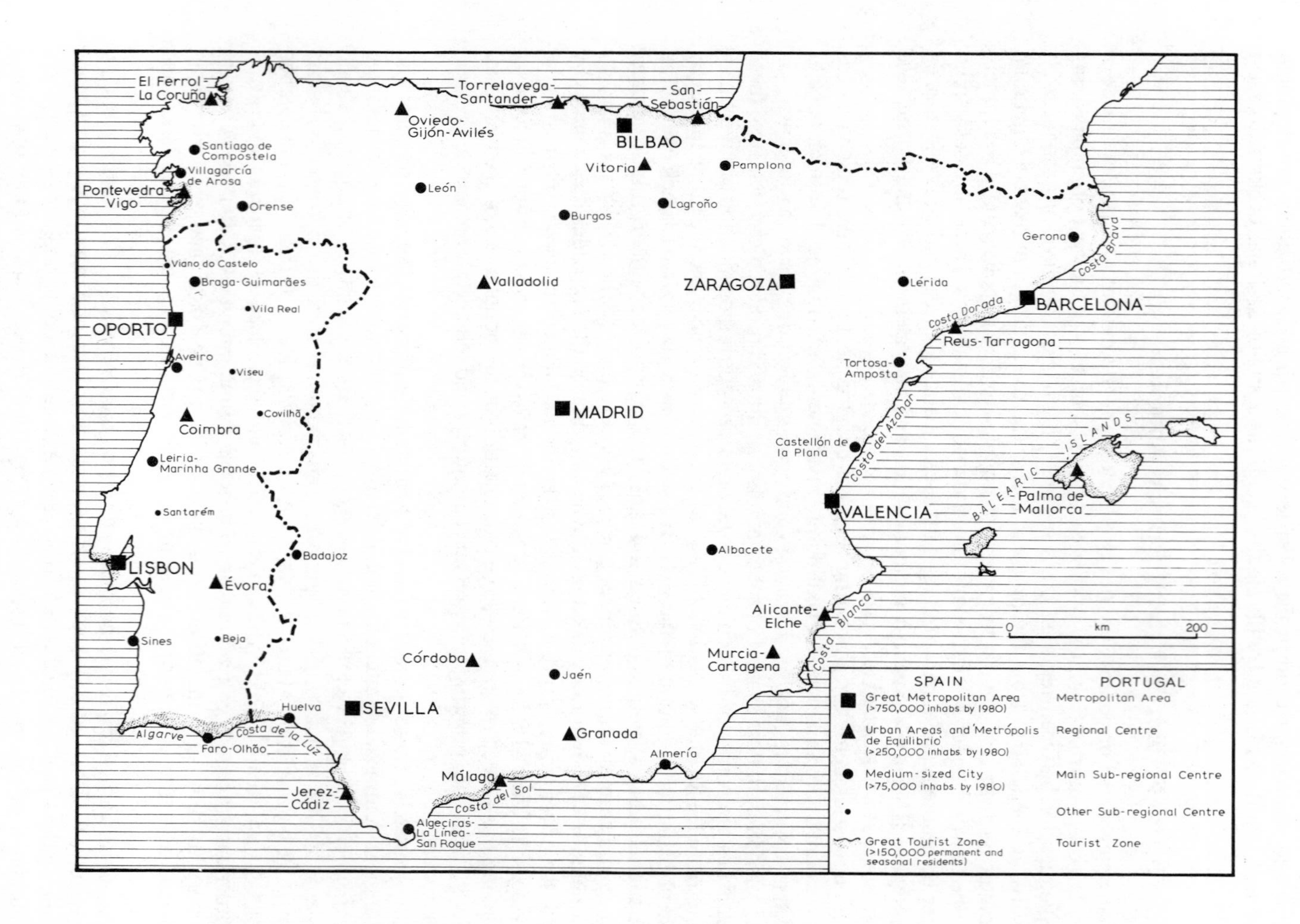

Figure 15.5. Iberia: urban centres and tourism.

are envisaged in the new Ley de Administración Local; but whether, in fact, Spain's present rulers would be prepared to take the political risks attached to such reforms is very doubtful. A similar political dilemma faces the stronger local initiatives called for in the Spanish III and IV Plans: the emergence of strong regional organizations, such as might become the foci of local separatist aspirations, would be anathema to the Madrid government, which would see the spectres of Catalan and Basque autonomy. A third difficulty involves finance. The most effective way of invigorating local authorities would be to allow them to collect their own taxes and to transfer to them larger amounts of state tax income, thus providing them with real funds for local investment and credit. In Spain at least, such moves—although also envisaged in the new Ley de Administración Local—would certainly meet with opposition from the Ministry of Finance, which provides the worst instance of the castration of local endeavour and the maximization of autonomy by a branch of the 'Administración Periférica'. Purely local sources of revenue, collected by local authorities, have gradually been replaced in Spain by taxes collected on their behalf by the Finance Ministry, which has thus assumed the role of paymaster, dictating to the local authorities how monies shall be spent rather than leaving them free to raise and spend money in accordance with their own conceptions of local needs.

The question therefore arises, at any rate in Spain, of whether the 'local initiatives' element in regional planning will not remain a technocrat's ideal. Such has been the experience of other more democratic countries, not least France, where the devolution of authority to the regions was introduced in the administrative reform of 1964. Critics of French planning practice claim that despite what is said about granting greater autonomy to the regions, the French government—fearing separatism—is really against the idea of such autonomy; and although the French national budget is now regionalized, the percentage of investment planned for the regions is often not achieved.

Conclusion

Even without treading on such thorny ground as regional autonomy, there are several things which central governments can do to ameliorate conditions in the backward regions of Iberia. The modernization of agriculture is still needed, but is of course no answer in itself. Heavy investments in economic and social overhead capital are promised under the latest development plans, but these may be beyond the national means of Spain and Portugal. Unfortunately, neither country can as yet draw upon E.E.C. regional development funds, as Italy has been able to do. However, appropriate fiscal policies on the part of the Iberian governments themselves could make an impact, especially a modernized fiscal system incorporating a more equitable distribution of taxes and the elimination of fraud. A sound and proportionate property tax would be an effective way of raising revenues in such regions as Andalusia, Extremadura and the Alentejo, as would an obligation upon firms operating in the poorer regions to invest part of their profits there. Firms controlled by the state could be obliged to locate a given percentage of their activity in the backward regions, as IRI is required to do in southern Italy. The state could also ensure that industry in the provinces benefited from legislation on 'industries of national interest', tax exemptions, credits, power and transport tariff reductions, subsidies for infrastructures, machinery and equipment, and guarantees against initial losses. In the Spanish case, such action might persuade part of the increasing amount of business investment now going abroad—for instance to southern France—to find profitable fields at home.

Without decided measures by the state, no rapid improvements can be expected in Iberia's backward regions. Many of the Spanish provinces which have had development plans are relatively no better off in the 1970s than they were when these plans began; some are relatively poorer. In Portugal, a good deal of regional planning exists solely on paper, and the nation must simultaneously face the imperative demands born of the 1974 revolution and the dissolution of the empire. In the meantime, the most industrialized Spanish and Portuguese regions—Catalonia, Vizcaya, Valencia, Navarra, Asturias, Greater Lisbon and Oporto—are not standing still and waiting for the poorer areas to catch up. They are planning their own large expansion programmes to face the challenges which will come when and if Spain and Portugal join the European Community. These plans have a vitality and a logic which efforts in the underdeveloped regions still lack.

References

Anon., 1973, Regional Policy in Spain, *O.E.C.D.Observer,* **64**, 8–12.

Bradshaw, R. P., 1972, Internal migration in Spain, *Iberian Studies,* **1**, 68–75.

Bradshaw, R. P., 1975, The development of regional planning in Spain, in A. D. M. Phillips and B. J. Turton (Eds.), *Environment, Man and Economic Change,* Longmans, London.

Lasuén, J. R. 1962, Regional income inequalities and the problems of growth in Spain, *Papers and Proceedings, Hague Congress 1961,* Regional Science Association, **8**, 169–188.

Linz, J. J. and A. de Miguel, 1966, Within-nation differences and comparisons: the eight Spains, in R. L. Merritt and S. Rokkan (Eds.), *Comparing Nations,* Yale University Press, New Haven, 267–319.

Medhurst, K. N., 1973, The central–local axis in Spain, *Iberian Studies,* **2**, 81–87.

Medhurst, K. N., 1973, *Government in Spain: The Executive at Work,* Pergamon, Oxford.

Mendes Espada, J., 1970, A Análise do Processo de Preparação e Avaliação dos Projectos de Hidráulica Agricola, *Planeamento e Integraçáo Economica, Boletim do Secretariado Tecnico,* Presidência do Conselho, Lisbon.

Ministério das Obras Páblicas, Direcção-General dos Serviços Hidráulicos, Direcção dos Servicos de Aproveitamentos Hidráulicos, 1965, *Plano de Valorizaçáo do Alentejo; Rego de 170 000 Hectares,* Lisbon.

Naylon, J., 1959, Land consolidation in Spain, *Annals of the Association of American Geographers,* **49**, 361–73.

Naylon, J., 1961, Progress in land consolidation in Spain, *Annals of the Association of American Geographers,* **51**, 335–338.

Naylon, J., 1966, The Badajoz Plan; an example of land settlement and regional development in Spain, *Erdkunde,* **20**, 44–60.

Naylon, J., 1967, Irrigation and internal colonization in Spain, *Geographical Journal,* **133**, 178–191.

Naylon, J., 1973, An appraisement of Spanish irrigation and land-settlement policy since 1939, *Iberian Studies,* **2**, 12–18.

Naylon, J., 1974, *Andalusia,* Oxford University Press, Oxford.

Pintado, A. and E. Berrenechea, 1972, *La Raya de Portugal; La Frontera del Subdesarrollo,* Editorial Cuadernos para el Diálogo, S.A., Madrid.

Presidencia del Gobierno, Comisaría del Plan de Desarrollo Económico y Social, 1963–75, *Plan de Desarrollo Economico y Social 1964–1967; II Plan de Desarrollo Económico y Social 1968–1971; III Plan de Desarrollo Económico y Social 1972–1975; IV Plan Nacional de Desarrollo 1975–78,* Madrid.

Presidência do Conselho 1968, *III Plano de Fomento para 1968–1973, Planeamento Regional,* Lisbon.

Presidência do Conselho, 1974, *IV Plano de Fomento 1974–1979, Tomo I, Metropole,* Secretariado Técnico, 1969–71, 1973, *Política de Ordenamento do Território; Anexo I. Ordenamento Urbano. A Rede Urbana do Continente. Hierarquia e Funcionamento; Anexo II. Política de Ordenamento Industrial do Território; Anexo III. Política de Ordenamento Rural do Território; Ordenamento do*

Territorio. Relatório do Grupo de Trabalho das Areas Integradas; Análise dos Desequilíbrios Regionais, Lisbon.

Richardson, H. W., 1971, Regional development policy in Spain, *Urban Studies,* **8**, 39–53.

Spanish Delegation, 1961, National regional and economic planning in Spain, in W. Isard and J. H. Cumberland (Eds.), *Regional Economic Planning: Techniques of Analysis,* European Productivity Agency of the Organization for European Economic Cooperation, Paris, 65–80.

16

Unity in Diversity?

Hugh D. Clout

Complexity, Disparity and Regional Development

'Diversity' is undoubtedly the appropriate word to apply to the spatial problems and regional development policies discussed in this book. Such an impression is due partly to the differing interests of the eleven authors, the themes that they have selected to discuss, and the emphasis that they have placed on particular issues. Nevertheless, this diversity is most certainly not just a reflection of variations in presentation but penetrates right to the heart of regional development issues in the sixteen countries that have been examined. Certain similarities in strategy are, of course, being applied in a number of states, and many of Western Europe's regional planners hang their thoughts and actions on a common intellectual frame. However, the haggling and bitter debate over a common regional policy in the E.E.C. emphasizes that the 'meeting of minds' theme should not be overstressed or accepted too readily.

Although starting from a basic recognition that spatial variations exist in 'wealth' and 'poverty', no matter how broadly defined, and a realization that action should be taken to modify these variations, national approaches to regional development then diverge one from another for three main groups of reasons. First, Western Europe's regional problems *are* in themselves highly varied and, more importantly, have been *perceived* in different ways by government institutions and planning bodies. Without entering the rarefied realms of methodology, suffice it to contrast the social and economic problems of the 'empty' lands of Northern Europe, with the need to revitalize industrial and mining areas such as the Belgian Borinage or the Ruhr, the task of decongesting Paris or Randstad Holland, and the concern of the Alpine states for conserving their tourist environment, in order to illustrate this point. Second, national systems of administration, government and planning vary enormously at the institutional level. Comprehensive machinery for regional development was established much more readily in centralized states than in federal nations (like Austria, Switzerland and West Germany) or countries such as Norway where much political power still resides in peripheral, depopulating regions. Third, national schemes for regional management vary in their stage of evolution. Well-established regional-planning systems in France, the Netherlands and the United Kingdom stand in contrast with programmes that are still in their infancy in the Alpine states, Denmark and Iberia.

The pages of this book have mirrored varying degrees of governmental intervention in economic and social life but the difficulty, if not impossibility, of trying to guage the precise impact of regional development policies must be recalled, since one may never know how regions *might* have fared without intervention. One must also remember that upward fluctuations in national and international economic 'health' may produce enormous regional effects and may be much more effective than decades of regional development schemes. Economic depression, likewise, has a devastating regional impact. Nevertheless, one may predict with certainty that governmental intervention in regional economic affairs will increase in Western Europe during the remainder of the twentieth century. National governments will exercise a larger say in each country, and supranational powers will probably intervene more strongly in the Nine. It would be unwise to venture beyond the latter general and guarded statement since the acrimony of recent political debate in the chambers of Brussels has raised new qualms about the innumerable and painful political compromises that will be necessary to bring the ideal of a 'United Europe' even a step or two closer to reality.

As far as the Nine are concerned, the quest for defining, refining, financing and operating the common regional policy will revolve around the following key questions that have been identified by Holland and Drewer (1974, p.116). How may labour and capital resources be allocated effectively between regions and nations with capitalist or 'mixed' economies? What are the appropriate roles of the state and the E.E.C. as regional policy makers? What are the current variations in existing regional problems and what new problems are likely to arise following greater European integration? To what degree will Community institutions be able to cope successfully with such problems? How effective will Community instruments be, including the common regional policy and the regional fund? What options will be open to member states if they find that Community instruments are inadequate or present obstacles to effective regional development in the national (rather than supranational) perspective? As yet there are few, if any, answers to such questions, but echoes of the on-going debates in Brussels and in national capitals will have to be monitored scrupulously if one hopes to trace, and in a modest way to comprehend, the changing geography of Western Europe in the next quarter century.

Alternative Spatial Scenarios

Various scenarios have been outlined which suggest how the pattern might evolve and the processes of economic and social change behave. In fact, such visions of the future, in association with the current economic downswing, raise questions and challenges by the score but offer few answers. For example, to what extent will free movement of labour under the Treaty of Rome reduce existing regional problems in the Nine, or will it contribute rather more powerfully to new ones? Professor Rochefort (1961) has argued that the organization of harmonious inter-regional migration of population should be developed as a vital component in the task of building a united Europe. Will the location of new employment facilities be planned and directed more closely than before, and in order to meet which objectives? Will the role of national states in patterning economic activities decline to be replaced by supranational authorities and great international commercial corporations? How will the current energy dilemma evolve, and what will the likely implications be with respect to moving goods and people and to operating regional development programmes? These and many other crucial issues need to be discussed but they cannot, of course, be answered at this stage. Nevertheless, it would seem to be essential that a range of policy options be kept open to accommodate the fundamental economic changes that will undoubtedly hit Western Europe in the depressing and challenging years of the immediate future.

Two contrasting themes have been projected from recent socio-economic trends in order to develop scenarios for the future patterning of Western Europe. The first of these has elaborated the theme of spatial differentiation (or divergence) with rich, urbanized parts of the Continent with dense communications networks continuing to attract the greater share of economic growth, new employment and, incidentally, environmental problems in the future. Boudeville (1974) first outlines the existing 'top-heavy hourglass' (from midland England through the Rhinelands and the 'wasp waist' of Switzerland to northern Italy) and then depicts an alternative 'two-tailed arrow' of possible growth (Figure 16.1). This second model links together affluent, currently rapidly-growing areas which may well display very high *per capita* incomes by the mid-1980s. The longer of the two 'tails' begins in Catalonia, running northward through Mediterranean France and the Rhône–Saône corridor to the Rhinelands where it is joined by the second 'tail' pushing through the Brenner Pass from northern Italy. Then the single shaft of the arrow continues northward to pierce Hamburg, Copenhagen and southern Sweden. A third divergent model would be roughly cruciform in design and more futuristic in conception, emphasizing major lines of road, rail and water communication running north–south along the Rhône axis into the Rhinelands and north Germany and being intersected by roughly east–west lines from southeastern England through northern France and the Low Countries (with a subsidiary Seine–Paris–Lyons link) to the Rhineland core and then eastward along the Rhine–Main–Danube waterway. Needless to say, there are considerable gaps in this idealized transport network, but many may be plugged by 1985. In each of these scenarios areas beyond the main axes of movement would certainly continue to experience industrial and tertiary growth but this would be limited in scale. The greater part of such areas would represent Western Europe's rural residuum, accommodating the space needs of agriculture, forestry and recreation.

The second group of scenarios builds on schemes for decongesting economic activities from over-developed city regions that have been at the heart of many national schemes for regional development in the last two decades. 'Growth foci' of all shapes and sizes, from large provincial cities to small rural 'holding points', are crucial elements in such a regional 'convergence' model which not only envisages affluence and employment opportunities being spread more liberally through Western Europe in the future but also presupposes adequate planning control being implemented to ensure that the kind of environmental degradation that has been experienced in urbanized regions should not be allowed to erode the quality of less-developed areas.

One is tempted to ask if regional 'losers' (namely 'the provinces') or 'winners' (in the form of capital cities, major ports and transport axes) will be backed in future schemes. In an epoch of economic difficulty will Western Europe be able to afford to help the 'losers' or will it have to fight for economic survival along established axes of development? To what extent will coalmining areas revive in the light of rising oil prices, and what will be the regional and broader impact of North Sea oil exploitation after 1980? Will economic decline continue and weaken such 'growth industries' as the car industry? Might this reduced potential for industrial employment reduce the demand for labour in city regions and retain population in the provinces and on the land, perhaps working farms in a rather less 'efficient' but more ecological way? Will rising oil prices and resultant costs of transport radically distort the state of high mobility and widespread urbanization that was being reached in the early 1970s? What kind of society and political organization will exist in Western Europe in the future? Ultimately, will her component countries and regions pull together in adversity or push further apart?

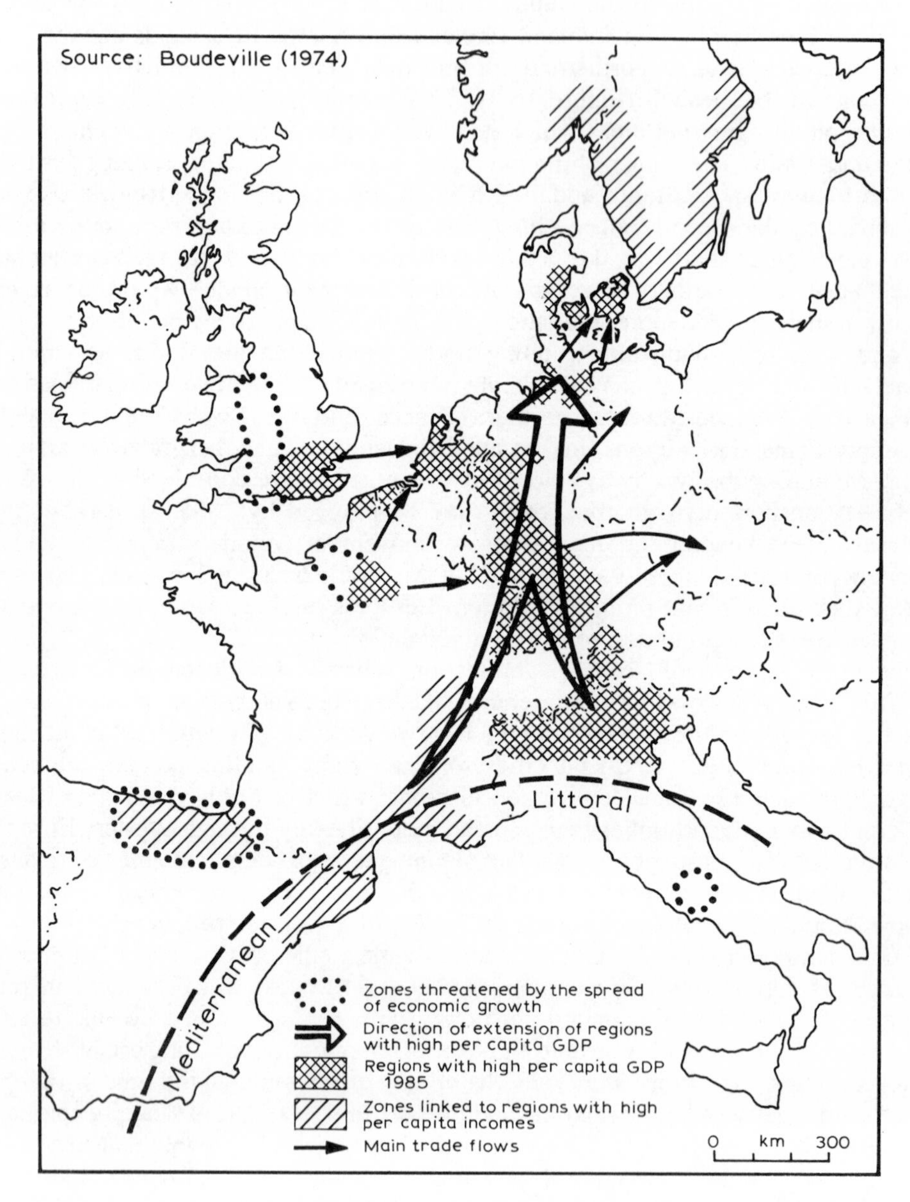

Figure 16.1. Axial development in Western Europe; forecasts for the 1980s (Reproduced by permission of Dr. M. Sant and the University of East Anglia)

References

Atlas of Europe: a Profile of Western Europe, 1974, Bartholomew, Edinburgh and Warne, London.

Boudeville, J. R., 1974, European integration, urban regions and medium-sized towns, in M.E.C. Sant (Ed.), *Regional Policy and Planning for Europe,* Saxon House, Farnborough, 129–156.

Holland, S. and S. Drewer, 1974, Regions versus Europe: an analysis of regional problems and policies in the European Community, in B. Reed *et al., The European Economic Community: Work and Home,* Open University Press, Milton Keynes, 115–128.

Kormoss, I. B., 1974, *The European Community in Maps,* Commission of the European Communities, Brussels.

Mayne, R. (Ed.), 1972, *Europe Tomorrow,* Fontana, London.

Plan Europe 2000, 1972a, *The Future is Tomorrow,* (2 vols.), M. Nijhoff, The Hague.

Plan Europe 2000, 1972b, *Fears and Hopes for European Urbanization,* M. Nijhoff, The Hague.

Rochefort, R., 1961, *Le Travail en Sicile: étude de géographie sociale,* Presses Universitaires de France, Paris.

Index